Springer Climate

Series Editor

John Dodson ⓘ, Institute of Earth Environment, Chinese Academy of Sciences, Xian, Shaanxi, China

Springer Climate is an interdisciplinary book series dedicated to climate research. This includes climatology, climate change impacts, climate change management, climate change policy, regional climate studies, climate monitoring and modeling, palaeoclimatology etc. The series publishes high quality research for scientists, researchers, students and policy makers. An author/editor questionnaire, instructions for authors and a book proposal form can be obtained from the Publishing Editor.

Now indexed in Scopus® !

More information about this series at https://link.springer.com/bookseries/11741

Simone Lucatello
Editor

Towards an Emissions Trading System in Mexico: Rationale, Design and Connections with the Global Climate Agenda

Outlook on the first ETS in Latin-America and Exploration of the Way Forward

On behalf of:

Deutsche Gesellschaft
für Internationale
Zusammenarbeit (GIZ) GmbH

Federal Ministry
for the Environment, Nature Conservation
and Nuclear Safety

of the Federal Republic of Germany

 Springer

Editor
Simone Lucatello
Instituto Mora
Mexico City, Mexico

ISSN 2352-0698   ISSN 2352-0701 (electronic)
Springer Climate
ISBN 978-3-030-82761-8   ISBN 978-3-030-82759-5 (eBook)
https://doi.org/10.1007/978-3-030-82759-5

This Springer imprint is published by the registered company Springer Nature Switzerland AG
The registered company address is: Gewerbestrasse 11, 6330 Cham, Switzerland

Foreword by William Acworth

2020 is a year that will not soon be forgotten. While the world was gripped by a global pandemic, bushfires raged, temperatures soared, and hurricanes and cyclones had devastating impacts on coastal communities—all indicative of the fact that the world is now 1.1° warmer than pre-industrial levels. Yet with close to 40% of global emissions now falling under net-zero commitments, we may still look back on 2020 as the year we finally changed course and made serious progress in addressing global climate change.

Mexico is leading the way by becoming the first emerging economy to implement an emissions trading system. For those familiar with Mexican climate policy, this should come as no surprise. Mexico was one of the first emerging economies to submit its Intended Nationally Determined Contribution in 2016 under the Paris Agreement, and its pioneering climate change law was passed unanimously by the Mexican Congress in 2012, which already committed the country's ambitious emission reduction goals.

Yet the significance of the launch of the Mexican emissions trading system must not be understated. Although emissions trading systems have proven an effective and economically sensible choice for reducing greenhouse gas emissions, they have to date been confined mostly to advanced industrialized economies. But this trend is now changing. The next wave of policies is set to be rolled out in emerging economies that face many different climate mitigation challenges.

In this context, *Towards an Emissions Trading System in Mexico: Rationale, Design and Connections With the Global Climate Agenda* could not be more important. Dr Simone Lucatello has curated a diverse set of contributions from Mexico's leading thinkers who provide their own insights and insider knowledge into how the Mexican ETS came to be. Through a peer-reviewed interdisciplinary and an approach that includes the application of state-of-the-art research methods, the authors unpack the design of the system in the context of Mexico's broader energy and climate policy, its legal basis and fundamental building blocks as well as how the system connects to the wider sustainability and development agenda. Lucatello and the authors combine both theory and practice to provide a complete perspective on the development of emissions trading in Mexico.

With only three decades remaining to reach net-zero global emissions, it is more crucial than ever that we take stock of recent policy developments, to distill and effectively communicate the key lessons learned. Only through this powerful form of collaboration can we hope to succeed in our global efforts to control climate change. With this urgency in mind, I compel anyone with a stake in meeting the challenge of the climate crisis to digest this necessary book.

Berlin, Germany   International Carbon Action Partnership
December 2020   Secretariat

Foreword by Dr. Dirk Weinreich

Our planet is inching towards the point of no return. As global temperatures continue to rise, the need to implement effective and efficient climate policies is becoming more critical than ever. Countries must ramp up their ambition so that the goals of the Paris Agreement can be met in time. Against this backdrop, the BMU stands by its commitment to financing international climate and biodiversity projects through the International Climate Initiative (IKI) and offering expertise and support when and where it can. *Towards an Emissions Trading System in Mexico: Rationale, Design and Connections with the Global Climate Agenda* is one part of this commitment come to fruition and serves as an invaluable resource for anyone interested in solving the climate crisis.

The meticulous, targeted, and efficient use of resources for achieving climate targets is particularly relevant for developing and emerging economies—and we are convinced that carbon markets allow for the most efficient form of climate action. In fact, market-based instruments also have their explicit place within the Paris Agreement. Article 6 lays the foundations for countries to create an international market, and many countries are already pursuing carbon pricing as part of their Nationally Determined Contributions. Together with our international partners, we are therefore actively pursuing a spectrum of activities to facilitate the development of market-based instruments for climate change mitigation. These include our long-standing engagement in the International Carbon Action Partnership (ICAP), our support for Partnership for Market Readiness, our "Capacity Building Programm Emissions Trading" as well as various bilateral cooperation projects under the International Climate Initiative of the BMU among a host of others.

Mexico is an important partner country for Germany in this endeavour, and the two countries have a long history of climate cooperation. Currently through the International Climate Initiative, we are supporting Mexico in implementing its NDC on climate change mitigation and adaption. Mexico shares with Germany the conviction that market-based mechanisms, such as Emissions Trading Systems (ETS), are cost-effective instruments that can catalyse the realisation of NDCs. This conviction is reflected in direct action: the 2018 reform to the Mexican General Climate Change Law established a mandatory national ETS. Mexico and Germany are extending their cooperation regarding ETS through the project "Preparation of an Emissions Trading System (ETS) in Mexico" (SiCEM), which began implementation in 2017 through the Deutsche Gesellschaft für Internationale Zusammenarbeit (GIZ) GmbH.

SiCEM has been providing technical support to the Mexican Ministry of Environment (SEMARNAT) in the development of an ETS through three main courses of action: the provision of scientific analyses and policy recommendations, capacity-building for stakeholders, and the facilitation of international dialogue. These collaborative efforts coalesced in SEMARNAT's publication of the regulation for the Mexican ETS Pilot Programme on October 1st, 2019. The Programme began implementation in January of 2020. With this, Mexico became the first country in Latin America with an ETS in operation—a trailblazer not only in the region but also in the global climate policy arena.

Towards an Emissions Trading System in Mexico is a timely and significant component of the broader support we provide to the design and implementation of the Mexican ETS. This book is the culmination of many conversations and seminars held in late 2019 and early 2020 between SiCEM and Mexican researchers, regarding the challenges and opportunities for ETS-related research in Mexico. The independence and scientific rigor of academic research are essential to an instrument whose cost-effectiveness is maximized if its design is suited specifically to its national context. Academia has long provided valuable insights into and detected opportunities for refinement in the design of ETSs around the world, including the European Union ETS. In the same vein, we have no doubt that Mexican researchers play a key role in strengthening the Mexican national ETS, whether through science-based recommendations to adjust the instrument design or by training the ETS practitioners of tomorrow.

We are weathering not only the climate crisis but now also a global social, economic, and public health upheaval brought about by the Covid-19 pandemic. There is no question that the challenges of today will impact our collective future. Whether and how we can pull together now to shape this impact will dictate how the future unfolds. Our recovery from Covid-19 must steer our economies away from a reliance on fossil fuels, our "new normal" must be a *green* normal. This is an extraordinary window of opportunity—and things look promising. The number of countries around the world setting net zero targets is on the rise. The European Green Deal is paving the way as an enormous and forward-looking policy package that promises to leave no one behind. China has announced its long-term targets and launched the largest carbon market in the world. The United States re-joined the Paris Agreement and announced the submission of its Nationally Determined Contributions for spring this year.

The climate crisis knows no borders. International cooperation is more important than ever, and Germany and Mexico will continue to work alongside each other to meet the greatest challenge of our time.

For their invaluable contributions to this book, I would like to thank Dr. Simone Lucatello of the Instituto de Investigaciones, Dr. José María Luis Mora-Conacyt as the technical editor and of course the researchers who gave their time and expertise to this collaborative effort. I would also like to thank our Mexican counterpart, SEMARNAT, for the long-standing trustful cooperation. Finally, I extend my gratitude to all those at the GIZ's SiCEM project for their continued efforts to strengthen the dialogues that only grow in importance each day. I also acknowlegde the work of William Acworth at ICAP Secretariat for supporting this project with great enthusiasm.

Berlin, Germany
April 2021

Dr. Dirk Weinreich
Head of Working Group Legal Issues Climate
Policy and Energy, Climate Legislation;
Emissions Trading. Federal Ministry for the
Environment, Nature Conservation and Nuclear
Safety

Introduction: Setting the Stage for the Emission Trading System in Mexico

Simone Lucatello

Abstract The introductory chapter describes the overall reason(s) and rational for the implementation of a Mexican ETS by setting the stage for analysing the content and structure of the book. The new emissions trading system will cover greenhouse gas emissions (mostly CO_2) and will add innovative features to the overall Mexican climate policy. Concerning the chapter, background information on the ETS will be provided to formulate the basis on which the book content is designed to address scientific and practical questions about the Mexican ETS and its application. Relevant information on the book is cited and explained in brief: the experimental design of the present study is described, in brief, highlighting the results and their significance and implications in understanding the broader issue and making the case for a Mexican ETS. The book draws upon a meticulous study of background documents and fieldwork from different authors to tell the story of how a Mexican ETS, the first of its kind in Latin America, can be set up in a major emerging world economy. The book also examines how the ETS can be designed and implemented building from previous experiences and lessons learned. Innovative paths of Mexico's research in mitigation will be explored and linkages with the global climate agenda will also be considered. The chapter ends with some takeaways about the importance of the ETS in Mexico.

Keywords ETS · Mexico · Mitigation · Carbon tax · Policy

Mexico and the ETS

Over the past few years and under the Paris Agreement reduction pledges, many countries in the world have adopted carbon pricing initiatives and climate policy tools in order to increase their emissions ambitions, declaring a "climate emergency" status for the planet. According to recent figures from the World Bank in the *2020 Report on States and Trend of Carbon markets*, there are currently 61 carbon pricing initiatives in place or scheduled for implementation, consisting of 31 Emission Trading Systems

(ETSs) and 30 carbon taxes covering 12 Giga-tons of carbon dioxide equivalent (GtCO2e) or about 22% of global GHG emissions. Among the Latin American countries, Mexico started the ETS implementation in 2020.

Mexico is the world's 10th largest emitter of greenhouse gases (GHG) and the 11th world economy. It is expected to be the world's seventh largest economy by 2050 (PWC 2019). Mexico total GHG emissions in 2019, excluding carbon sequestration, were 789tCO2e, approximately 1.8% of total global emissions. The most important contributors to GHG in Mexico are transportation, electricity generation and industry: major emitters, along with the private sectors, are government owned companies like Pemex, the National Oil Industry and the CFE, the Federal Commission for Electricity (Climate Tracker 2020).

In 2012, Mexico's Congress passed the General Law on Climate Change (LGCC), which entered into force that same year. The law and other national mitigation instruments like the National Program for Climate Change (PECC), set a target for a 30% reduction in GHG emissions below business as usual (BAU) by 2020, and a 50% reduction below 2000 levels by 2050. Since the law inception in 2012, Mexico has been evolving in its commitment to climate change mitigation and adaptation policies by introducing new internal governance mechanisms like the climate planning instruments, which broadly include the following: the National Climate Change Strategy (ENCC) 10-20-40, which provides a short, medium and long term vision on climate action and includes economic instruments. Second is the Special Climate Change Program 2014-2018, which provides a framework that links development and other national priority targets with mitigation and adaptation goals at federal level but also at state and municipal level (IETA, 2018).

Mexico was also the first developing economy to submit its Intended Nationally Determined Contribution (INDC) to the UNFCCC in 2016. Mexico's INDC recalls the need to improve regional or bilateral market-based mechanisms to achieve rapid and cost-effective mitigation. Mexico signed the Paris Agreement in April 2016, and the country is now committed to its non-conditional target of 22% GHG emission reductions, compared to Business As Usual (BAU) scenario, and the conditional target of a 36% reduction by 2030 under the NDC (SEMARNAT 2019).

Under the current political administration (2018–2024) led by President Andrés Manuel López Obrador, the LGCC mandates that the government develop an updated Climate Change Program and introduce new carbon price mechanisms. The government is currently defining well-suited policies and measures for achieving these targets. Through the 2018 reform to the general climate change law, the Mexican Ministry of Environment has been given the mandate to establish an Emissions Trading System (ETS) in the country. The regulation for the ETS pilot program was published in October 2019 and the program started operation in January 2020. It will last until December 2021, before entering a one-year transition period while the formal phase of its functioning and implementation is set to begin in January 2023.

In July 2020, Mexico launched the National Environmental Sectoral Plan 2020–2024 that is the main political instrument for governing environmental public policies, including climate change mitigation and adaptation efforts. The plan is made up of 5 main goals, among which the second one states that the plan will promote *the*

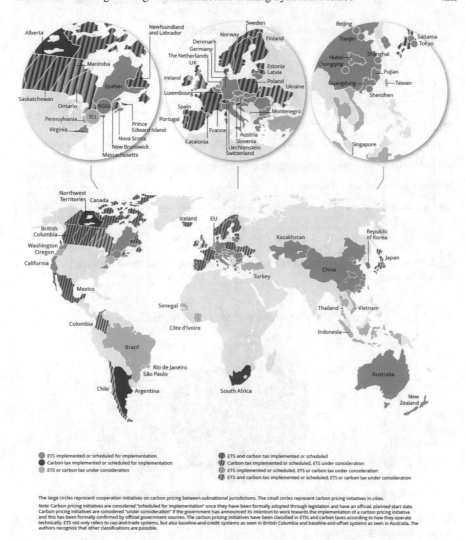

Fig. 1 Carbon pricing initiatives implemented, scheduled for implementation and under consideration (ETS and carbon tax). *Source* World Bank. 2020. *State and Trends of Carbon Pricing 2020*. Washington, DC: World Bank. © World Bank. https://openknowledge.worldbank.org/handle/10986/33809 License: CC BY 3.0 IGO

implementation and strengthening of climate actions aiming to reach a low-carbon economy and resilient societies, ecosystems and infrastructure based on scientific knowledge, local traditions and available technologies (SEMARNAT 2020). Under this pillar, climate initiatives such as the ETS find a legal basis to continue with previous decade-long efforts to tackle climate change in Mexico and worldwide.

In this brief overview of policy instruments, it is also worth keeping in mind that significant changes in Mexican regulatory policy outputs for the energy sector

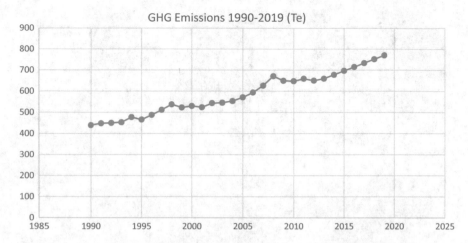

Fig. 2 Mexico GHG emissions growth 1990–2019. *Source* own elaboration based on the Mexican National Emission Registry

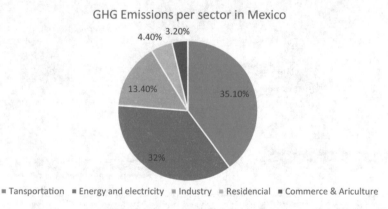

Fig. 3 GHG emission per sector in Mexico (2019). *Source* own elaboration based on the Mexican National Emission Registry

occurred during the period 2013–2018 under President Peña Nieto. Negotiations over the 2013 Energy Constitutional Reform Bill went fast, being approved by both legislative chambers at the end of December 2013. The bill was the subject of extensive support from international organizations such as the OECD and the US government, which recognized the relevance of the reform to its own energy security and fossil fuel exploitation (Echeverria Victoria and Banuelos-Ramirez 2017; Vargas 2015). For example, it included a mandate for Congress to approve a legal framework to establish the basis for achieving environmental protection through the incorporation of criteria and best practices regarding energy efficiency, GHG reductions and lower carbon footprints in energy-related processes (Cámara de diputados 2018). However, the bill's main aim was to increase the production of oil and gas through private investment, something potentially inimical to climate change mitigation, since meeting the goals of the Paris Agreement implies that significant portions of the world's fossil fuel reserves remain undeveloped. Furthermore, investing in the production of oil and gas will have a lock-in effect on investments in high-emissions infrastructure, thus slowing down the transition toward low-carbon development (Piggot et al. 2018). Hence, in contrast with previous laws, the 2014 hydrocarbon law does not include provisions to reduce gas flaring and venting during oil and gas extraction, even though this is the main source of CO_2 emissions associated with offshore oil extraction (García-Lucatello 2020).

With the new administration of President Lopez Obrador, which clearly intends to take some distance from previous neo-liberal economic administrations, important reviews of the main energy law and other mechanisms is under profound scrutiny. For example, new, recently approved measures impose a number of limitations and tests on new clean energy projects and give the National Center for Energy Control, known as CENACE, the power to reject new plant study requests and prioritize the state utility CFE. Without the tests, new plants would not be able to come online (Bloomberg 2020). It remains to be seen how the current legislative changes by the government and the above highlighted controversial issues may affect the settings for the Mexican ETS implementation and its continuity.

The ETS and Its Rationale for Mexico

An Emission Trading System is a market-based instrument that is based upon the principle of "Cap and Trade" (C&T) and its main purpose is to reduce emissions at the least possible cost. C&T is a common term for a government regulatory program designed to limit (cap) the total level of emissions of certain gases, particularly carbon dioxide, as a result of industrial activity (IETA 2018). There are two main forms of an ETS: the above-mentioned C&T and baseline-and-credit. In the case of a baseline-and-credit system, baselines are set for regulated emitters. When emissions are above their designated baseline, emitters need to "surrender" credits for emission above their baseline (WB 2020). Emitters that have reduced their emissions below their baseline can obtain credits for emission reductions and can sell them to other emitters.

Proponents of ETS argue that this is an important alternative to a carbon tax (IETA 2016): both measures are attempts to reduce environmental damage—mostly polluting emissions—without causing economic impacts to the industry. It is worth mentioning that in the absence of an ETS, the only domestic policy tool that currently sends an explicit price signal to the Mexican economy is the carbon tax. This was introduced in the country in 2014 and has since been adjusted annually for inflation. The tax rate was capped at 3% of the sales price of fuel and proved to be an important collector of revenues over time.

Unfortunately, as happens with carbon taxes examples around the world, much criticism accompanies the implementation of such instruments. In the case of Mexico, revenues coming from this source are not necessarily used for specific environmental purposes but instead they broadly contribute to general state revenues (Reynoso-Montes 2016).

Concerning the functioning of an ETS, a limit of overall emissions in one or more economic sectors is defined and reduced each year (allocation of permits). Within this limit, companies can decide to buy and sell emissions rights, as these are needed in order to comply with the limit. Through this mechanism, companies have the flexibility to minimise their emissions in the most cost-efficient way. The instrument therefore contributes to changing the behaviour of producers, consumers and investors so as to reduce emissions, while the inherent market mechanism provides flexibility on who takes which actions and when (WB 2016).

Implementation of market-based instruments, such as Emissions Trading Systems (ETS), are desired when the variation of the marginal pollution abatement costs across sectors is significant (Baumol and Oates 1988). In such cases, an ETS helps to achieve emissions reductions in the most cost-effective way (Ellerman et al. 2003). Some recent literature also points to the fact the costs of climate change mitigation can be reduced when other policy instruments, besides the ETS, are also set in place (Kreibich et al. 2019). Additionally, perspectives of linking ETS at international level as well as implementing national climate programs in its different versions, provide new insights for analysing and evaluating how non-ETS climate policy instruments, such as carbon taxes or green certificate trading schemes, could serve as a basis for establishing robust and new ETS like the Mexican one (Borghesi 2019).

The promise of cost-effective emissions reductions was incorporated into the Mexican General Climate Change Law (GLCC) enacted in 2012, which allowed the Mexican Ministry of Environment and Natural Resources (SEMARNAT) to establish a voluntary ETS in order to incentivise "emission reductions achieved at the lowest possible cost" (GLCC 2012). In 2018, amendments to the GLCC transformed the voluntary nature of the ETS into a mandatory instrument and added a consideration to protect the regulated sectors' competitiveness in international markets (Art. 94, GLCC 2018). The amendments also dictated the establishment of a 3-year pilot programme "without economic impacts" (Trans Art. 2, GLCC, 2018). On 1 October 2019, SEMARNAT published the Preliminary Bases for the Emissions Trading System Pilot Programme (the pilot programme's regulation), which began operation on 1 January 2020.

The Mexican ETS marks the beginning of ETS implementation in Latin America and is thus far one of few systems in emerging economies (ICAP 2020). In line with the GLCC, the Preliminary Bases state that the pilot programme aims at supporting the achievement of the Mexican climate targets and incentivizing emission reductions at the lower possible cost. However, the regulation also explicitly acknowledges the 'work in progress' nature of the pilot programme, as additional objectives include generating better quality data, building capacity, and testing the functioning of the system.

For an ETS to achieve its promise of cost-effective emissions reductions, its design must aim for a binding cap (aligned with the country's level of ambition), efficient allowance allocation, a functional trading market, long-term certainty and flexibility for regulated participants (ETS Handbook 2016).

Therefore, a key concern is to understand whether these conditions will be given during the pilot programme; alternatively, whether the Mexican ETS can be strengthened during the pilot to ensure these conditions are increasingly incorporated into the operational phase.

As the book will explain in some of the chapters, international experience has shown that historical, facility-based, reliable GHG emissions data has been deficient in most jurisdictions during the initial phase of any ETS implementation. This, coupled with pressures from regulated sectors, may hamper a government's capacity for designing a sufficiently binding cap. *Additionally, certainty about emissions reductions associated with a cap is also reliant on the institutional capacity to establish a "credible enforcement regime with appropriate penalties"* (ETS Handbook 2016).

If, following international experience, the cap for the Mexican ETS pilot programme were overestimated, the pilot programme still offers an opportunity to collect further historical data and provide the conditions to design a cap that effectively achieves emissions reductions at a low cost. This period can also be used to develop an effective sanction regime, building on the existing design which, although it does not consider economic sanctions, does include a penalty for noncompliant entities, to which two allowances will be deducted for every missing allowance in the free allocation of the operational phase.

Achieving the Mexican ETS objectives—cost-effective emission reductions coupled with protection of vulnerable sectors—also requires a well-balanced allocation mix. To support the transition of Mexican companies to an ETS, the preferred allocation method during the pilot programme is free allowance allocation through grandfathering, which means allowances are allocated in direct relation to the installations' historical emissions (Diario Oficial 2019). However, auctioning is the most appropriate method for encouraging efficient emissions abatement (GIZ 2018). Therefore, the pilot programme also presents a great opportunity to start building the conditions to introduce other allocations methodologies (e.g. free allocation through benchmarking) or building an auctioning scheme.

A functional allowance trading market in the Mexican ETS requires enough supply and demand levels to have market liquidity, as well as a clear definition of the legal nature of the allowance and its fiscal treatment. The trading behaviour observed during the pilot programme will give the necessary arguments to reinforce the ETS design for the operational phase, not the least including the price predictability or cost containment mechanisms necessary for long-term price certainty. The observed behaviour may also reveal a lack or an excess of flexibility in the use of offset credits for compliance. In this respect, the pilot programme regulation allows for up to 10% of the obligation to be met with offset credits.

However, given that offset protocols are yet to be published by SEMARNAT, and offset projects take at least a year to begin operation, it is unclear whether their use in the pilot programme will be sufficiently widespread so as to be able to conclude anything about its impact in the market.

This is only a handful of the issues that remain to be solved for the Mexican ETS to keep its promises. As the pilot program begins operation, building institutional capacities within the regulated entities as well as the ministries involved, additional questions arise: Are the legal provisions for a formal ETS sufficient? Can the ETS's cost-effectiveness in achieving emissions reductions help envisage more ambitious climate targets? Could the ETS be broadened to include more sectors (e.g. forestry)? What are the major obstacles for establishing a well-functioning ETS in Mexico? Who are the major stakeholders in the ETS and what are their capacities?

Book Objectives and Content

This book guide provides detailed information about the incoming Mexican Emissions Trading System, including an analysis on why the system was implemented, how the system was designed, how it operates, how it could work and how it could be strengthened by 2023 when it will be formally launched. This document is aimed at those who want to understand how an ETS can operate in an emerging economy. Although it has been written for experts and non-experts, this book does not provide the underlying theory of market-based instruments and emissions trading systems in general. The book can be read from start to finish, but can also be used as a reference for specific components of regional ETSs.

The book draws upon a meticulous study of background documents and fieldwork from different authors to tell the story of how a Mexican ETS, the first of its kind in Latin America, can be set up in the country. It also examines how the ETS can be designed and implemented building from previous experiences and lessons learned. Innovative paths of research will be explored in connection with the broader climate agenda.

The book consists of three parts. Part I, *Emissions Trading and Mexican Climate Policy: National and International Perspectives* deals with an overview of key policy design and theoretical environmental economics principles that underpin the concept of emission trading systems (ETS) as a policy approach to address climate change. It presents the technical basis for the broader discussion that this book as a whole presents on the launch of the pilot phase of the Mexican ETS on 1 January 2020. It also deepens the understanding of international experiences, international cooperation and best practices to ensure Mexico's success in the transition to an operational ETS compliance system. The four articles presented give an idea that the decision to adopt an ETS relies not only on specific characteristics of the instrument, but also on institutional constraints and political considerations. Factors involved in the deployment of the ETS pilot project include distinctive and shared characteristics in line with international experiences and domestic Mexican climate policy evolution. Experiences from North America, Europe and other geographical areas are considered as they can help to understand what can be improved for the Mexican ETS.

Chapter 1, by Blas Pérez Henriquez, presents a brief overview of the policy design and theoretical environmental economics principles that underpin the concept of emission trading systems (ETS) as a policy approach to address climate change. It discusses basic environmental economics principles pertinent to the development of market-based solutions to mitigate greenhouse gas (GHG) and co-pollutants. The chapter serves as the technical basis for the broader discussion that this book as a whole presents on the launch of the pilot phase of the Mexican ETS on 1 January 2020. The chapter also outlines a set of key policy lessons and design parameters to support the transition from the pilot Mexican ETS to an operational compliance phase in a socially just, environmentally sound, and cost-effective manner.

Chapter 2, by Alejandra Elizondo, analyses the decision to deploy carbon markets and their interaction with other instruments in Mexico's climate policy. Instrument selection has been thoroughly explored in the regulation and public policy literature but its application to carbon markets is mainly focused on environments such as Europe, the US and, more recently, China. The author analyses considerations involved in the deployment of the ETS pilot project, looking at its distinctive characteristics and those it shares with other available instruments, as well as the requirements for its implementation.

In Chap. 3, Daniela Stevens argues that the design of carbon pricing policies takes place as a sequential, negotiated process whereby specific constituencies have privileged access to shape policy design because they have high stakes in regulations. Using theory-guided process tracing and the policy stages heuristics framework, the author empirically analyses and explores the policymaking process of the Mexican pilot emission trading system and discusses key features of its design.

Chapter 4, by Neydi Cruz and Mirelle Meneses, explores how Mexico has participated in different international climate initiatives and has benefited from international collaboration in climate change mitigation efforts. Mexico played a key role in international carbon pricing initiatives, and in the technical sphere, the country benefited from peer-to-peer international experiences and knowledge. The chapter analyses those initiatives and their contribution to continue broadening collaboration toward a carbon market in the country. It explores how the recent changes to the environmental agenda, adopted in 2018 by the new federal administration, could hinder the implementation of the market mechanism.

Part II, titled *Legal Frameworks and Design Perspectives for a Mexican ETS. the Building Blocks,* looks at the different legal and technical provisions for the implementation of the ETS in Mexico. Authors provide general guidance to identify key components of an emissions trading system, the so-called 'ETS building blocks' (Kreibich et al. 2019). One example of such a block is the emission cap of the trading system, which defines the maximum quantity of emission allowances issued by the regulator and is thereby one of the key determinants of the mitigation impact of the future Mexican ETS. Legal provisions are also explored under national and international current frameworks (from the UNFCCC to the National Climate Change Law).

Authors with a legal background analyse the role that legal principles can play, including cap definition and allocation of allowances. As a developing country, Mexico needs technical and economic support from developed nations and global international carbon funds. The lack of infrastructure, capacity-building, transfer of technology and finance are the principal barriers for proper ETS implementation. Perspectives related to carbon taxes, carbon finance and other issues are also taken into account in this section. Findings from this section also provide an outlook on how the development of the ETS could be supported by considering existing instruments such as the carbon tax, the CDM and building up on the new ETS for completing the transition towards a low-carbon economy. Examples from other recent ETS set up like New Zealand are brought into the discussion. The section concludes with an overview of the offsetting programs worldwide and analyses why problematic issues impact offset programs, especially in the Mexican program linked to the nascent emissions trading system. A final concern in this section is about governance mechanisms. The creation of an ETS in Mexico as an answer to the international policy on climate change forces government and corporations to create new activities and responsibilities to address this topic. Additionally, it's important to know who is going to be the decision-maker and who is in charge of the institutional work (representation and negotiation). The main objective of this chapter is to point out who are the stakeholders involved into the design, implementation, evaluation and transparency of the system, according to the national regulatory framework and international experiences.

Part two begins with Chap. 5, by Alicia Gutierrez, who gives an overview of the international influence of the Emissions Trading System (ETS) in Mexico. It briefly examines both the international Climate Change regime through the description of such instruments as the 1997 Kyoto Protocol and the 2015 Paris Agreement, and the national regime by reviewing such instruments as the 2012 General Law on Climate Change (*LGCC*) and the National Emissions Registry (*RENE*) and its Regulations. The chapter also analyses the legal framework of the pilot phase of the ETS in Mexico (under the cap and trade principle) which seeks to reduce only carbon dioxide emissions (CO_2) in the energy and industry sectors whose emissions are greater than 100 thousand direct tons of CO_2. In doing so, it also explains the relevance of implementing an ETS as a cost-effective mitigation measure to achieve the Nationally Determined Contributions (NDCs) in order to reduce 22% greenhouse gas (GHG) emissions by 2030 (increasing to 36% if there is international support and financing) and 50% by 2050 as a developing country.

In Chap. 6, Rosalia Ibarra explores the legal basis for mandatory regulation of the emissions trading system in Mexico. Those bases are derived from the primary international instruments on climate change: the United Nations Framework Convention on Climate Change (UNFCCC) and its ambitious objective, the quantifiable commitment of the Kyoto Protocol and its tie to economic instruments. The Paris Agreement, the Nationally Determined Contributions (NDCs) and the market mechanisms regulated in Article 6, the implementation of which is essential to achieve the Agreement's objectives are also part of this broad system. The author looks at the core aspects of this system, which are highlighted from a national regulatory analysis, with special emphasis on the importance of a limited cap and its future reduction, as well as the legal nature of allowances that are assigned by the public administration to the regulated industries facilities.

Chapter 7, by Juan Carlos Belausteguigoitia, Vidal Romero and Alberto Simpser, analyses the political economy of the introduction of a carbon tax in Mexico in 2013 with the objective of learning from that process in order to facilitate the eventual implementation of an effective cap-and-trade system in Mexico. Many of the lessons in Mexico are likely to be applicable elsewhere. As countries struggle to meet the goals of international environmental agreements, it is of the utmost importance to understand the conditions under which it is feasible to implement policies that reduce carbon emissions.

Simone Lucatello and Eduardo Tovar, in Chap. 8, look at how general lessons from the past decade in climate policy and carbon initiatives such as the CDM or NAMAs in Mexico can help to better understand how carbon finance experiences can improve the new ETS in Mexico. Who is going to finance the starting process for allocating emissions, financing bonds and other design issues for the implementation of the Mexican ETS? Who will be financing and offering technical cooperation to the follow up of eligible projects for the ETS and who will be supporting education and information activities about the ETS implementation? Those and other questions are explored in the article, in the light of international and regional experiences.

Chapter 9, by Benjamin Rontard and Humberto Reyes Hernandez, deals with the Emissions Trading System in New Zealand (NZ ETS) as the only case of an ETS integrating forestry as a mandatory actor. This is the result of prolonged political discussions and the characteristics of New Zealand forestry. In Mexico, the implementation of an Emission Trading System in 2020 is evidence of the country's commitment to controlling domestic emission under the Paris Agreement. Nevertheless, for now, the forestry sector is not involved as a liable actor. The author argues that the potential impacts of this integration are both positive and negative. The economic impacts would be highly favourable for forest landowners if market volatility were controlled, but there is a potential loss of public revenue for the state. Finally, carbon forestry has the potential to cause conflict between economic sectors involved in land use and among participating communities.

In Chap. 10, Marcela Lopez Vallejo, provides an overview of how Mexico will utilize an emissions trading system as one of its carbon pricing instruments and in particular she stresses the importance of the "offsets". Offsets allow for market participants to compensate their emissions through mitigation projects. However, offset projects usually present various challenges to quality. The chapter also looks at the offset programs worldwide and identifies non-additionality, overestimated supply, and double counting as their three most pressing problems. This analysis sheds light on the nascent Mexican system and its offset program.

Chapter 11, by María Concepción Martínez Rodríguez, Catherine Nieto Moreno, Mariana Marcelino Aranda, points out who are, or should be, the stakeholders involved in the design, implementation, evaluation and transparency of the Mexican ETS, according to the national regulatory framework and international calls. The chapter also analyses the mechanics and information provided by the system and how it helps to make environmental policy clear to reduce emissions. Emphasizing the potential of training spaces as a place of transformation and developing a learning framework whose relevance relies on the focus on emergent strategies, the chapter also explores the environmental integrity and the conditions for the country's competence in the international context.

Part III of the book, titled *Mexican ETS Connected Issues with the Broader Climate Agenda*, pays attention to several spillover effects and linking issues to the nascent ETS. An important dimension is related to increasing environmental justice concerns. Environmental justice is achieved when no group is disproportionately affected by an environmental policy or environmental phenomenon. The Mexican ETS has the potential to reduce the country's GHG emissions but it should not come at the cost of increasing environmental injustice throughout the country. On the other hand, Mexico is a country surrounded by oceans. Very powerful nature-based solutions for climate change mitigation responses are required. Blue carbon can be a novel mechanism to promote communication and cooperation between the investor, the government, the users and beneficiaries of the environmental services of these ecosystems, creating public-private-social partnerships through mechanisms such as payment for environmental services, credits or the voluntary carbon market. Another important concern of this section is the relation between the ETS and the SDGs agenda.

In Chap. 12, Danaé Hernandez Cortes andErick Rosas López, estimate the pollution burden faced by marginalized neighbourhoods in Mexico. This is relevant for Mexico given the beginning of the pilot program of the Mexican emissions trading system and the country's history of income inequality and poverty. Using linear regression and two-way fixed effects methods, authors found that the highest emitters regulated under the ETS are located near poor populations. They estimated a 5% CO_2 emissions reduction scenario corresponding to national targets and associated NO_2 emissions to that scenario. The chapter also discusses other potential sources of environmental injustice that could result after the beginning of the ETS and the potential to address them.

Chapter 13, by Antonina Ivanova and Alfredo Bermudez-Contreras, deals with the possibilities of incorporating blue carbon in emissions markets. Authors explore the huge potential of Mexico's blue carbon to sequester CO_2. They then analyse the new market instrument that allows countries to sell or transfer mitigation results internationally: The Sustainable Development Mechanism (SDM), established in the Paris Agreement. Secondly, they present the progress of the Commission for Environmental Cooperation (CEC), to standardize the methodologies to assess their stock and determine the magnitude of the blue carbon sinks. Thirdly, as an opportunity for Mexico, the collaboration with California cap-and-trade program is analysed. They conclude that the blue carbon is a very important mitigation tool to be included in the compensation schemes on a regional and global level, including in the Mexican

ETS. Additionally, mangrove protection is an excellent example of the relationship mitigation-adaptation-sustainable development, as well as fostering of governance by the inclusion of the coastal communities in decision-making and incomes.

Finally, in Chap. 14, Gustavo Sosa seeks to explain the relation between the ETS and the SDG. The importance of emission trading systems can be observed when assessing their relevance for achieving the 2030 Agenda for Sustainable Development. This is the case of the goals related to energy, economic growth, inclusive industrialization, sustainable cities, sustainable production and consumption patterns, marine and land life, as well as climate itself. In this context, the author contributes to assessing the manner in which this relationship takes place in the global fora and in Mexico. A key argument is that there should be participation of a wider set of sectors and actors.

Key Take-Aways from the Book

- In 2020, Mexico launched the first ETS in Latin America with a pilot program. The three-year pilot will test the ETS design, covering 37% of national emissions, before transitioning to an operational trading system in 2023.
- As in other jurisdictions, the decision to adopt an ETS in Mexico relied not only on specific characteristics of the instrument, but also on institutional constraints and political considerations.
- A Mexican Cap and Trade requires robust emissions monitoring, reporting and verification—essential for any climate policy to preserve integrity. Allowing for the use of offsets, which lowers compliance costs, can help involve non-covered sectors in the fight against climate change.
- When establishing a cap and trade policy, the Mexican government should try to establish the correct cap on the producers of emissions. A cap that is too high may lead to even higher emissions, while a cap that is too low would be seen as a burden on the industry and a cost that would be passed on to consumers.
- The combination of an absolute cap on the level of emissions permitted and the carbon price signal from trading helps firms identify low-cost methods of reducing emissions on site, such as investing in energy efficiency—which can lead to a further reduction in overheads. This helps make business more sustainable for the future (IETA, 2016).
- The development of the Mexican ETS can be supported by considering existing national instruments such as the carbon tax, the CDM and NAMAs, as well as by learning from international experiences through international cooperation.
- Operation of the Mexican ETS pilot program will allow for building institutional capacities within the government and regulated entities, as well as determine whether the legal and technical provisions can be strengthened for the formal phase.
- The discussion around the ETS in Mexico must incorporate environmental justice and wider sustainability concerns.

- The Mexican Government and stakeholders must engage with de-carbonization strategies through international cooperation in order to reduce emissions and comply with the Paris Agreement.

References

Baumol P, Oates Y (1988) The theory of environmental policy. Cambridge University Press, New York

Borghesi S, Montini M, Barreca A (2016) The European emission trading system and its followers: Comparative analysis and linking perspectives/Borghesi S. Springer

Compromisos de mitigación y adaptación ante el cambio climático para el periodo 2020–2030. National Institute of Ecology and Climate Change, 2015. http://www.inecc.gob.mx/descargas/adaptacion/2015_indc_esp.pdf

Cámara de Diputados, Ley General sobre Cambio Climático, Ciudad de México, disponible en: http://www.diputados.gob.mx/LeyesBiblio/pdf/LGCC_130718.pdf

Diario Oficial de la Federación, Ley General sobre Cambio Climático, Ciudad de México, disponible en: http://www.diputados.gob.mx/LeyesBiblio/ref/lgcc/LGCC_orig_06jun12.pdf

Diario Oficial de la Federación (2019) Acuerdo por el que se establecen las bases preliminares de del Programa de Prueba del Sistema de Comercio de Emisiones, Ciudad de México, disponible en: https://www.dof.gob.mx/nota_detalle.php?codigo=5573934&fecha=01/10/2019

Echeverria V, Martin and Reyna Banuelos-Ramirez. (2017) The Pro-Government Bias as Frame Fidelity: Media Coverage of the 2013 Energy Reform in Mexico. *Revista Mexicana de Ciencias Politicas y Sociales* 229–254

Ellerman et al (2003) Emissions trading in the U.S., experience, lessons and considerations for greenhouse gases, pew center on global climate change. Arlington, disponible en: http://citeserx.ist.psu.edu/viewdoc/download?doi=10.1.1.180.2487&rep=rep1&type=pdf

EU-ETS Handbook, (2016). https://ec.europa.eu/clima/sites/clima/files/docs/ets_handbook_en.pdf

García A, Lucatello S (2020) *Between policy entrepreneurs and policy windows: explaining climate policy integration in Mexico from 1997 to 2018. Forthcoming.*

Healy et al (2018) Distributing allowances in the mexican emissions trading system: Indicative allocation scenarios, SEMARNAT and Deutsche Gesellschaft für Internationale Zusammenarbeit, Ciudad de México, disponible en: https://www.gob.mx/cms/uploads/attachment/file/505767/Distributing_Allowances_in_the_Mexican_ETS.pdf

Huesca Reynoso L, López Montes A (2016) *Carbon Tax in Mexico and Progressivity: an analytical review.* Economía Informa, Volume 398, May–June 2016, pp. 23–39. https://doi.org/10.1016/j.ecin.2016.04.003

IETA, (2018) Mexico: A market based climate policy case study. https://www.ieta.org/resources/Resources/Case_Studies_Worlds_Carbon_Markets/2018/Mexico-Case-Study-Jan2018.pdf

ICAP, (2021). International Carbon Action Partnership ETS Detailed Information, Mexico.https://icapcarbonaction.com/en/?option=com_etsmap&task=export&format=pdf&layout=list&systems[]=59

Kreibich N, Butzengeiger-Geyer S, Obergassel W, Phylipsen D, Hoch S, Michaelowa A., Friedmann V, Dransfeld B, Wang-Helmreich H. How can existing national climate policy instruments contribute to ETS development? Climate Change 11/2019. https://www.perspectives.cc/fileadmin/Publications/Kreibich_et_al._2019_-_How_can_existing_national.pdf

La Hoz et al. (2020) Emissions trading worldwide, international carbon action partnership. Berlin, disponible en: https://icapcarbonaction.com/en/?option=com_attach&task=download&id=677

Mexico carbon tax: Semarnat presentation. 22 March 2017. https://www.thepmr.org/system/files/documents/Mexico%20Carbon%20Tax_PMR_march_2017.pdf7

Partnership for Market Readiness (PMR) and International Carbon Action Partnership (ICAP) (2016) https://icapcarbonaction.com/en/?option=com_attach&task=download&id=364

Practice: A Handbook on Design and Implementation, The World Bank, Washington D. C., XX pp, disponible en: https://icapcarbonaction.com/en/?option=com_attach&task=download&id=364

Piggot G, Erickson P, Harro van Asselt, Lazarus M. (2018) Swimming upstream: Addressing fossil fuel supply under the UNFCCC. *Climate policy,* 1189–1202 doi: 10.1080/14693062.2018.1494535

PWC (2017) The Long View: How will the global economic order change by 2050? https://www.pwc.com/gx/en/world-2050/assets/pwc-the-world-in-2050-full-reportfeb-2017.pdf

Semarnat (2020) Programa sectorial de Medio Ambiente. https://www.gob.mx/profepa/acciones-y-programas/programa-sectorial-de-medio-ambiente-y-recursos-naturales-promarnat-2020-2024

Semarnat, Mexican government. http://www.gob.mx/cms/uploads/attachment/file/145709/ReporteAnual2015_Retos2016.pdf

World Bank (2016) Emissions trading in practice: A handbook on design and implementation

World Bank. Mexico overview (2020) https://www.worldbank.org/en/country/mexico/overview

Other sources consulted

BLOOMBERG GREEN; *AMLO's Clash With Business Spreads to Mexico's Renewables Sector*. Available at: https://www.bloomberg.com/news/articles/2020-05-18/amlo-s-clash-with-business-spreads-to-mexico-s-renewables-sector

Contents

Editor and Contributors

About the Editor

Simone Lucatello is a full-time researcher and professor at Instituto Mora, a Public research Centre belonging to the Mexican National Agency for Science and Technology (CONACYT) in Mexico City, Mexico.

He is one the leading authors of the IPCC, Intergovernmental Panel for Climate Change, Working Group II, which deals with the impacts, adaptation and vulnerability to climate change for the next IPCC Sixth Assessment Report (AR6, due in 2021), and coordinating leading author of the AR6 North American Chapter.

His research interests include climate change (both mitigation and adaptation) sustainability in the global south and disaster risk reduction in Latin America and the Caribbean. He served as a consultant to several international organizations such as the Inter-American Development Bank (IDB), UNEP, UNIDO, OCHA, the European Union (Europe Aid) in the Balkans, Central America and Mexico. He is member of several international academic Networks, and he is also actively engaged in international research projects across the Americas, Europe and Asia.

He holds a Master's Degree in International relations from the London School of Economics (LSE) and a Ph.D. in Governance for Sustainable development from the Venice International University, Italy. He is member of the Mexican research system (SNI), level 2.

Contributors

Mariana Marcelino Aranda SEPI-UPIICSA, Instituto Politécnico Nacional. Av. Té 950 Granjas México, Ciudad de México, México

Juan Carlos Belausteguigoitia Centro ITAM de Energía y Recursos Naturales, Ciudad de México, Mexico

Alfredo Bermudez-Contreras Autonomous University of Baja California Sur (UABCS), La Paz, Mexico

Neydi Cruz Instituto Mora-CONACYT, Mexico City, Mexico

Alejandra Elizondo Centro de Investigación y Docencia Económicas, A.C. (CIDE), Carretera México-Toluca 3655, Col. Lomas de Santa Fe., Distrito Federal, México

José Eduardo Tovar Flores Eosol Energy de México, S.A.P.I. de C.V., Mexico City, Mexico

Alicia Gutierrez González Facultad de Estudios Globales de la Universidad Anáhuac México, Estado de México, México

Danae Hernandez-Cortes School for the Future of Innovation in Society, School of Sustainability, Arizona State University, Tempe, AZ, USA

Humberto Reyes Hernandez Facultad de Ciencias Sociales Y Humanidades, Universidad Autónoma de San Luis Potosí, San Luis Potosí. S. L. P., México

Rosalía Ibarra Sarlat Institute of Legal Research (IIJ) of National Autonomous Universityof Mexico (UNAM), CDMX, México City, Mexico

Antonina Ivanova Boncheva Department of Economics and Director of APEC Studies Center, Universidad Autonoma de Baja California Sur, La Paz, Mexico

Simone Lucatello Instituto Mora/CONACYT, Mexico City, Mexico

Marcela López-Vallejo Departamento de Estudios del Pacífico (CUCSH), Centro de Estudios Sobre América del Norte, Universidad de Guadalajara, Zapopan, Jalisco, Mexico

Mireille Meneses Instituto Mora-CONACYT, Mexico City, Mexico

Catherine Nieto Moreno Centro Interdisciplinario de Investigaciones y Estudios sobre Medio Ambiente y Desarrollo, Instituto Politécnico Nacional, Ciudad de México, México

Blas L. Pérez Henríquez Precourt Institute for Energy, Stanford University, Stanford, CA, USA

María Concepción Martínez Rodríguez Centro Interdisciplinario de Investigaciones y Estudios sobre Medio Ambiente y Desarrollo, Instituto Politécnico Nacional, Ciudad de México, México

Vidal Romero Department of Political Science, ITAM, Mexico City CDMX, Mexico

Benjamin Rontard Programa Multidisciplinario de Posgrado en Ciencias Ambientales (PMPCA), Universidad Autonoma de San Luis Potosi (UASLP), San Luis Potosí, SLP, México

Erick Rosas-López Coordination for Climate Change Mitigation, Instituto Nacional de Ecología y Cambio Climático, Mexcio City, Mexico

Alberto Simpser Department of Political Science, ITAM, Mexico City CDMX, Mexico

Gustavo Sosa-Nunez Instituto Mora-CONACYT, Mexico City, Mexico

Daniela Stevens División de Estudios Internacionales, Centro de Investigación y Docencia Económicas, Ciudad de México, CDMX, Mexico

Acronyms

AAU	Assigned Amount Unit
AB32	Assembly Bill 32 (California State Law)
ADB	Asian Development Bank
AEC	Agreement on Environmental Cooperation
AfDB	African Development Bank
AFOLU	Agriculture, Forestry and Other Land Use
AGS	Afforestation Grant Scheme (New Zealand)
AIJ	Activities implemented jointly
BANCOMEXT	Mexican National Bank for Foreing Trade
BAU	Business as Usual
BMU	German Federal Ministry for the Environment, Nature Conservation and Nuclear Safety
BMV	Mexican Stock Exchange
C&T	Cap and Trade
CAR	Climate Action Reserve
CARB	California Air Resources Board
CDM	Clean Development Mechanism
CEC	Clean Energy Certificates
CEC	Commission for Environmental Cooperation between Canada, Mexico and the United States
CEL	Clean Energy Certificates (accronym in Spanish)
CEMS	Continous Emission Monitoring System
CENACE	Mexican National Center of Energy Control
CER	Certified emission reduction
CFE	Mexican Federal Electricity Commission
CH_4	Methane
CIF	Climate Investment Funds
CMA	Conference of the Parties serving as the meeting of the Parties to the Paris Agreement
CN	Black carbon
CO_2	Carbon dioxide

CO_{2eq}	Carbon dioxide equivalent
COA	Annual operating certificate
COA-WEB	Platform for reporting emissions and transfer for pollutants to obtain the annual operating certificate
CONABIO	Mexican National Commission for the Knowledge and Use of Biodiversity
CONAFOR	Mexican National Forestry Commission
CONAPO	Mexican National Population Council
CONAVI	Mexican National Housing Commission
CONCAMIN	National Chamber of the Transformation Industries
COP	UN Climate Change Conference of the Parties
CORSIA	Carbon Offsetting and Reduction Scheme for International Aviation
COVID19	Coronavirus disease 2019
CPEUM	Political Constitution of the United Mexican States
CPLC	Carbon Pricing Leadership Coalition
CPPs	Carbon Pricing Policies
DAC	Development Assistance Committee
DOE	Designated Operational Entity accredited by CDM Executive Board
DOF	Official Gazette of the Federation
EBRD	European Bank for Reconstruction and Development
ECLAC	United Nations Economic Commission for Latin America and the Caribbean
EDF	Environmental Defense Fund
EIB	European Investment Bank
ENAREDD+	Mexican National Strategy for Reducing Emissions from Deforestation and forest Degradation
ENCC	Mexican National Climate Change Strategy
ERUs	Emission Reduction Units
ETS	Emissions Trading System
EU ETS	European Union Emissions Trading System
EU	European Union
FFA	Farm Forestry Association (New Zealand)
FOA	Forest Owners Association (New Zealand)
FOMECAR	Mexican Carbon Fund
GDP	Gross domestic product
GEEREF	EU Global Energy Efficiency and Renewable Energy Fund
GHG	Greenhouse gases
GIZ	Deutsche Gesellschaft für Internationale Zusammenarbeit
GS	Gold Standard
HFC	Hydrofluorocarbons
HIS	High-interest stakeholders
HPS	High-power stakeholders
IADB	Interamerican Development Bank

ICAO	International Civil Aviation Organization
ICAP	International Carbon Action Partership
IDB	Inter-American Development Bank
IEPS	Mexican Special Tax upon Services and Production
IETA	International Emissions Trading Association
IFAD	International Fund for Agricultural Development
IFC	International Finance Corporation
IMF	International Monetary Fund
INDC	Intended Nationally Determined Contribution
INECC	Mexican National Institute of Ecology and Climate Change
INEGI	Mexican National Institute of Statistics and Geography
IPCC	Intergovernmental Panel on Climate Change
ITMO	Internationally transferred mitigation outcome
JI	Joint Implementation
KP	Kyoto Protocol
LCFS	Low carbon fuel standard (California)
LDC	Least Developed Countries
LET	Mexican Law for Energy Transition
LGCC	Mexican General Law on Climate Change
LGEEPA	Mexican General Law of Ecological Balance and Protection of the Environment
LPG	Liquefied petroleum gas
LRF	Linear Reduction Factor
LULUCF	Land Use, Land Use Change and Forestry
MDB	Multilateral Development Bank
MDG	Millennium Development Goals
MOP	Members of the Protocol (Kyoto Protocol)
MoU	Memorandum of understanding
MRV	Monitoring, reporting and verification
MSR	Market Stability Reserve
MW	Megawatt
N_2O	Nitrous oxide
NACLD	American Climate Leadership Dialogue
NAMA	National Appropriate Mitigation Actions
NCS	Natural climate solutions
NDC	Nationally determined contributions
NGO	Non-governmental organization
NO_2	Nitrogen dioxide
NO_x	Nitrogen oxides
NPA	Natural protected areas
NZ ETS	New Zealand Emissions Trading System
ODA	Official Development Assistance
OECD	Organisation for Economic Co-operation and Development
PECC	Mexican Special Plan for Climate Change
PEMEX	Petróleos Mexicanos, state-owned oil company

PFC	Perfluorocarbons
PFSI	Permanent Forest Sink Initiative
PM	Particulate matter
$PM_{2.5}$	Particulate matter with a diameter less than 2.5 micrometres
PMI	Partnership for Market Implementation
PMR	Partnership for Market Readiness
PRI	Mexican Institutional Revolutionary Party
PROFEPA	Mexican Federal Attorney for Environmental Protection
PSA	Payment for environmental services
R&D	Research and development
REDD+	Reducing Emissions from Deforestation and Forest Degradation
RENE	Mexican National Emissions Registry
RETC	Mexican Emissions and Transfer of Pollutants Registry
RGGI	Regional Greenhouse Gas Initiative
RPS	Renewable Portfolio Standard of California
SCC	Social Cost of Carbon
SDG	Sustainable Development Goals
SDM	Sustainable Development Mechanism
SEDESOL	Mexican Ministry of Inclusion and Social Welfare
SEDUE	Mexican Ministry of Ecology and Urban Development
SEMARNAP	Mexican Ministry of the Environment, Natural Resources and Fisheries
SEMARNAT	Mexican Ministry of Environment and Natural Resources
SENER	Mexican Ministry of Energy
SF6	Sulfur hexafluoride
SHCP	Mexican Ministry of Finance and Public Credit
SiCEM	GIZ Project "Preparation of an Emissions Trading System in Mexico"
SO_2	Sulfur dioxide
SO_x	Sulfur oxides
SPCC	Special Program of Climate Change
UN	United Nations
UNCTAD	United Nations Conference on Trade and Development
UNDP	United Nations Development Programme
UNFCCC	United Nations Framework Convention on Climate Change
UNWCED	United Nations World Commission on Environment and Development
US	United States
USD	US dollar
USMCA	United States-Mexico-Canada Free Trade Agreement
VCS	Verified Carbon Standard
VER	Verified Emissions Reduction
WB	World Bank
WBCSD	World Business Council for Sustainable Development
WCI	Western Climate Initiative

Part I
Emissions Trading and Mexican Climate Policy: National and International Perspectives

Chapter 1
Key Theoretical, Policy, and Implementation Experience Considerations for the Mexican ETS: Toward an Equitable and Cost-Effective Compliance Phase

Blas L. Pérez Henríquez

Abstract This chapter presents a brief overview of the policy design and theoretical environmental economic principles that underpin the concept of emissions trading systems (ETS) as a policy approach to address climate change. It discusses basic environmental economic principles pertinent to the development of market-based solutions to mitigate greenhouse gas (GHG) and co-pollutants. The chapter serves as the technical basis for the broader discussion that this book as a whole presents on the launch of the pilot phase of the Mexican ETS on January 1, 2020. Understanding international program design experiences, theoretical principles, and implementing best practices is key to ensuring Mexico's success in the transition from the pilot or learning phase to an operational ETS compliance system. This will ensure Mexico fulfills its national climate policy goals and nationally determined contributions (NDC) under the Paris Agreement in a cost-effective manner, while also providing compliance flexibility to the industrial sectors covered under the program. A well-designed ETS ultimately provides the right incentives for industrial carbon emission reductions to drive cost-effective abatement and clean innovation. Secondly, this chapter presents a more in-depth review of policy developments focusing specifically on key implementation lessons from the two most advanced ETS systems in operation to date: (1) the European Union ETS and (2) California's cap-and-trade program. In short, this chapter outlines a set of key policy lessons and design parameters to support the transition from the pilot Mexican ETS to an operational compliance phase in a socially just, environmentally sound, and cost-effective manner.

Keywords Mexico climate policy · Emissions trading · Carbon market · Policy design

B. L. Pérez Henríquez (✉)
Precourt Institute for Energy, Stanford University, Stanford | Mexico Clean Economy 2050, 473 Via Ortega, Suite 324, Stanford, CA, USA
e-mail: blph@stanford.edu

© The Author(s) 2022
S. Lucatello (ed.), *Towards an Emissions Trading System in Mexico: Rationale, Design and Connections With the Global Climate Agenda*, Springer Climate,
https://doi.org/10.1007/978-3-030-82759-5_1

Environmental Regulation: Conventional and Alternative Market-Based Approaches

Conventional Environmental Regulation

The most common practice in environmental regulation is to impose a limit or quantity control on pollution, which provides certainty about the policy objective but does not create incentives to reduce total demand. This approach is also known as command-and-control regulation. This non-market approach traditionally sets a maximum allowable emissions quantity (e.g., standard) for easily identifiable point sources (e.g., smokestack). The regulator can also prescribe installing specific abatement technology (e.g., scrubbers). Because the cost of reducing emissions varies among sources, a one-size-fits-all governmental rule (or command) to control air pollution emissions, while effective, is in practice an economically inefficient approach. In addition, by picking and mandating the installation of specific cleaning technologies, regulators limit the innovation cycle and disincentivize the use of ingenuity to introduce novel production processes (i.e., clean innovation) or managerial solutions to achieve emission reductions at a lower cost to the emitter.

Direct regulations for environmental control require high levels of enforcement and inspection. The centralized standard-setting process is conducted by regulators with little knowledge of the universe of production systems on site and across industrial supply chains to be covered by the new regulation. Moreover, regulators lack information about the marginal cost of abatement for each regulated facility. Obtaining this information can be a costly and difficult endeavor for the regulatory agency, and industry has no incentive to provide it. The regulator wants to control pollution, businesses want to minimize regulatory costs. The asymmetry of information between regulated entities (e.g., production processes) and regulators' aim to reduce emissions creates a misalignment of incentives and adds costs to sharing information necessary to improve the quality of environmental protection programs (Tirole 1988, 4). Voluntary approaches, such as industry self-regulating systems, can have some positive effects but tend to be lax, inconsistent, and also require costly certification and public verification efforts to produce tangible environmental outcomes.

Market-Based Environmental Policy

An alternative policy approach to address pollution externalities is the use of market incentives (Hahn and Stavins 1992). Markets drive the efficient allocation of resources in the economy. The theoretical challenge is that there is no market for environmental resources or environmental services. There is incomplete information about prices for environmental goods. Economists would point to reasons such as non-exclusion, non-rival consumption, asymmetric information, and non-convexities that make assumed economic market behavior not hold as a function of supply and demand in this case (Hanley et al. 2016). Such market failures make efficient allocations of environmental services difficult to achieve. Resolving these failures has been an ongoing quest for public economic theorists.

The environmental economics paradigm calls for comparing the benefits of pollution reduction with the cost of pollution control for the regulator to set controls at the level where the marginal benefits equal the marginal costs of control (Hahn and Stavins 1992). The interdependence of ecosystems makes a full estimate of impacts elusive. In the real world, we do not have complete assessments of environmental damages, a key consideration for governance and policy design. To approximate price and make predictions, economists can elicit information through surveys to get the willingness to pay for non-market goods (i.e., contingent valuation), like a clean environment (Hanemann 1994). Prices can also be estimated through the valuation of other related goods, for instance, housing in an area with clean air, or public health expenditures (a disutility) because of air pollution, as proxy measures. These methodologies, in an indirect manner, can help economists estimate costs needed to "patch" the price system. The technological, time, and human resource requirements to consistently adapt the price to market conditions, however, makes price-setting an ongoing onerous demand on regulator resources. In short, imperfect information limits efficient outcomes. Therefore, there is a role for governmental intervention to minimize both transaction costs through careful policy design, as well as to decide on the right level of governmental supervision or control (Perez Henriquez 2013, p. 12).

Addressing global commons issues such as climate change has increasingly been on the agenda of multilateral negotiations since the mainstreaming of the concept of sustainable development at the United Nations' World Commission on Environment and Development (UNWCED) in 1987, which was aimed at avoiding compromising the natural planetary capital endowment and welfare of future generations. Crocker (1966) was the first to introduce the idea of using a market to address atmospheric pollution control systems. In 1997, the United Nations Conference on Trade and Development (UNCTAD) launched the GHG Emissions Trading Policy Forum. At the inaugural multi-stakeholder event, Maurice Strong, Chairman of the Earth Council and Secretary-General of the 1992 Earth Summit, argued that "an international market for greenhouse gases bridges the gap between the environmental objective of lowering emissions and industry's need for flexible, economical paths for achieving this objective, while encouraging new investments in sustainable development" (UNCTAD 1997). These discussions jumpstarted the debate on the use of emissions trading as a policy approach within the United Nations climate negotiations process. Comparing alternative institutional arrangements or policy approaches is key to evaluating options and achieving policy outcomes in a more efficient manner. Distributional equity is an important criterion that cannot be dismissed, particularly in the context of international climate policy (Baumol and Oates 1988).

Carbon Pricing, Markets, and Innovation

There is a high degree of certainty and consensus among the international scientific community that the heat-trapping effect of GHG emissions (or carbon) accumulation in the atmosphere due to human activities over the last 150 years is the main cause of global warming, threatening biodiversity and human existence. A price on carbon is the most practical policy solution to decarbonize our global economy. Whether through taxes (Pigou 1932) or cap-and-trade programs (Coase 1960; Dales 1968), a price on carbon transfers the social cost of climate change to emitters and requires them to choose how to address this cost. It therefore promotes emission reductions and investment in smart, clean energy solutions and low-carbon economic development. In the real world, policy analysts must consider how market-based policy instruments are constrained by political, administrative, and other institutional factors, as well as by self-interested actors (Pérez Henríquez 2013).

As suggested by Nordhaus (2017), the social cost of carbon (SCC) is a central concept for understanding and implementing climate change policies. The SCC is the monetary value of environmental damage from emitting an additional ton of carbon into the atmosphere. Avoiding these negative impacts on the economic welfare of nations makes carbon mitigation and adaptation investments worthwhile. For example, if you are a country highly vulnerable to climate stressors and risk as in the case of Mexico, water scarcity in surface and groundwater recharge from extended droughts along with reduced hydropower generation could negatively impact the delicate water-energy-food nexus balance of these interdependent shared-resource systems. Further, hurricanes and superstorms are expected to affect critical infrastructure (e.g., power stations and dams), disrupting housing, physical and electronic communications, as well as supply chains and economic activity in general due to more frequent, damaging, and costly extreme weather events (United States Agency for International Development 2017).

As noted in Burke et al. (2016), researchers have dramatically advanced our understanding of the physical science of climate change. Implications of this knowledge for society remain limited, but some progress has been achieved in formalizing climate-economy linkages. Refining the SCC and enhancing our capacity to assess the economic impacts of alternative policy approaches, in particular for developing countries, are current topics in the environmental economics research agenda. Emitting one unit of carbon, in a city, industrial park, or by wildfires, contributes the same to climate change regardless of where it is emitted. Consequently, the social damage caused is the same. Stern (2007) provoked a debate on how best to estimate the benefits of carbon emission mitigation in order to avoid costly climatic effects in the future. Dietz and Stern (2008) later emphasized that adaptation investment strategies play a key role in minimizing the costs and maximizing the benefits of planetary warming. Both carbon pricing mechanisms as described below create a new source of revenue for governments. The revenue raised can reduce the social cost of these programs, for instance, by eliminating distortionary taxes (Goulder 1995). However, the use of revenue may be controversial and politically contentious because of its potential impact on the overall cost of market-based approaches and their distributional effects. Revenue can go, for instance, to the general government treasury, or support environmental and social remediation, or climate investment and low-carbon development initiatives and adaptation projects.

Despite calls for urgent climate action from scientists, environmental groups, and young voices of the next generation, the Paris Agreement mitigation and climate action plans need enhanced ambition and are behind schedule to stabilize emissions and avoid catastrophic warming by the end of this century. Moreover, the levels of climate investment funds necessary for a transition to a global clean economy, are not readily available. While in the short term, policy makers can reap low-hanging fruit benefits from stated governmental mitigation programs (i.e., NDCs) and supplementary actions, long-term goals of transitioning to a net-zero carbon future by 2050 will require deeper, more costly carbon emission reductions. Some industrial sectors have a heavier lift ahead in their decarbonization process (e.g., hydrocarbon industry, cement, steel, and aluminum). Fossil fuel use continues to grow at the global level, in particular in large developing countries, including Mexico. To effectively address this situation, scientists and expert research groups tracking global carbon emissions such as the Global Carbon Project state that "we need accelerated energy efficiency improvements and reduced consumption, rapid deployment of electric vehicles, carbon capture and storage technologies, and a decarbonized electricity grid, with new renewable capacities replacing fossil fuels, not supplementing them. Stronger global commitments and carbon pricing would help implement such policies at scale and in time" (Jackson et al. 2019). Moreover, climate investment in innovative carbon management technologies can provide cost-effective solutions in the long run (Gillingham 2019).

Experience shows that meaningful carbon prices create a strong business incentive to invest in low-carbon technology and clean innovation. As reported by the World Economic Forum (2017) based on the European carbon market experience, at certain carbon price levels, companies will file more patents in the new clean economy sectors described above. Recent policy fine-tuning of the EU ETS is delivering enhanced incentives for innovation and adoption of low-carbon technologies (Teixidó et al. 2019). Accelerating and de-risking deployment is central to this process. How we choosing to control carbon is key to achieve the desired policy outcomes (Goulder and Parry 2008). Theoretically, under a full-information scenario, tradable allowances (or quotas) and taxes are equivalent regulatory instruments (Montgomery 1972; Weitzman 1974). However, as established above, this is not the case in practice given the uncertain level of environmental damages. Hybrid policy designs with overlapping policies may be appropriate for cost-effective and equitable climate action, but this requires a high degree of regulatory coordination by government (Pérez Henríquez 2013).

The two main carbon pricing mechanisms are as follows.

Carbon Tax

From the environmental policy implementation perspective, the key aspect of a carbon tax is that it provides price certainty to regulated entities. This is why this approach is generally preferred by industry. A carbon tax creates a uniform price on emissions irrespective of the source. This allows the cost to regulated entities of reducing their emissions by one unit at the margin as defined by economists to be equalized across all facilities and sectors of the economy, thus making it technically appealing (Pizer 1999; Nordhaus 2005).

However, a significant concern, particularly for some environmental groups, is that there is no certainty on the amount of emissions reduced in a certain period. While considered by many a straightforward proposition, the technically elegant global carbon tax has also not yet passed the political feasibility test across most national and regional policy processes around the world. Passing a new, significant green tax is a tough sell for any politician. As reported by *The Guardian*, the French government's ongoing experience with the *Gilets Jaunes* or "yellow vests" movement against eco-taxes in the transportation sector, with support of 70% of the general population, illustrates this point (Willsher 2018).This is despite France's strong support for the Paris Agreement, its own "ecological transition" national objectives, and its overall global leadership on climate policy. In Europe, in general, "pollution and resource taxes account for a very small portion of the tax revenue" vis-à-vis energy taxes (Willsher 2018). Instead of a carbon tax, therefore, an ETS was the system that the European Union adopted to address GHG emission reductions.

Likewise, several carbon legislation proposals have been in wait to survive the policy process in the United States Congress, but without much potential to become federal policy. However, the significant federal revenue-raising potential from a carbon tax is often cited as an incentive to decisionmakers to favor the passing of carbon tax legislation. This may become a salient political feasibility consideration, particularly in the face of the potential budgetary shortfalls that the Covid19 public health crisis may create across all levels of government in the near future.

Emissions Trading Systems

The key policy aspect of implementing a cap-and-trade approach is that it establishes a clear environmental goal. This type of ETS ideally sets a science-based limit, but in practice, a politically agreed-upon maximum allowable limit on emissions from all program participants by a central agent (i.e., regulator), and grants flexibility to program participants to search for the most cost-effective emission reductions through investments in pollution removal systems, technological innovation, or by applying managerial ingenuity to mitigate emissions. Some environmental groups tend to favor this approach because it focuses on achieving a clearly defined environmental policy objective (i.e., the cap).

In a cap-and-trade ETS, regulated entities are required to meet a universal target (or cap) on all emissions in the economy or in a specific sector (e.g., electricity generation), but without directly prescribing how to achieve the emission reductions (Tietenberg 2005, 2006). Emission permits or allowances (quasi-legal property rights) based on the total cap are allocated among participants (or purchased through annual auctions). Those who can reduce emissions in the most cost-effective manner will have a surplus of emission credits to sell in such an environmental commodity market. Those with relatively more expensive mitigation costs will have to buy a number of permits needed to meet their annual quota and compensate for added pollution. Cost minimization opportunities in a market-based system arise as the differences in the cost of pollution abatement across sources increases (Newell and Stavins 2003). GHG sources are diverse and heterogeneous. Thus, heterogeneity and market scale determine the amount of such opportunities, while the system's gains in economic efficiency are enhanced as the size of the ETS expands, particularly across jurisdictions regionally or at a global level.

Approaches focused on setting maximum caps on targeted emissions with allowance trading provide incentives to mitigate emissions per unit of production output. The theoretical promise is that these policy approaches are relatively cost-effective to both the regulator and the emitter. Prices are not set by the regulator beyond potential floors and ceilings, but are determined by the relationship between the supply of excess emission credits from those entities that emit less than the cap, and the demand for credits from those entities that have not met their caps. The financial and technical context of each regulated entity determines whether they decide to invest in improved emission controls, fostering technological innovation and novel managerial approaches, or in purchasing excess emission allowances from those entities that have exceeded their requirements.

From the perspective of the environmental administrator, theoretically, this frees the public regulator from having to track and analyze all of this exchange data. The necessary caveats to this approach, however, include the need to pay close attention to distribution changes in the policy's environmental benefits, as it can potentially exacerbate existing public health disparities if the heaviest polluters decide to purchase credits instead of reducing emissions when they are located in already over-polluted areas. Additionally, the market has to be closely monitored for manipulation, particularly if the underlying regulated entities are part of a sector with monopolistic characteristics (electricity distribution, etc.). However, modern information and communication technologies, through sensors, mobile monitoring, continuous emission monitoring systems (CEMS), and online operation tracking, reduce administrative and transactional costs in the implementation and oversight of environmental commodity markets (Pérez Henríquez 2004).

More than 30 years of implementation experience with cap-and-trade programs offers a wealth of lessons on how economic efficiency was gradually achieved through policy design improvement and fine-tuning. This experience has also been informing the process of developing new ETS programs around the world to address climate change (Schmalensee and Stavins 2017). However, a smart policy approach can easily become inefficient and ineffective if its design and implementation processes are compromised or if it lacks the flexibility to iteratively adapt to evolving political, economic, environmental, and social conditions. Political concessions during the policy process and exogenous factors can diminish the economic efficiency of an ETS (Perez Henriquez 2013).

There are several lessons from this implementation history with cap-and-trade systems that are key to the future success of growing carbon markets and emerging ETS policies.

Cap: Perhaps the most significant lesson to date is the importance of establishing a well-defined, transparent, and sufficiently strict initial emissions cap. Such a cap will ensure both the environmental and market credibility of the program and, through scarcity, trigger the necessary demand for emission credits to foster an emissions market. Accurate *ex ante* emission data is required to establish a cap that is stringent enough to produce scarcity in the market at the start (with the expectation of increasing ambition as needed in the future).

Carbon pricing and policy continuity: A meaningful allowance price (and/or price floor), along with a credible government commitment to long-term carbon market policy continuity toward a zero-emission future, creates confidence in the effectiveness and legal certainty of emissions trading markets.

Allocation: (1) Direct: Allowance distribution is provided directly to regulated entities, utility rate payers, and other special sectors to smooth the transition to a low-carbon economy, as well as to minimize emission leakage (i.e., emission increases outside of the ETS-covered industry or jurisdiction) while maintaining local production. Free allowance allocations enhance the political feasibility of the program at the onset and support economic development and competitiveness. (2) Auctions: Alternatively, auction sales for all market participants provide access to allowances. Emission auctions jumpstart the market through price signals. They also provide new industry entrants with allowances and allow ETS participants in general to buy permits as they plan ahead in balancing their production goal with environmental compliance.

Climate investment: As mentioned above, carbon pricing can also provide revenue for climate investment (e.g., allowance auctions, carbon price floors, and carbon taxes), foster low-carbon development, or provide direct dividends to society from the decarbonization effort.

Cost Containment and Competitiveness: Cost containment measures such as credits outside the cap (e.g., carbon offsets from an array of approved projects), as well as price caps and allowance reserves for market stabilization, can also provide flexibility and price certainty to maintain jurisdictional economic competitiveness during the decarbonization process.

Penalties: Meaningful penalties for failure to meet annual emission reduction goals—through emission reduction or purchase of allowances—are also key to making the system effective.

Emission clustering and environmental justice: Air pollutants produced along with GHGs from hydrocarbon combustion frequently result in illness and environmental hazards to communities living in urban areas, near industrial facilities, or transport hubs like port facilities. Localized emissions (or hotspots) exacerbate asthma and morbidity rates among at-risk populations and, in most cases, in economically and socially disadvantaged areas. Thus, climate action plans that strongly link market-based carbon mitigation with air pollution protection from co-pollutants will have more equitable, just outcomes.

Creating Global Environmental Commodity Markets

The use of carbon pricing policy by governments around the world is increasing as a means to address climate change by stabilizing GHG emissions in a cost-effective manner and limiting global temperature rise to below 2 °C within this century. The World Bank (2020) has developed a Carbon Pricing Dashboard which reports that, to date, 61 regional, national, and subnational programs (e.g., carbon tax, ETS, and hybrid systems) are operating or scheduled for implementation globally. These carbon pricing initiatives include 46 national and 32 subnational jurisdictional mitigation efforts, representing about 22.3% of global GHG emissions in 2020. The European Union ETS and the California cap-and-trade carbon market operating jointly with the Canadian province of Quebec as part of the Western Climate Initiative represent a collaborative effort toward a future North American ETS. Both regional carbon trading programs represent the state-of-the-art design and implementation of this market-based policy approach. However, important developments in other regions are in place, including ETS programs launched in China in 2017, South Korea in 2015, and Kazakhstan in 2013, representing key Asian developments toward a future global carbon market.

It is true that many countries, subnational governments, and regions around the world have introduced cap-and-trade emission trading systems to mitigate carbon emissions. However, as attested by the negotiations under the United Nations Climate Framework Convention on Change Climate (UNFCCC) Paris Agreement over Article 6, at the multilateral level, the idea of establishing a global carbon market remains contentious. Even the use of the word "market" is questioned by some stakeholders. Some environmental organizations still consider it unethical to "trade" environmental goods or to use the natural capital of indigenous communities to compensate for emissions from large industrial emitters. Other civil society and non-governmental groups, think tanks, and advocates who see the benefits of carbon pricing as a policy tool want assurances that a future global carbon market will avoid double-counting (e.g., NDC goals vs. an international offset program), forbid carryovers of units derived from older systems such as the Kyoto Protocol, and demand the inclusion of more holistic social and environmental safeguards that protect human rights, ensure effective mitigation of global emissions, and achieve sustainable development goals (SDGs) (Climate Action Network 2019). Industry lobbies want to ensure the minimization of compliance costs and achieve overall climate goals in a more efficient manner. The savings potential in the implementation of NDCs could reach ~$250 billion per year in 2030, or alternatively "facilitate the removal of 50 percent more emissions (~5 $GtCO_2$ per year), at no additional cost" (Edmonds et al. 2019). All this is contingent on the implementation of a well-designed, operational global carbon market.

UN diplomatic negotiations over Article 6 have agreed on terms such as "cooperative approaches" to develop internationally transferred mitigation outcomes (ITMOS) in order to avoid using the term "carbon market". Under these future rules, countries would allow the exchange of ITMOS and linkages among systems. Decisions on setting the rules for such a system at the multilateral level are highly contentious. Heterogeneity in policy objectives and approaches among national, regional, and local governments and their interaction with the NDC system has become a hurdle to advancing rules and aspects of implementation. Demonstrating that credible and technically sound convergence in accounting, environmental integrity, and transparency policy design features is feasible between different national approaches among treaty signatories can be the first step. Such experiences can provide valuable implementation lessons that can serve as trust-building foundations to support cost-effective global collective climate action while at the same time providing policy design protocols for others to formally join these efforts. A well-functioning global carbon market could allow for cost-savings to be invested in the transition to a clean economy and to foster low-carbon development around the world.

According to the International Carbon Action Partnership (ICAP) in its most recent report on ETS developments around the world (2020), one-sixth of the global population lives under an ETS. These jurisdictions represent 42% of global gross domestic product (GDP)—up from 37% a year ago—and the systems cover 9% of GHG emissions worldwide. By 2021, ICAP estimates that 14% of global emissions will come under an ETS as more systems come online, including China. Launched in 2017, the China ETS will be approximately twice the size of the EU ETS and almost nine times the size of the California cap-and-trade system making it the largest in the world (Stavins 2018). Carbon market implementation lessons will soon emerge from Asian manufacturing and export-oriented peers to Mexico. However, at this time, the two most pertinent experiences to date that can inform Mexico's ETS implementation process are the European Union (EU) ETS and the California experience.

EU ETS

The first full-fledged multi-national ETS developed to address climate change in the world was the European Union ETS. In tandem with a series of supplementary measures and programs, this cap-and-trade system was introduced by the EU Commission as its main policy for GHG mitigation in the region. It covers approximately 45% of EU emissions from the power, industrial, and internal aviation sectors. More recently, the European Green Deal has set an ambitious goal for EU members to become climate neutral by 2050. For instance, it aims to reach clean energy and energy efficiency levels of 32 and 32.5%, respectively, by 2030, and periodically revises these goals upwards. An important effort to strengthen the EU ETS is to improve the integration of monitoring, reporting, and verification (MRV) rules to meet its own policy objectives and international commitments under the Paris Agreement. The EU ETS is entering into its fourth phase and implementing its most recent fine-tuning adjustment. Unfortunately, the UK, where the program was first piloted for the benefit of the EU, formally retired from this association in 2019. There is some market uncertainty triggered by this process. However, in January 2020, the EU ETS successfully linked up with the Swiss carbon market.

As noted above, the EU ETS is not an economy-wide cap-and-trade system. It has been gradually implemented, under a sector-based approach focusing on already highly regulated, energy-intensive sectors such as electric power generators and large, stationary industrial sources. Coverage has been expanding throughout the years. Initially, it allowed member nations to develop internal capacities and inform the system independently. However, the EU experience demonstrates the importance of a central coordinating agent that ensures the environmental integrity of the system.

Launched in 2005, the EU ETS has faced some significant learning experiences such as over-allocation of allowances and exogenous shocks such as the deacceleration of the economy due to the 2008 financial crisis. As a result of over-allocation and reduced demand, the EU ETS found it necessary to introduce a cap adjustment to trigger some scarcity of allowances in the market through a policy known as backloading. This temporarily reduced the auction volumes during the 2014–2016 period. The amount of reduced allowances (900 million total) was reintroduced into the system in 2019 and 2020. This caused price instability and some strategic market speculation with these environmental commodities. There was also an oversupply of offsets, which forced the EU to close the door to such cost-containment instruments, de facto halting the clean development mechanism process. The size of the EU ETS and insufficient cybersecurity for emissions registries produced large fraud and tax avoidance schemes. After adapting to these challenges through fine-tuning and learning by doing, the EU ETS has managed to gain some relative price stability while helping to decarbonize the EU region.

Phase IV of the EU ETS is currently setting the new market rules for the next decade (i.e., post 2020). According to the European Commission (2017), in order "to achieve the EU's overall greenhouse gas emissions reduction target for 2030, the sectors covered by the EU Emissions Trading System (EU ETS) must reduce their emissions by 43% compared to 2005 levels. The revised EU ETS Directive, which will apply for the 2021–2030 period, will enable this through a mix of interlinked measures." Below is a summary of key adjustments to the program:

Adjusting the Cap: A key factor signaling to market participants that there will be gradual reductions in the availability of allowances is the "Linear Reduction Factor (LRF)" that adjusts the EU ETS cap starting in 2021 to a 2.2% annual decrease in total allowances from the previous 1.74%. This gradual cap reduction will result in 43% fewer allowances available in the market by 2030 compared to the start of the program in 2005. Free allocations will now target emission intensity improvement over absolute mitigation, and will have a schedule of two allocation periods—2021–2025 and 2026–2030. Allocations will be based on real data in the process of collection using a benchmarking system set by the 10% best-performing facilities in each sector. A new measure to avoid over-allocation is to provide more flexibility in addressing capacity changes or activity levels, stopping operations, mergers and acquisitions, etc. based on a certain percentage of reduced or increased operations to determine free allocation adjustments.

Market Stability Reserve (MSR): The EU established the MSR to reduce the surplus of emission allowances in the system and improve EU ETS resilience to future market shocks. To strengthen its function, "between 2019 and 2023, the amount of allowances put in the reserve will double to 24% of the allowances in circulation. The regular feeding rate of 12% will be restored as of 2024. As a long-term measure to improve the functioning of the EU ETS, and unless otherwise decided in the first review of the MSR in 2021, from 2023 onwards the number of allowances held in the reserve will be limited to the auction volume of the previous year. Holdings above that amount will lose their validity." (European Commission, 2017)

Carbon Leakage Risk: The system of free allocation will be prolonged for another decade and has been revised to focus on sectors at the highest risk of relocating their production outside of the EU. These sectors will receive 100% of their allocation for free. For less-exposed sectors, free allocation is foreseen to be phased out after 2026 from a maximum of 30% to 0 at the end of phase 4 (2030). A considerable number of free allowances will be set aside for new and growing installations. This number consists of allowances that were not allocated from the total amount available for free allocation by the end of phase 3 (2020) and 200 million allowances from the MSR. A series of additional flexibility rules aim to better align the free allocation process with actual production levels while minimizing carbon leakage risk.

Climate Investment: The EU ETS has decided to fund directly from their auctions in order to "de-risk" innovation projects and accelerate decarbonization in the region while supporting research to market projects. Through the Innovation Fund, the EU will fund demonstration projects of cutting-edge technology both on a large and small scales. This could include novel clean energy systems, energy storage, and a boost to carbon capturing, utilization, and storage projects. The EU ETS revenue will also provide for a modernization fund that will provide additional funding to continued efforts to modernize energy system transformation in the 10 lowest income EU members. No funding will be directed to coal-fired electric generation and minimal finance will be provided to improve natural gas power plants.

Offsets: Offsets will continue to be prohibited under Phase IV to maintain a strict cap on regional emissions. However, the EU Commission has signaled that once an agreement on Article 6 of the Paris Agreement is reached, ITMOs will be part of the system. No Kyoto offset instruments will be allowed in the EU ETS beyond 2030.

Perhaps the most important lesson from the implementation of the EU ETS is that addressing the climate challenge requires an array of policies and measures

(Delbeke and Vis 2019). However, overlapping policies can impact the environmental and economic performance of these programs. An ETS relies on the market to identify the least-cost carbon mitigation opportunities. Complementary policies can disrupt these market incentives. Careful policy design, implementation, and review to balance policy objectives and other co-benefits are required. The trade-offs between policies and objectives need to continuously be reviewed to improve the overall cost-effectiveness, environmental integrity, and equitable implementation of the program toward a carbon–neutral economy. This is particularly relevant in the face of global trends toward deep decarbonization efforts to achieve net-zero emission transformation of key economic sectors such as energy and transport systems.

North American Developments

In North America, California and Quebec have taken the lead implementing ETS programs, and developing workable linking protocols for both parties. Both subnational jurisdictions see the electrification of the economy as a key infrastructure transformation challenge that, when supported by clean innovation solutions, will enable a net-zero carbon future. Quebec is strengthening its institutional setup around climate by creating a climate change advisory committee and restructuring its climate investment program through a revamped Green Fund. Both subnational jurisdictions have energy sectors with a high supply of renewable power.

Political cycles have affected the level of support for carbon pricing policy in North America. Both the US and Canada have been in and out of the UNFCCC climate process. Canada withdrew from the Kyoto Protocol in December 2012 but is now an active member of the Paris Agreement. Under the Trudeau premiership, Canada returned to action, introducing a national carbon price floor and allowing provinces to choose their own policy approach to meet Canada's climate objectives. Under the Canadian federal backstop system, provincial governments can opt between (1) An output-based pricing system and (2) A carbon levy (or tax). This applies to provinces and territories having no carbon pricing systems validated by the central government in place. In 2016, this pan-Canadian carbon pricing system was introduced and set a minimum price of CAN $10 per ton of carbon in 2018, increasing to CAN $50 in 2022. Alternatively, provinces may choose option 1 and establish an equivalently scaled ETS. Experts believe welfare and implementation costs are "manageable" and the system will create an important source of revenue (Parry and Mylonas 2017). Carbon pricing in Canada has been an evolutionary process. For instance, since 2008 British Columbia has demonstrated that it can use a neutral carbon tax, granting carbon rebates and spurring energy-efficient processes without slowing down economic performance. The principle of neutrality applies to the revenue raised from the central backstop system, as it is designed to be returned to those sectors and households that paid the charge. On the other hand, as mentioned above, elections matter. At one point, Ontario was in line to link with California and Quebec's carbon market, but a political cycle canceled that process on July 3, 2018, as a conservative government opposing the policy got elected. Constitutional challenges by some provinces resisting the implementation of the backstop system are being litigated in courts.

In the United States, climate action is being mostly led by local governments and communities around the country. Under the Obama administration, much support was given to global climate action. However, meaningful federal climate laws to support executive actions (e.g., Clean Power Plan) have been dismantled by the Trump administration. Delivering on a campaign promise, on January 1, 2017, the Trump administration announced it would withdraw from the Paris Agreement.

In the case of Mexico, its government has been an active participant in the UNFCCC process. As host of the COP13 in Cancun, its diplomatic skill and political leadership at that summit were key to boosting the weakened multilateral process. A member of the OECD and the G20 multi-national groupings, it has historically had access to and participates in high-level discussions about how to best address climate change while advancing its sustainable development objectives. In 2012, Mexico became the first developing country, and second in the world after the United Kingdom, to pass a federal climate bill—the General Law of Climate Change (LGCC). Also, in 2014, it passed a carbon tax at a very low level on fossil fuels with the intention of raising awareness on carbon emissions, but with limited effect on emission reductions. However, this prompted the debate around the use of compensatory measures through emission offsets, mainly by the private sector. In January 2020, the pilot phase of the Mexican ETS was launched. If fully implemented, it will become the only federal ETS in North America, and the first national cap-and-trade system in the Americas. This process occurs at the start of a new presidential administration with a different perspective in terms of decarbonization, and with new directions for the energy sector that emphasize hydrocarbons, which will make the implementation of Mexico's climate action plan challenging.

California

Climate risk is a reality for all Californians. Like Mexico, California is highly vulnerable to global warming and its economic impacts as demonstrated by recent periods of long-term droughts, followed by superstorms and destructive firestorms. California is recognized as a global subnational leader in climate action contributing to the global efforts under the Paris Agreement. In 2006, the trailblazing Global Warming Solutions Act or AB32 was passed into law. While not a country party to the UNFCCC process, California represents the fifth largest economy in the world. Californians have been able to sustain economic growth while becoming less carbon-intensive, as seen in Fig. 1.1 below. This has demonstrated that clean tech innovation can be an engine for economic growth while transitioning to a smarter, inclusive, resilient, and cleaner economic development model.

With no climate policy coming out of Washington, DC, the State has taken the lead on climate action in the United States. This has not occurred without contention

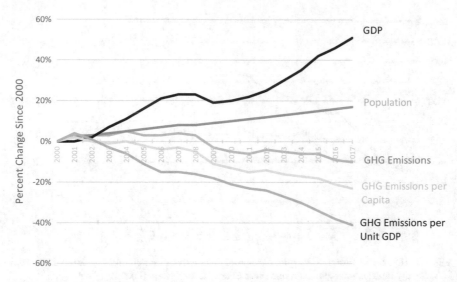

Fig. 1.1 Decoupled GHG emissions, GDP, and population trends in California from 2000 to 2017. Data from CARB (2019)

in recent years. While in the past, California had been the darling of federal environmental policy makers, winning regulatory exceptions based on its record of developing innovative environmental programs, more recently the state has been challenged by the Trump administration. Trump's administration has sued California because of its international cooperation with the Canadian province of Quebec in linking carbon markets, as well as for de facto setting national fuel efficiency standards to vehicles, among many other legal disputes. Under the Trump administration, as summarized by *National Geographic*, there has been an unwinding of important federal environmental regulations across the board to reduce regulation and control regulatory costs (Gibbens 2019).

AB 32 initially set a GHG emission target to reduce emissions to 1990 levels by 2020. California surpassed its 2020 climate targets in 2016 (see Fig. 1.4 below), while the economy continues to grow. Through its Climate Change Scoping Plan, the state periodically outlines and updates a series of programs and policies to decarbonize its economy and move toward a clean economy. These include reducing the carbon footprint of the state through carbon pricing, resource efficiency, and the deployment of clean energy and sustainable transport solutions. This portfolio approach aims both at addressing global climate change and state adaptation needs, while also ensuring that all Californians are able to enjoy their rights to clean air, clean water, and a healthy and safe environment.

Among the most innovative sectoral programs implemented in California to become a resource-efficient, clean, and climate responsible economy are the following.

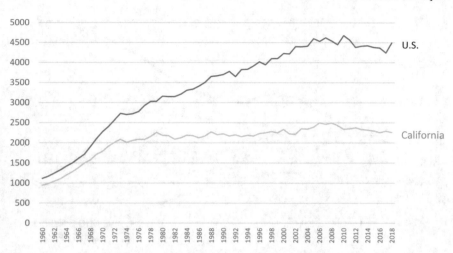

Fig. 1.2 U.S. and California per capita residential electricity consumption 1960–2018. Data from U.S. Energy Information Administration State Energy Data System (2020a)

Energy efficiency standards

California launched Appliance Energy Efficiency Standards in 1977 and the Building Energy Efficiency Standards in 1978 in part to respond to the energy crisis of the 1970s. These standards have been credited as a significant reason why California was able to delink its continued economic growth from growth in energy demand, although rising fossil fuel prices from 1973 to 1981 likely triggered efficiency innovation beyond regulatory requirements. Figure 1.2 below shows how California has been able to become an energy resource-efficient economy vis-à-vis the United States in general.

Renewable portfolio standards (RPS)

California instituted an RPS in 2002 under Senate Bill 1078 with the original aim of sourcing 20% of retail electricity sold from renewable sources by 2017. This was increased to 50% of electricity sales by 2030 under SB 250 in 2015. SB 100, signed into law in 2018, increased this requirement to 60% by 2030 and 100% by 2045 and specified the renewable sources must be carbon-free. By 2017, a majority of retail sellers of electricity had met or exceeded their updated interim 2017 target of 27% renewable sourcing (Figs. 1.3, 1.4).

Low-carbon fuel standards (LCFS)

In 2002, with the implementation of AB 1493, known as the Pavley Regulations, California instituted corporate average fuel economy standards stricter than those of the U.S. EPA. After an extended battle with automakers and trade associations, and eventually negotiations between California, the federal government and automakers, the EPA granted California a pre-emption waiver under the Federal Clean Air Act in 2009, which would allow California to set higher standards. Because California is

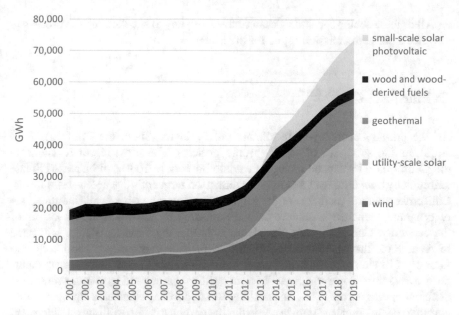

Fig. 1.3 Renewable electricity generation in California by resource type. Data from USEIA (2020b)

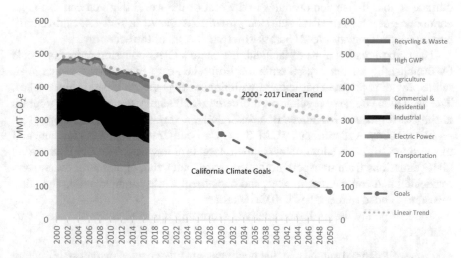

Fig. 1.4 California GHG emissions 2000–2017, linear trend to 2050, and climate goals. Data from CARB (2019)

the single largest market for automobiles in the U.S., California's standards became the de facto national standards, as automakers all retooled their production lines to meet these higher standards. However, in 2019, the Trump administration announced that it would revoke California's exemption.

All of these sector-specific measures have been key to accelerating the process to advance a net-zero emission future for the State.

California ETS

As the primary economy-wide climate policy in the State, the California ETS augments the successes of sector-specific policies. The most recent revisions to the California ETS expand its legal mandate horizon to 2030 and increase climate ambition by now aiming at an emission reduction target of 40% below 1990 levels. California's carbon market is at the center of a comprehensive, well-integrated set of programs launched to achieve carbon neutrality by 2045 across key sectors of the economy. One contingent consideration is that the State legislature will need to reauthorize the cap-and-trade program to provide continuity beyond 2030. So far, one of its strengths has been trans-administrational political commitment continuity across three governors from different political parties and legislature control by different parties across time. This continuous support has signaled a strong credible commitment to climate action, in particular the carbon market, to all stakeholders and economic agents in California and elsewhere. Californians in general support strong climate action. It also demonstrates to the rest of the world that you can decouple economic growth from GHG emission growth. As shown in Fig. 1.4, mid-century carbon neutrality goals are set across different sectors of the economy.

The carbon market aims to facilitate the State's GHG emission reduction goals by establishing a strict overall emission limit (i.e., cap) that decreases each year, while providing program participants with flexibility in their mitigation approaches. The agency in charge of regulatory enforcement and administration of this program is the California Air Resources Board (CARB), a very powerful and independent regulatory body. As described by CARB, the Cap-and-Trade Program is fundamental to meeting California's long-range climate targets at low cost. The CA ETS covers GHG emissions from transportation, electricity, industrial, agricultural, waste, and residential and commercial sources, and caps them while complementing the other measures needed to meet the 2030 GHG target.

The key policy design lessons derived from this implementation to date are as follows:

(1)   A well-defined annual cap that declines annually ensures market stability and investor certainty.
(2)   Robust measurement/monitoring, reporting and verification systems (MRV) based on accurate emission data ensures effective tracking of compliance and market function.
(3)   Auctions and a hybrid price collar system (i.e., price floor, akin to a carbon tax when market prices are low, and a price ceiling, akin to a carbon subsidy when market prices are high) address potential market volatility.

(4)   A strong regulatory link with existing air quality standards for criteria (e.g., NOx, SOx, and volatile organic components) and toxic air pollutants protects against increases in distributional inequities in GHG co-pollutant exposures as a result of the ETS, supported by strong default penalties under the State's health and safety code.

(5)   High penalties for non-compliance.

Additional supplementary considerations

Climate Investment Program and Equity: The revenue generated by auctions of allowances from the California ETS provides direct financial benefits to electricity rate payers (i.e., climate rebate), but more importantly funds low-carbon investment projects. As of March 2020, California's climate investments totalled $12.7 billion dollars, supporting among other projects energy-efficient affordable housing, zero-emission vehicles and transport infrastructure (e.g., high-speed rail), and land use and urban forest initiatives. A key element of such climate investment fund is to provide solutions and benefits to disadvantaged and low-income communities. So far, $3.5 billion has been allocated to benefit priority populations (CARB 2020).

Market Supervision: There is a market advisory committee supported by leading academics and environmental finance experts that assesses market performance. However, CARB has acquired its own internal capacity through staff to independently verify and assess potential market manipulation and the overall market performance of its cap-and-trade system.

Equity, Access, and Environmental Justice: It is important to devise environmental justice safeguards and compensatory programs to protect disadvantaged communities from the harmful effects to human health and welfare of localized exposures to GHG co-pollutants (criteria pollutants) emitted by facilities participating in the ETS and other pricing mechanisms. A quarter of the climate investment revenue goes to address the needs of disadvantaged communities. CARB's Environmental Justice Advisory Committee has also submitted recommendations for CARB to implement AB 197 which requires prioritizing emission reductions at the largest GHG sources and those specifically in disadvantaged communities. AB 197 also requires consideration of the social cost of GHG emissions, which includes the public health costs of GHG co-pollutants.

Linking: From the start, California made sure that the system to be implemented would be robust enough to meet its intended goals. CARB actually delayed its original start date to make sure that that was the case. As part of its international cooperation activities, it has advised many countries and regions around the world. There are four legal requirements for linking market-based mechanisms with other jurisdictions, as follows:

1.   The jurisdiction with which the state agency proposes to link has adopted equivalent or stricter program requirements for greenhouse gas reductions, including, but not limited to, offset requirements.

2. Under the proposed linkage, the State of California is able to enforce statutes against any entity subject to regulation under those statutes and against any entity located within the linking jurisdiction to the maximum extent permitted under the United States and California Constitutions.
3. The proposed linkage provides for the enforcement of applicable laws by the state agency or by the linking jurisdiction of program requirements that are equivalent to or stricter than those required by California law.
4. The proposed linkage and any related participation of the State of California in the Western Climate Initiative shall not impose any significant liability on the state or any state agency for any failure associated with the linkage.

Climate action in California constitutes a comprehensive, holistic approach to achieve ambitious GHG emission reduction goals in order to reach mid-century carbon neutrality goals across different sectors of the economy. Central among a suite of complementary policies, carbon pricing in the form of a cap-and-trade system is helping minimize the costs of advancing toward a smarter, innovative, clean economy model. The California ETS is an economy-wide program, setting a limit on approximately 85 percent of the state's GHG emissions. It complements and buttresses sectoral polices such as RPS and the LCFS. Clean technological innovation, resource efficiency, and cost-effective emission reductions are central to making a net-zero carbon California a reality by mid-century. As in the case of the EU ETS, California requires continuous adaptive management of its cap-and-trade system as part of a portfolio approach to sustain cost-effective carbon mitigation efforts, while delivering other co-benefits to its residents to underpin a prosperous, healthy, and climate-resilient future to communities across the state.

Mexico

For decades now, Mexico has been following and learning from the international experience designing and implementing ETS through its participation in the UNFCCC, G20, OECD, and other multilateral organization support groups like the Work Bank Program for Market Preparedness (PMR). The General Law of Climate Change (LGCC) of 2012, listed the concept of an ETS as a possible approach to reduce carbon emissions in order to achieve national climate policy goals, but without legally mandating its implementation. The need to reform the LGCC to align it with the Paris Agreement became clear after the participation of a group of key Mexican legislators in the Mexican delegation to the UNFCCC COP 23 in November 2017 and internal GOM and expert consultations at the summit. On December 12, 2017, the chamber of deputies sent its proposed amendments to the senate. Among these, the reforms to Article 94 mandated the implementation of an ETS and assigned authority to the Ministry of Environmental and Natural Resources (SEMARNAT). The law was ultimately passed in July 2018. An important aspect of these amendments to the LGCC is their emphasis on establishing a "consultation and representation" mechanism for

ETS participating industrial sectors. Also, like past experiences in California and the EU, the ETS will include "progressive and gradual implementation" and minimize competitiveness impacts to industry vis-à-vis international markets. The same month, a new president was elected in the country. The new administration, with some delays and revisions, pushed forward with the pilot ETS (LGCC 2018).

The Mexican ETS: From Learning-Phase to Compliance Mechanism

On January 10, 2019, SEMARNAT published the agreement to establish the preliminary basis for the implementation of a "test" period for the Mexican ETS, toward developing a formal operational phase as per Article 94 of the LGCC. The program is being developed based on historical emission data (2016–2019) provided to the National Emissions Registry (RENE). In the last few years, the government has launched an effort to improve reporting and true-up reported emissions. The agreement establishes an emissions cap, a transactions tracking system of emission allowances and offsets (the exchange market), and introduces flexible compliance mechanisms (e.g., offsets). The 3-year learning phase has two periods: (1) From January 1, 2020, through December 31, 2022, and (2) the last year, 2022, which will be a transition toward the operational phase or compliance system where accurate reporting will be demanded by regulators. The objectives of the pilot period are to.

- Advance the achievement of Mexico's emission reduction goals;
- Promote emission reductions at the lowest cost possible in a measurable, reportable, and verifiable manner;
- Test the functioning of the ETS in the Mexican context, and educate participating sectors while developing the internal institutional capacity building for its successful implementation;
- Identify areas of improvement to fine-tune the system toward its operational phase;
- Generate robust and high-quality information;
- Create value for the emission permits and the compensation credits.

In this stage, no economic penalties will be imposed and allocations will be granted free and independent of reserve allowances.

Program scope: The Mexican ETS is not an economy-wide system, it includes only the industrial and energy sectors outlined as follows.

Energy Sector:

- exploitation, production, transport, and distribution of hydrocarbons; and
- the generation, transmission, and distribution of electricity.

Industrial Sector:

- Automotive;
- Cement;

- Chemical;
- Food and beverages;
- Glass;
- Steel;
- Metallurgic;
- Mining;
- Petrochemical;
- Pulp and paper;
- Other subsectors that emit direct GHG from fixed sources.

Participation threshold: 100,000 tCO_2 or above of direct GHG emissions in any of the years between 2016 and 2019. All facilities reaching this emission level will have to participate for the duration of the pilot period, independently of reaching lower emission levels at some point during the implementation of the program.

Program participants: 300 facilities that represent more than 90% of the total emissions reported to RENE.

Regulated substance: Direct CO_2 emissions from fixed sources as part of industrial processes and fuel burning in preparation for the compliance and operational phase of the program. Other gases and components with carbon equivalent properties will not be included during the pilot period.

Reserve: SEMARNAT can transfer credits from the general reserve to new entrants and those with increases in emissions because of expansion in production to maintain the "environmental integrity" of the program. (Art 21).

New entrants: Facilities that reach the 100,000 tCO_2 threshold and above starting in 2020 and beyond.

Auctions: Can be used at the start of the second year of the learning phase derived from the allowances deposited in the reserve contingent on the ETS market performance.

Reconciliation period: On 1 November of each calendar year, participants and new entrants to the program will have to hold a number of credits equivalent to the emissions reported from the immediate previous year and verified according to a preestablished submission schedule for the pilot period (i.e., Annex 1 of the market rules). On-time and fully compliant reporting will allow participants to trade excess allowances in their accounts or to be used to comply with future requirements within the test period. Those failing to submit on time and not remaining fully compliant before 15 November of this account reconciliation period will not be able to trade allowances, and each excess allowance in this stage will be discounted at a double rate (2 for each non-compliant allowance) during the first allocation at the start of the operational phase of Mexican ETS compliance. All allowances issued for the pilot period will be canceled at the end of 2023.

Facility retirement: If a fixed source is permanently closed, the allowances allocated to such facility will be returned to SEMARNAT and will follow a preset schedule and proportional return depending on the date of the closure. These allowances will be retired from the program.

Market transactions: These operations will be conducted among program participants through the tracking system and exchange platform of the Mexican ETS. They will only be considered fully executed if they are registered in the aforementioned tracking system.

Market supervision: SEMARNAT will maintain the environmental integrity of the system by setting maximum amounts of allowances to be purchased per participant. If allowance hoarding, market manipulation, and perceived monopolistic behavior are detected in accordance with the Federal Law of Economic Competition, it will inform the corresponding agency (i.e., Federal Commission of Economic Competition) to take legal action against such actors.

Electronic Exchange Platform: The Mexican ETS is expected to provide an electronic system to issue, transact, and cancel allowances and compensatory credits (e.g., offsets). It will provide a means to account for valid allowances and compensatory credits as part of the systems and to create a directory of registered program participants and their accounts. This electronic system will make it possible to validate and register all transactions as well as the regulatory action taken by SEMARNAT, such as:

- Allocation;
- Buying and selling;
- Acquisitions through auctions;
- Compliance allowance submissions;
- Cancelation of allowances;
- Maintenance throughout compliance periods (preset schedule);
- Creating and keeping up accounts;
- Registering the number of emission products for the verification process.

In short, the Mexican ETS platform will be the repository of transactions, accounts, and all aspects relevant to its operation while following legal transparency requirements regarding public information and the protection of participants' confidential information.

Precautionary Measures: If SEMARNAT detects activity contrary to the environmental integrity of the system, such as gaming or intervening in the system, or abusing the ETS by any other means, it will suspend culprit accounts. Accounts will have 15 days to clarify actions and SEMARNAT will then decide on the continuation of suspension or restoring account rights.

As for crediting, clarification is required in terms of early action and compensatory measures. An important element of the learning phase is the information derived from the actual implementation experience. Article 10 requires comprehensive reporting regarding.

- Price behavior;
- Emission reductions achieved;
- Percentage of emissions that account for total national emissions;
- Actual administrative costs as well as the operational costs of the tracking systems.

Based on this information and fine-tuning of the system, SEMARNAT and the Mexican ETS Advisory Committee will review the achievements of the pilot period and progress toward meeting national emission reduction goals. Analytical support from the National Institute of Climate Change (INECC) should supplement this review. In particular, these institutions will evaluate the potential to achieve additional emission reductions, the benefits to the population in general and to program participants, mitigation costs, and the administrative and MRV functioning of the program. Additionally, from the start, some industries, for instance, cement, were very concerned about the advantages and disadvantages vis-à-vis non-participating facilities and public information requirements. Moreover, the assessment of the pilot period is expected to include an analysis of the impact of the ETS on other GHG mitigation policy instruments in Mexico's climate action plan. Finally, the assessment process will evaluate if new sectors of the economy should be covered under the Mexican ETS.

Implementation Challenges Ahead

The pilot phase of the Mexico ETS will inform the operational compliance phase. Challenges remain ahead, and exogenous factors threaten its full-fledged implementation as a flexible, cost-effective compliance mechanism. These range from political risk and funding to institutional capacity and technical aspects such as high-quality information on emission and transparency measures. Some of these include the following.

Budget Shortfalls and Policy Continuity Risk: The Obrador Administration's planned "budget savings" and the process of shrinking of government agencies, along with the expected impact of the COVID19 global health crisis on the economy, will further limit the resources available for environmental protection in Mexico. Moreover, the ongoing institutional reorganization of SEMARNAT and the changes in leadership at the helm of the agency, including three ministerial appointments since December 2018, threaten policy commitment and continuity for this policy approach. The Mexican ETS does require some key immediate investments that will face budgetary constraints, such as hiring staff with appropriate expertise at SEMARNAT and setting up the electronic platform for the program. The IT systems of the Mexican ETS will also have to invest in cybersecurity safeguards and other emerging solutions such as blockchain, as they mature.

Institutional Frameworks and Inter-Agency Coordination: SEMARNAT will also have to develop methodologies to assess the environmental integrity of the ETS and its key components, for instance, in the case of compensatory measures and the offset system to provide additive cost containment mechanisms to emitters. This includes policy alignment with the existing carbon tax, which will require a high degree of inter-agency cooperation and collaboration, for instance, with the Ministry of Finance and the National Forestry Commission. It will also require open consultation processes with stakeholders to developing the appropriate institutions to sustain carbon pricing revenue, environmental commodities trading, and climate investment mechanisms. Mexico has great potential to develop natural climate solutions (NCS) or nature-based solutions (NBS) through agricultural, forestry, and other land uses (AFOLU) as a means to conserve natural capital and support communities in these areas. This in turn will require collaboration with the communities potentially hosting these projects, capacity building, legal certainty, and strong institutional and political support for the creation of such environmental financial products. Additionally, NCS will require clarification vis-à-vis NDCs to avoid double counting. The Ministry of Energy is no longer prioritizing decarbonization of the sector and deployment of clean energy at this time. Policy-constructive engagement from energy regulators is key to effective climate action.

Inventory and MRV systems: The process of improving the quality of the information provided to the emissions registry (i.e., RENE) should continue. Also, there is room for improvement on regulatory emission reporting procedures, as the conventional Environmental Certificate of Operation (COA) and CO_2 direct emission reporting may be too burdensome in addition to the GHG monitoring plans. Thus, there is an opportunity to harmonize and streamline the reporting system. RENE's data is critical to setting a transparent, stringent cap. MRV errors may result in over-allocation of allowances at the start of the compliance phase. The new industry of GHG certification entities is working on adapting their services to provide more seamless reporting products to regulated entities.

Mexico has the opportunity to deepen its collaborative work with the rest of the Americas in expanding carbon pricing, markets, and clean innovation opportunities. It has been working, for instance, with Colombia and Chile. Looking north, California has already served as a blueprint for the Mexican ETS and continues to regularly advise and provide technical assistance. This is important. Aspiring to meet the highest standards in the design and implementation of national ETS around the world can foster certainty, integrity, and credibility in the implementation path of a workable and cost-effective global carbon market. Getting it right from the start in the implementation of the Mexican ETS will no doubt contribute toward the institutional build-up of the North American carbon market as a major step toward the development of a global system. This will demonstrate that it is possible to align accounting and policy designs between developed and developing nations. Mexico, like California, Quebec, and the European Union, is a responsible actor contributing to the global effort to stabilize the accumulation of GHG emissions in our atmosphere. The Mexican government should also aim to maximize the low-carbon development opportunities that the transition to a cleaner, more resource-efficient economy presents. That process is being put on hold because of new directions in Mexico's federal energy policy. A resource-efficient nation is a more competitive one. Energy policy is climate policy. Alignment and coordination between resource, energy, and environmental federal policy goals are sorely needed.

Concluding Remarks

The road ahead for the implementation of the Mexican ETS is a rocky one. There is a high level of political risk given current uncertainties in the disconnect between energy and climate policy in Mexico. The UNFCCC still has to clarify important aspects of Article 6 of the Paris Agreement. To succeed, policy continuity and full-institutional endorsement of this policy approach at the federal level must be achieved during the transition between the learning phase and the ETS operational compliance. This would signal a credible commitment from the federal government to the Mexican carbon market. A well-integrated, workable cap-and-trade system for carbon in Mexico will take into account the idiosyncrasies of the Mexican implementation context that will emerge from its learning phase, but should also aim to implement the highest international policy design standards for its compliance phase to ensure future connectivity and linking with international markets.

Gradual ratcheting up of ambition for cost-effective carbon mitigation, strong accounting standards, and MRV measures, using auctions and revenue proceeds for climate investment and clean innovation, along with strong penalties for non-compliance are central to the economic and environmental performance of an ETS. These objectives need to be balanced with other national policy objectives for a more prosperous, just, and clean economy in Mexico. The U.N. SDGs provide an array of examples on how to achieve this in a sustainable manner. Moreover, linking the ETS and offset mechanisms with strong regulations on GHG-co-pollutants which effect public health will be key to avoiding growing disparities in exposures across the country (i.e., environment justice). Additionally, care needs to be taken in the design of any offset program in the country. NCS projects' social dimension consideration is important, for instance, and should be designed in close partnership with the communities conserving Mexico's natural capital assets to ensure they are equal negotiators in these projects and they benefit from these climate investment flows. On the other hand, Mexico will have to work on providing legal certainty, institutional support, and permanence to such environmental products.

Getting it right from the start of the compliance period based on the pilot period experience is important. Periodically fine-tuning the system based on new scientific, technical, and accumulated implementation experience domestically, as well as from other carbon market developments from around the world, will be necessary. As pointed out above, the necessary interactions between social, institutional, and political forces bear on the quality of an emissions trading mechanism's design (Perez Henriquez 2013). Regulators will need to remain vigilant in the transition to the operational phase of the Mexican ETS, particularly by avoiding the granting of too many political concessions to all regulated entities or by not considering interaction with other governmental policies and measures that would reduce the environmental integrity and the overall cost-effectiveness of the program. A high degree of institutional coordination and alignment is needed within government agencies to prepare Mexico for the new clean economy. Ultimately, uncoordinated decision-making tends to result in resource use that is socially inefficient.

References

Baumol WJ, Oates WE (1988) The theory of environmental policy. Cambridge University Press

Burke M, Craxton M, Kolstad CD, Onda C, Allcott H, Baker E, Barrage L, Carson R, Gillingham K, Graff-Zivin J, Greenstone M (2016) Opportunities for advances in climate change economics. Sci 352(6283):292–293

California Air Resources Board (CARB) (2017) Climate change scoping plan: the strategy for achieving California's 2030 greenhouse gas target

California Air Resources Board (CARB) (2019) California Greenhouse gas 2000–2017 emissions trends and indicators report

California Air Resources Board (CARB) (2020) Cap-and-trade auction proceeds: annual report to the legislature

Climate Action Network (2019) COP25. Eco—NGO Newsletter. Winter 2019. Madrid, Spain

Coase R (1960) The problem of social cost. J Law Econ 3(1):1–44

Crocker T (1966) Structuring of atmospheric pollution control systems. In: Wolozin H (ed) The economics of air pollution. W.W. Norton, New York

Dale JH (1968) Pollution, property, and prices: an essay in policy-making. University of Toronto Press, Toronto

Delbeke J, Vis, P (eds) (2019) Towards a climate-neutral Europe: curbing the trend. Routledge

Dietz S, Stern N (2008) Why economic analysis supports strong action on climate change: a response to the Stern Review's critics. Rev Environ Econ Policy 2(1):94–113

Edmonds J, Forrister D, Clarke L, de Clara S, Munnings C (2019) The economic potential of article 6 of the Paris Agreement and implementation challenges. International Emissions Trading Association, University of Maryland, and Carbon Pricing Leadership Coalition, Washington, D.C.. World Bank. Retrieved from: https://openknowledge.worldbank.org/handle/10986/33523 License: CC BY 3.0 IGO

European Commission (2020) Environmental taxes in the EU. Eurostat, statistics explained. January 2020. Retrieved from: https://ec.europa.eu/eurostat/statistics-explained/index.php/Environme ntal_tax_statistics#Environmental_taxes_in_the_EU

European Commission (2017) Emissions trading system (EU ETS) revision for phase 4 (2021–2030). Retrieved from: https://ec.europa.eu/clima/policies/ets/revision_en

Gibbens S (2019) 15 ways the Trump administration has changed environmental policies. National Geographic. Retrieved from: https://www.nationalgeographic.com/environment/2019/02/15-ways-trump-administration-impacted-environment/

Gillingham K (2019) Carbon calculus: for deep greenhouse gas emission reductions, a long-term perspective on costs is essential. Financ Dev 56(004)

Goulder LH (1995) Environmental taxation and the double dividend: a reader's guide. Int Tax Public finance 2(2):157–183

Goulder L, Parry I (2008) Instrument choice in environmental policy. Rev Environ Econ Policy 2(2):152–174

Hahn RW, Stavins RN (1992) Economic incentives for environmental protection: integrating theory and practice. Am Econ Rev 82(2):464–468

Hanemann WM (1994) Valuing the environment through contingent valuation. J Econ Perspect 8(4):19–43

Hanley N, Shogren JF, White B (2016) Environmental economics: in theory and practice. Macmillan international higher education

International Carbon Action Partnership (ICAP) (2020) Emissions trading worldwide: status report 2020. International Carbon Action Partnership. Berlin, Germany. Retrieved from: https://icapca rbonaction.com/en/icap-status-report-2020

Intergovernmental Panel on Climate Change (IPCC) (2007) Summary for policymakers. In: Solomon S, Qin D, Manning M, Chen Z, Marquis M, Averyt KB, Tignor M, Miller HL (eds) Climate change 2007: the physical science basis. Contribution of working group I to the fourth assessment report of the intergovernmental panel on climate change. Cambridge University Press, Cambridge, United Kingdom and New York, NY, USA

Jackson RB, Friedlingstein P, Andrew RM Canadell JG, Le Quéré C, Peters GP (2019) Persistent fossil fuel growth threatens the Paris agreement and planetary health. Environ Res Lett 14(12):121001

Newell RG, Stavins RN (2003) Cost heterogeneity and the potential savings from market-based policies. J Regul Econ 23:43–59

Nordhaus WD (2017) Revisiting the social cost of carbon. Proc Natl Acad Sci 114(7):1518–1523

Nordhaus, W.D. (2005). Life after Kyoto: alternative approaches to global warming. National Bureau of Economic Research (NBER) Working Papers, (11889)

Parry IW, Mylonas V (2017) Canada's Carbon Price Floor. Nat Tax J 70(4):879–900

Pérez Henríquez B (2004). Information technologies the unsung hero of environmental markets. Resources Fall/Winter, 9–12

Pérez Henríquez B (2013) Environmental commodities markets, towards a low carbon future. Resource for the Future Press, Washington, DC.

Pigou AC (1932) The economics of welfare, 4th edn. Macmillan and Co., London

Pizer W (1999) Optimal choice of policy instrument and stringency under uncertainty: the case of climate change. Resour Energy Econ 21:255–287

Schmalensee R, Stavins R (2017) Lessons learned from three decades of experience with cap and trade. Rev Environ Econ Policy 11(1):59–79

Secretaría de Medio Ambiente y Recursos Naturales (SEMARNAT) (2019) Acuerdo por el que establecen las bases prelimnares del Programa de Prueba de Sistema de Comercio de Emisiones. Diario Oficial de la Federación: México, Enero 10, 2019. Retrieved from: https://www.dof.gob. mx/nota_detalle.php?codigo=5573934&fecha=01/10/2019

Stavins R (2018) What should we maker of China's announcement of a national CO_2 trading system. An economic view of the environment a blog by robert stavins. Retrieved from: http://www.rob ertstavinsblog.org/2018/01/07/make-chinas-announcement-national-co2-trading-system/

Stern N (2007) The economics of climate change: the stern review. Cambridge University Press, Cambridge

Teixidó J, Verde SF, Nicolli F (2019) The impact of the EU emissions trading system on low-carbon technological change: The empirical evidence. Ecol Econ 164:106347

Tietenberg TH (2005) Tradable permits in principle and practice. Penn St Envtl L Rev 14:251

Tietenberg TH (2006) Emissions trading: principles and practice. Resources for the Future (RFF Press), Washington, D.C

United Nations Conference on Trade and Development (UNCTAD) (1997) Greenhouse gas emissions trading on the global climate agenda. Retrieved from: https://unctad.org/en/Pages/PressR eleaseArchive.aspx?ReferenceDocId=3277

United States Agency for International Development (USAID) (2017) Climate change risk profile for Mexico. Retrieved from: https://www.climatelinks.org/resources/climate-change-risk-pro file-mexico

United States Energy Information Administration (USEIA) (2020a) Data dashboard. https://www. eia.gov/beta/states/data/dashboard/renewables

United States Energy Information Administration (USEIA) (2020b) State energy data system. https://www.eia.gov/state/seds/sep_sum/html/xls/use_es_capita.xlsx

United Nations World Commission on Environment and Development (UNWCED) (1987) Report of the world commission on environment and development: our common future. United Nations General Assembly Document A/42/427

Willsher K (2018) Gilets Jaunes protesters threaten to bring France to a standstill. The Guardian, Nov 16, 2018. Retrieved from: https://www.theguardian.com/world/2018/nov/16/gilet-jaunes-yellow-jackets-protesters-france-standstill

World Bank (2015) Escalating call to put a price on carbon. Retrieved from: https://www.worldbank.org/en/news/feature/2015/12/01/escalating-calls-to-put-a-price-on-carbon

World Bank (2020) Carbon pricing dashboard. Retrieved from: https://carbonpricingdashboard.worldbank.org/map_data

Chapter 2
Bringing Emissions Trading Schemes into Mexican Climate Policy

Alejandra Elizondo

Abstract Emissions trading schemes (ETS) have become popular as a policy instrument to tackle climate change. This chapter analyses the decision to deploy carbon markets and their interaction with other instruments in Mexico's climate policy. Instrument selection has been thoroughly explored in the regulation and public policy literature (Kern et al. in Res Policy 48, 2019; Capano and Lippi in Policy Sci 50(2):269–293, 2016; Wurzel et al. in German Policy Studies 9:21–48, 2013; Harker et al. in Climate Policy 17(4):485–500, 2017; Baldwin et al. in Understanding regulation, Oxford University Press, 2012; Jordan et al. in Policy instruments in practice. Oxford handbooks online 536–549, 2011), but its application to carbon markets is mainly focused on environments such as Europe, the US and, more recently, China. The decision to adopt an ETS relies not only on specific characteristics of each instrument but also on institutional constraints and messy political considerations. A combination of preferences and institutional factors affect the choice of instruments, and the ultimate decision must be legitimate and instrumental for each context. I analyse the considerations involved in the deployment of the ETS pilot project, looking at its distinctive characteristics and those it shares with other available instruments, as well as the requirements for its implementation.

Keywords ETS · Climate change · Mexico · Policy instruments · Policy tools · Patterns of choice

Introduction

The appeal of emissions trading systems (ETSs) is in part due to their regulatory logic for government and industry, as well as a belief that environmental measures are not necessarily expensive (Bailey and Maresh 2009). Their emergence also responds to an emphasis on higher levels of dynamic and allocative efficiencies compared to

A. Elizondo (✉)
Centro de Investigación y Docencia Económicas, A.C. (CIDE), Carretera México-Toluca 3655, Col. Lomas de Santa Fe., CP 01210 Distrito Federal, México
e-mail: alejandra.elizondo@cide.edu

© The Author(s) 2022
S. Lucatello (ed.), *Towards an Emissions Trading System in Mexico: Rationale, Design and Connections With the Global Climate Agenda*, Springer Climate,
https://doi.org/10.1007/978-3-030-82759-5_2

other regulations. In this chapter, I will look into the introduction of ETS in Mexican climate policy and the factors that influence their entry.

In general, economic instruments are more efficient than other regulations, leaving decisions about technology, operations, and plant life to agents participating in the market (Isser 2016). Less flexible regulatory instruments set uniform standards and control targets, or specify processes and technologies to use. Nonetheless, costs usually vary greatly among firms, technologies, and strategies, making uniform regulations more expensive for industry and society. Additionally, conventional regulatory instruments require more information and have a heavier regulatory burden when emissions sources are diverse. In short, they seem to result in larger costs for society.

Market instruments internalize the cost of externalities by taking into consideration the social cost of emissions when choosing activity levels. Economic instruments that tackle externalities are divided into two groups: (1) fiscal policies, such as carbon taxes; and (2) the creation of markets, such as emissions trading systems (ETSs).

This chapter is organized as follows. First, I briefly describe ETS experiences around the world as inputs to the Mexican decision to adopt one later on. Next, I focus on explaining the events, stakeholders, and decisions that shaped what we know today as the ETS pilot program in Mexico. Then I describe the analytical framework and its applicability to the market. Finally, I offer some concluding remarks.

Background

The market that initially inspired carbon market deployments worldwide was the US Acid Rain Program, a permit system for sulphur dioxide emissions created in the mid-nineties that led to markets of sulphur dioxide and nitrous dioxides. This experience was taken as evidence that markets for pollution could work effectively, encouraging technological innovation and reducing the cost of pollution abatement (Isser 2016). Since then, more than 60 entities—including national governments and sub-national jurisdictions—have implemented carbon pricing instruments such as carbon markets and taxes. Some of them have been influential in the design of the Mexican carbon market. Specifically, the Regional Greenhouse Gas Initiative (RGGI), the European Union (EU), California and, more recently, Chinese pilot projects have provided knowledge on potential routes for Mexico and will continue to do so.

The Regional Greenhouse Gas Initiative (RGGI)

The RGGI pioneered a mandatory ETS covering emissions from the power sector. It covers 10 US states (Connecticut, Delaware, Maine, Maryland, Massachusetts, New Hampshire, New Jersey, New York, Rhode Island, and Vermont) and will soon cover two more (Pennsylvania and Virginia). The market has been operating since 2009 based on general rules (ICAP 2020a, b) and specific CO_2 budget trading programs, caps, and design adjustments for each participant. It is expected to reduce emissions by 30% compared to 2020 between 2021 and 2030 (more than 65% below the RGGI cap in 2009). An interesting feature of this ETS is that an emissions containment reserve will come into action in 2021, automatically adjusting the cap downward if there are lower than expected costs.

The European Union (EU)

The EU ETS is distinctive because it cedes state power to a supranational climate agency (Bailey and Maresh 2009) and is the first transnational arrangement. Various factors facilitated its creation. An entrepreneur (the Commission) received support from Member States and business groups, and a set of pioneers led the way in its implementation (the United Kingdom, the Netherlands, Denmark, and Sweden) (Jordan et al. 2011a, b). Institutional factors and the potential to buy off political opponents through their right to distribute permits facilitated agreement from all the parties involved. An initial decentralized design was followed by a more centralized decision-making process as the European Union strengthened its authority over other economic actors (Bailey and Maresh 2009).

California

Learning from the EU market (EU ETS) and the RGGI, California then designed its own market. Participants accounted for 85% of California's total greenhouse gas emissions. Its initial target in 2006 was set at reaching 1990 emissions levels by 2020. In 2016, it passed legislation to change its target to 40% below 1990 levels by 2030 (EDF 2020). The distribution of allowances is a mixed system, with free allowances by industry and efficiency, and allowances purchased at auctions or via trade. The state not only reduces emissions but takes less carbon to grow the economy, creates benefits for local populations, and promotes clean energy jobs and local air-quality initiatives (EDF 2020). Flexibility mechanisms are provided through offsets, banking, and strategic reserves (California Cap and Trade 2020). The market is now linked to Quebec and Ontario, and a potential link to Mexico has been discussed.

China

China was the first developing country to implement a pilot ETS in 2013. Some of China's features, such as size and emission levels, made it a distinctive case, accounting for 1.2 billion tons of CO_2 across seven regions. The country had to deal with a lack of strong legal regulations, weak enforcement, the dominance of state-owned enterprises, and a shortage of available data, experiencing low levels of liquidity in all pilot projects as well as continuous interventions by regulators (Munnings et al. 2016). After launching seven pilot ETS, policymakers focused on discussing interprovincial trading and solving market fragmentation problems (Jiang et al. 2016).

The Definition of ETSs in Mexico

For the last decade, Mexico has been an active player in the international climate policy framework. It was the first developing country to have a general law of climate change (LGCC) mandating a long-term climate policy, and regulatory and market instruments. This was second only to the UK worldwide. The Mexican government has been at the forefront of international negotiations, participating in the United Nations Framework Convention on Climate Change (UNFCCC), the Kyoto Protocol, and the Paris Agreement, and hosting the 2010 United Nations Climate Change Conference in Cancun.

Policy implementation, on the other hand, has faced several challenges. In 2014, the Mexican congress passed a carbon tax bill designed to reduce the consumption of fossil fuels. The initial bill included differentiated rates for each fuel according to its carbon content, but it ended up with exemptions for natural gas and lower rates for certain fuels that had relatively high levels of emissions. Together with a generous reduction to the burden, the tax ended up having only a minor impact (if any) on the decisions of fossil fuel consumers.

After this difficult initial experience, an ETS, also mentioned in the law, then became of interest as a viable alternative. The fact that gasoline price increases are at odds with the current administration's interests further lowers the political viability of a carbon tax. While a tax was seen as controversial and politically challenging, the implementation of an ETS seemed more feasible to enact in Mexico. It was hoped that this new policy would then incorporate lessons learned from previous missteps. For instance, a carbon tax is a fiscal instrument administered by the Ministry of Finance. An ETS, by contrast, is classified as an environmental tool. The Ministry of Environment (SEMARNAT) is in charge of designing and managing its resources and collecting fines. It seemed a more suitable tool due to its single goal: decreasing CO_2 emissions relative to business as usual.

In April 2018, the Mexican Senate approved reforms to the LGCC by a vast majority (zero votes against and one abstention). The amendment included the targets listed in the Mexican Nationally Determined Contribution (NDC): a 22% reduction in emissions by 2030 and, if certain conditions regarding financial support were met, the reduction would amount to 36% compared to business as usual (BAU). The law also eliminated the word voluntary from the ETS, and the Second Transitory Article established the implementation of an emissions trading pilot program 10 months after the reform.

Removing a market's voluntariness may initially sound disturbing if not understood within the ETS context. From an economic point of view, a market is voluntary by definition. In this case, however, the government defines the regulations, technicalities, and an institutional framework for the ETS. Participants are then required to comply with a certain limit or cap, but they are free to enter or exit the market as they choose, as with any other market structure.

In 2019, an executive order was signed to launch the carbon market pilot program. It started in January 2020, requiring all industries generating more than 100,000 tons of CO_2 annually from direct and fixed emissions to participate in the market.

The initial market structure was designed by a number of groups. Think tanks organized dialogues with civil society organizations (CSOs) to obtain their input. The Environmental Defense Fund (EDF) then constructed a market simulator for the industry and other interested parties to foster their knowledge on the subject. The World Bank funded these efforts, and the Mexican-German Climate Change Alliance (GIZ) then facilitated the whole process by conducting technical studies. Together, governments, think tanks, and international organizations widened the understanding of the program and introduced interested parties to the ETS's language and technicalities.

In 2012, Mexico joined the World Bank's Partnership for Market Readiness (PMR), an association of parties interested in carbon pricing. Some of them act as fund providers and others as resource recipients. The PMR provides resources to Mexico for technical support, consulting services on technical topics, and training for both the public and private sectors. The initial focus was on developing Nationally Appropriate Mitigation Actions (NAMAs), and resources shifted towards creating the market after 2015. The PMR ends in 2020 and, if interested, Mexico must enter the Partnership for Market Implementation (PMI), starting in 2021.

In 2014, SEMARNAT launched the first tool in preparation for the market: the National Emissions Registry (RENE). This tool gathers emissions information from facilities emitting 25,000 tons of CO_2e or more, including sectors such as energy, transportation, and agriculture. Facilities must report emissions of carbon dioxide, methane, nitrous oxide, black carbon, and fluorinated gases, among others. RENE was the first tool implemented in preparation for the market and was in effect for over 3 years before the pilot project came into practice.

The government's links with other countries and regions with established markets began before they had any certainty about their own market. The Mexican ETS design was heavily influenced by two markets: the European ETS and the California market. First, Mexican policymakers have traditionally had strong ties to their Californian counterparts and California is a close ally of the federal government. "California has always behaved as a leader in this topic… Mexico has followed California's DNA", said one interviewee from a civil organization. "California pays attention to sectoral protocols. Rules for each sector, transparency, permanence… It is complex, but it has been carefully documented." The Mexican government also showed interest when the US drafted legislation for a national ETS, but the Waxman-Markey bill, as it is known, was never voted in by the Senate. "Officials from the Mexican Ministry of Finance were interested in being linked to US markets", mentioned a CSO executive.

"But as soon as the bill came to a halt, so did the Mexican government's interest in the ETS." Despite the Ministry of Finance's decreasing salience, environmental authorities have maintained their interest.

Considerable efforts were then expended to design and implement the Mexican market. In 2014, the government signed a Memorandum of Understanding (MOU) on climate change with California, with a special section devoted to the ETS. This initiative triggered dialogues between SEMARNAT and the California Air Resources Board (CARB) and the support of think tanks, which had previously helped other governments to design their markets. "We were looking at the possibility of linking the market and actually having a North American ETS", mentioned a former official from SEMARNAT. "The regulations in California mention that its market can be linked with others when the legal framework from its potential partner is equivalent in terms of soundness and requirements. In addition, it must have operated for three years with such regulations to be considered. Hence the importance of having a three-year pilot program in Mexico." The Mexican regulation for the carbon market was designed according to California's requirements.

Knowledge about Europe in general and Germany in particular came through assistance from the German government. The GIZ offered resources to analyse the entry of an ETS in topics such as carbon pricing policy mix, legal analyses, and the introduction of the market within a Mexican setting. The most recent support came through a 3-year project to assist in launching the pilot project with technical knowledge, legal and regulatory analyses, and direct support for the parties involved. A fraction of the assistance was devoted to creating and disseminating material for the Spanish-speaking audience. The Ministry of Environment worked closely with all development assistance offices in a coordinated effort. "Good communication among teams working on the topic was key to the process. The Ministry and the two assistance agencies [GIZ and PMR] shared information about their work, they closely coordinated their work … and gathered all the elements needed for the design and implementation of the market" (interview from an international assistance office).

A highly technical and intersectoral instrument such as a carbon market requires the participation of a wide set of institutions. Ministries and regulators from sectors such as finance and energy and state-owned enterprises like the Federal Commission of Electricity (CFE) and the oil and gas company Pemex were involved in the market's design and implementation. Up to now, only the CFE has been consistently present during the development of the new framework. Pemex was present during policy implementation, but the rotation of its personnel has been detrimental to its effectiveness. A key player, the Energy Ministry (SENER), was actively involved in past administrations. "We partnered in exploring possible implications of policies that share impacts but that differ in their origin, from the energy or environmental sectors", mentioned a former official. Together, both ministries addressed concerns from the industry about the interaction of Clean Energy Certificates (CELs) and the market, with technical analyses provided through international assistance. This type of partnership ceased with the new administration in 2018 and has therefore been absent in the initial phase of the pilot program. Financial institutions, such as the Ministry of Finance, the National Banking and Securities Commission, and the

Mexican Central Bank, have not been consistently active. For example, the Ministry of Finance worked closely with SEMARNAT on the subject when the officer in charge had previously worked in the environmental sector, but the ties were linked to the person and not the institution and so were lost after the officer left their post.

International experience reveals that the incentives of the various market participants determine their involvement in the market. In the EU, for instance, manufacturers saw the market as a compliance tool and only traded at the last minute, while energy utilities were highly active in the market, hedging their future positions (Bailey and Maresh 2009). Brokers, financial institutions, banks, and hedge funds entered the market to speculate or offer special services. The vision of market participants in Mexico is still a missing piece. Up to now, the intention to participate is unclear and the pilot program lacks incentives for potential participants to reveal their positions.

From 1 January 2020, emissions reported at the RENE have been considered for the pilot program. Another registry where allowances are generated and firms can insert their follow-up and register compliance is still lacking (the German government has offered its support once more here). The platform needs to be operating by October 2020, when the first allowances are distributed.

These elements, plus a continuous effort by SEMARNAT officials to remove political barriers, foster dialogues, and share information, resulted in negotiations with industry representatives. Authorities had learned their lesson about possible distortions and undesired results from the carbon tax and, in this case, 2 years of conferences, dialogue, travel, and discussions resulted in a much better working relationship between government and industry.

A New Relationship with Industry

Industry's position as regards participation in the market was initially entrenched, and organized industrial groups opposed the idea throughout the initial phase. The government, with the assistance of international experts, began a long process of negotiations with capacity-building elements, Q&As with experts, studies to solve common queries about the impact on competitiveness, and interactions with the various stakeholders.

The first and most important question that arose was how "obligated" parties would actually be required to participate in the market. The law lacked clarity because it declared that a voluntary ETS was an option for SEMARNAT. A change in the law was in order, and the word voluntary was eliminated in LGCC Article 94. Congress supported SCOs and the federal government in this process, but more objections emerged from industry at the same time. When the government first decided to take the ETS route and implemented the RENE, representatives showed their disagreement and government officials invited them to discuss queries and uncertainties. "The negotiation was very open and transparent. It overcame challenges from the private sector, and was conducive to having them on board in the project", said a manager from an international assistance institution.

Government officials held informal meetings with private firm representatives discussing possible implications for changing the law. One of the main concerns was whether SEMARNAT had the technical capacity and resources to implement the instrument. By then, Mexico had support from PMR and was in the middle of negotiations with Germany, requesting technical assistance and support for the ETS. "Then we did not depend on the government's budget in a period of constant budget cuts, or on how many carbon market experts our offices had", mentioned a former public official.

A new emissions market calls for particular skills that might be unfamiliar to some firms. This may mean the creation of new profiles, hiring specialized consultants, and developing new skills. There was an uneven level of knowledge among participating firms, and thus the government, SCOs, and international assistance agencies combined their efforts to educate them about the forthcoming market. According to one interviewee from an SCO, "some firms had a clearer idea of what was coming since their subsidiaries participated in carbon markets in the EU or California. Even if they haven't developed such competencies here, it is easier for them to import them." On the other hand, some firms had never had any experience with such instruments. "These firms saw the market as another burden coming from the government and an increasing cost to their businesses", said a former government employee.

The interaction that the ETS would have with other policies brought uneasiness to members of the private sector. The establishment of an ETS is part of a set of climate instruments, and its interaction with the rest of the toolkit can bring uncertainty on prices or increase the need to monitor credit veracity (Bailey and Maresh 2009). So far, the agreement is that CELs and the market work as separate entities. The use of CELs drives down emissions if they are managed well, and that is considered when defining the cap but CELs cannot be considered in participants' allowances. "They are working with renewable energy providers; we are working with those that haven't made technological changes yet", declared an international expert. The carbon tax, the other existing instrument, has had a minimal effect on the market, so it could not be taken as a reference or lower bound on the market price. "We discussed having the carbon tax as the floor price for the ETS, but then [those in charge of the pilot project design] thought the floor price could be higher", said a representative of an international development office. The lack of connection to carbon content in the tax rate means a less-expensive burden for coal than for any other fuel, and natural gas cannot be used as a benchmark reference as it is exempt.

Cooperation from industry, as in any other transaction/negotiation, involved trade-offs. Although the process advanced in the technical, political, and legal arenas, it was far from smooth. Some features of the market indicated that concessions needed to be made to keep the process going. As a result, the pilot project has no economic sanctions, and there is provision for a 5% increase over the cap, among other reserves. "In the end, the Ministry of Economy had to act as mediator in the negotiations between private sector and environmental authorities, for the last round of meetings drafting the law. They were held at the Ministry of Economy", comments an interviewee. "We had to make adjustments … It seemed worthy of consideration because this is only the pilot project with a start and an end date and it will only last for three years." The industry seems to perceive the regulation as not having had a significant impact so far. The next few years will facilitate a learning curve without risking profits, learning about trading, the implications of the registry, legal requirements, types of contracts, and financial operations.

A market without economic consequences.

The regulation for the pilot ETS entails a.

testing program with no economic effects, meaning that there won't be monetary sanctions, that initial allocation of allowances will be free of charge in a proportion equivalent to the emissions of the participants, independent of the allowances destined to the reserves.

(Art. 6, Bases Preliminares del Mercado)

Previous research in this area reveals that the industry's perspective of how an initial payment would damage its competitiveness may lead to the inclusion of grandfathering to gain its cooperation (Bailey and Maresh 2009). This is not unique to the Mexican case. Industries have convinced authorities to provide generous allocations in other countries. "The elimination of barriers from industry was a good call. There will be more negotiations to come and proper adjustments will be made", said an interviewee from an SCO.

For Mexico, it means that fines will not be applied during the pilot phase. However, if a firm does not comply, the Ministry can apply a fine in terms of future allowances equal to the magnitude of the non-compliance. That certainly has an economic effect. "In my opinion, when the system is in place and the operational phase includes economic sanctions, resistance from industry is going to re-emerge", declared an international advisor.

Conceptual Framework and Analysis

To structure and analyse the process of the ETS's initial design described above, I used an analytical framework developed by Capano and Lippi (2016) that integrates elements of legitimacy and instrumentality. When analysing (i) **instrumentality**, the focus is on the effectiveness and coherence of the instrument with the goal; when evaluating (ii) **legitimacy**, the choice of instrument must seem appropriate and related to values such as being just and lawful.

Proposing efficient or cost-effective regulations, or analysing their theoretical advantages, is not sufficient for their adoption. Contextual factors, institutional realities, and the ability of certain actors to interfere in the process also impact a policy's design. Looking at the literature, economists, public policy scholars, and political scientists have traditionally investigated these aspects separately. To overcome this constraint, the analytical framework adopted provides insights on both sides of the market: instrumentality and legitimacy. Policymakers focus on instrumentality when looking at the theoretical impact of a policy, and on legitimacy when their main drive is the pursuit of a suitable choice.

An instrument can be *specialized* or *generic*. If *specialized*, the instrument is considered original, non-substitutable, and a best practice to follow. Its definition is clear according to those who choose the instrument, and those involved must consider cognitive and legal implications. The creation of symbols, codes, and languages creates a border that divides insiders from outsiders. "Mention of legal factors is a reference to specific procedures or a characteristic regulation affecting the instrument, whereby everyone can recognize it within its legal framework" (Capano and Lippi 2016, p. 280).

On the other hand, policy instruments may be *generic*, broad, and flexible, with less-coercive use, allowing an increasing number of actors, problems, and situations. Generic instruments can encompass a broad range of problems, within and outside the policy field, leaving room for interpretation and reshaping. The instrument's regulations and technicalities are loosely defined, so many actors can converge and the tool can be applied to different situations and policy problems.

Internal legitimacy comes from insiders considering aspects rooted in the practice, legal framework, and moral background of the sector. Insiders "are the fundamental source of legitimation of the adoption of new instruments" (Capano and Lippi 2016, p. 276), and legitimacy is rooted in the values and arguments from a specific policy field, a legal framework, or a moral background of a policy sector. Such legitimacy is often taken for granted by policymakers.

An instrument faces *external* legitimacy when it comes from a different policy sector or political context but becomes fashionable and appealing. International organizations, the private sector, or other countries are often called upon to provide input in policy discussions. External legitimation can be a result of policy diffusion or transfer, with policy designers perceiving an instrument as a best practice that could work for them and deciding to transfer it to their own policy sector or environment.

Combining legitimacy and instrumentality, Table 2.1 shows that decisions about policy instruments can be classified as (i) routinization (internal and specialized), (ii) contamination (internal and generic), (iii) hybridization (external and specialized), or (iv) stratification (external and generic).

A policy instrument observes *routinization* when there is a continuous adoption of previous instruments, confirming paths and past behaviours. The instrument is perceived as clear and specific, so there is no need for new trials. Decision-makers are convinced that it is the best choice in terms of performance, or that there is no other choice available. The downside is that the instrument's effectiveness is not tested. It is characterized by path dependence and specialization, as well as preservation of the status quo.

In *contamination*, decision-makers adopt new tools in an unspecific way, looking for a change in perception. The intention is to appear capable of dealing with previous ineffective policies. In doing so, stakeholders must adapt their preferences to the requirements of the new tool. This novelty changes the set of adopted tools, with generic new instruments that are broadly designed to cover a wider range of situations.

Table 2.1 Classification of policy instruments

Instrumentality	Legitimacy	
	Internal	External
Specialized	Routinization: adoption of the same instruments in the same way. Specialized and path-dependent, popular and uncontested	Hybridization: innovation within a policy sector with a highly specialized tool. Re-framing of the existing set with a new policy mix with more actors and situations
Generic	Contamination: adoption of new tools in an unspecific way. Actors adopt new tools in a patching-up process	Stratification: introduction of new instruments in a generic way, readily accepted in other fields. Instruments may not be enforced in practice. No real impact on policy dynamics

Source Adapted from Capano and Lippi (2016)

The term *hybridization* comes from a biological concept explaining the inter-breeding of individuals from distinct populations (Capano and Lippi 2016, p. 284), and it reveals a pattern where innovation comes through the insertion of a specialized instrument. Local decision-makers innovate in order to gain external legitimacy. The resulting set of tools is a mix of policy principles, new and old, that may result in a decrease of not only congruence but also the integration of new actors and situations. It is observed in environmental policies, where civil society, corporations, and supranational actors may influence decision-makers into innovating through specific instruments.

Finally, the concept of *stratification* entails a decision to introduce a new instrument generically, juxtaposing it with existing instruments. Since it is adopted generically, it gains legitimacy through innovation and by not being a real threat to any stakeholder. However, the impact is not expected because it is adopted independently of other instruments and is not necessarily enforced.

To analyse the Mexican experience, I carried out a content analysis with publicly available information on the process and conducted semi-structured interviews with former officials, CSO representatives, and experts from international assistance offices. The interviews aimed to acknowledge the part that they and other stakeholders played in designing and shaping the market, as well as their vision of the process.

The design and implementation of the Mexican ETS support the idea that instrument selection is neither linear nor determined. A framework that bases its classification on those premises helps us understand how the carbon market was adopted in Mexico. Had Mexico improved both the design and implementation of the carbon tax, it would have followed a *routinization* process, continuing along the same path and keeping the same instruments with the firm belief that a carbon tax is the right instrument to internalize global externalities on the environment from industry. The tax embodies internal legitimacy, making sense to all policymakers involved through the internalization of the externality by means of a Pigouvian tax. At the same time, it is meant to be specialized following what is often termed as "best practice" in the financial, economic, and environmental fields. It did not, however, follow this course.

Instead, Mexico opted to design and create a new market. SEMARNAT policy-makers took the initiative several years ago and decided to join the World Bank's PMR. This may signal internal legitimacy. Nonetheless, their motivation was due to their links to international actors, even during the early stages of the process. California's government and the MOU, foreign CSOs' assistance, and the GIZ were instrumental in the decision to go ahead with the ETS. Supranational actors "placed the question on the political agenda and suggested the type of instrument to be adopted" (Capano and Lippi 2016, p. 285).

The ETS was considered best practice by local actors in charge of choosing the type of tool to implement, but it was not the obvious choice for all actors in the environmental field, not to mention other sectors. Cooperation and coordination with local CSOs resulted in approval from Congress and successful negotiations with private actors. In terms of legitimacy, the adoption of a market made sense for many.

The market's underlying regulations, the introduction of the RENE, the missing registries, and even the basic logic of the mechanism resulted in a highly *specialized* tool. The above description of its adoption exemplifies the technical knowledge that is going to be needed for the market to work properly. It is not only a matter of goodwill and general understanding but deep immersion in a new logic that comprises financial, technical, and intersectoral requirements for all actors involved. The definition of the pilot project and the accompanying regulations and prerequisites illustrate the nature of the market.

Both conditions together make the ETS an example of a hybridization process, according to Capano and Lippi's (2016) classification (Table 2.2). The creation of the market reflects decision-makers' determination to innovate through external legitimation and the adoption of a clearly defined tool. A long list of technical requisites was discussed in preparation for the market, and international assistance offices actively provided guidelines, lessons, and other materials from previous experiences. The registries to operate the market were also financed through international assistance.

Nevertheless, pressure from the private sector in the last stage of the negotiations reflected an effort to generalize the tool. The lack of economic consequences during the pilot phase of the project and the introduction of additional allowances above the cap to provide flexibility compromised its effectiveness. It remains to be seen whether the authorities can encapsulate these and other attempts into the pilot phase, and then use these concessions as part of the learning curve for all actors.

The Mexican ETS's design seems to allow more flexibility for the market to allocate emissions in the best possible way, leaving decisions about technology and operations to those that have the best knowledge and information: market participants (Isser 2016). As a market mechanism, it always allows subjects or participants certain leeway within which to choose actions to follow, since economic instruments neither prescribe nor prohibit the activities involved (Vedung 1998). As a regulation, an ETS must follow rules and directives, which may result in lower emissions and costs. The internalization of externalities will depend to a great extent on the reliability

Table 2.2 Classification of the Mexican carbon tax and ETS

Instrumentality	Legitimacy	
	Internal	External
Specialized	Routinization Carbon tax (alternative policy)	Hybridization Mexican Emissions Trading Scheme (ETS)
Generic	Contamination	Stratification

of information from RENE. The 3 years during which the system has been in place should have provided knowledge to improve its functionality.

The Mexican experience so far has some similarities with other experiences mentioned above. A regulation aligned with California's legal framework materializes the possibility of opening the market and expanding the scope of the Mexican ETS. The experiences of the RGGI and Europe, and California's linkages to Quebec and Ontario, support the hypothesis of further benefits of regional markets with similar rules adapted to their specific features.

The analyses and lessons from China as a developing country facing similar institutional challenges in this process may increase awareness of certain topics for Mexico, including market expansion and linkages. In fact, the pilot project resembles the Chinese market in the fact that, even without noticeably affecting emissions, the initial phase may change firms' behaviour and increase knowledge for its future implementation.

The negotiations with the private sector that resulted in concessions on reserves are not unique to the Mexican experience. In fact, one of the general advantages of the ETS mentioned above is the possibility of negotiating with potential opponents through the distribution of permits.

Conclusions

The development of events in recent years implies that the different agents participating in the market are satisfied with both their interaction and the road travelled. There are, however, common fears for the future of its implementation. The need for specialized personnel within the government, a continuous negotiation process with industry, the finalization of prerequisites, such as a system for tracking progress and delivering allowances, and the members of the advisory council are still missing pieces (at the time of writing). Institutional capacities are a constant challenge. When the market comes into full operation, more specialized roles will be needed within the government, both inside and outside the Ministry of Environment.

Other sectors' authorities should be incorporated into the process. A climate instrument not only involves the environmental ministry (even if the ministry is the responsible party). The financial sector, for instance, should be included in the process and start thinking about instruments and policies that should accompany the market. The energy sector must take responsibility for climate change, and not only through Clean Energy Certificates, whose destiny is still to be determined, or energy efficiency. Authorities from the energy sector and the environmental sector should realize that the continuation of parallel policies is not enough to actually flatten the curve. Most energy decisions and policies have a direct impact on emissions. The Mexican ETS must be further analysed in terms of the policy mix, focusing on both the form of policy instruments and the context in which they are implemented, and observing interactions between policy instruments, policy strategy, implementation settings, and target groups (Mavrot et al. 2019).

The characterization of the carbon market must include economic consequences, which requires the definition of fines, a well-established and vetted monitoring and enforcement system, and willingness to participate. We must remember the fate of the carbon tax, given industry's capacity to lobby, when preparing the next phases of the market. An appropriate initial design is necessary, but it is not sufficient to achieve the final goal of the ETS: to lower the level of GHG emissions. The policy needs further analyses of the market and its interactions with other regulations and instruments, along with potential consequences for market participants. The pilot program provides this opportunity.

The initial design and now implementation of the market are closely linked to international assistance. Inputs so far have been provided from outside sources. The public budget for the program is limited and clearly insufficient to administer the market. In a subsequent phase, the Mexican government must develop its own abilities, and that involves devoting resources to this policy. This is a new topic for the country, and it must continue to be a part of its national politics and policies.

The initial phase of the Mexican ETS provides early lessons for its future implementation. The creation of a carbon market is both a political and a technical decision that involves many sectors besides the environment. In the political arena, the core of the policy must remain untouched—that is, the decision to use the market to effectively drive down GHG emissions. As a specialized instrument, multiple stakeholders within the government have to be on board for the market to deliver, learning their share of responsibilities and increasing their specific knowledge of ETS technicalities to define rules that are aligned with the stated objectives. External actors have played a substantial role in making the market a reality, at least up to the pilot program. It is time for national authorities to take the reins. A previous attempt to lower emissions through an economic instrument demonstrated that the involvement of all stakeholders (inside and outside of government) is as important as a precise definition of the objective and its technical requirements.

References

Bailey I, Maresh S (2009) Scales and networks of neoliberal climate governance: the regulatory and territorial logics of European Union emissions trading. Trans Inst Br Geogr 34(4):445–461. https://doi.org/10.1111/j.1475-5661.2009.00355.x

Baldwin R, Cave M, Lodge M (2012) Understanding regulation. Oxford University Press
California Cap and Trade (2020). Center for climate and energy solutions. https://www.c2es.org/
    content/california-cap-and-trade/
Capano G, Lippi A (2016) How policy instruments are chosen: patterns of decision makers' choices.
    Policy Sci 50(2):269–293. https://doi.org/10.1007/s11077-016-9267-8
Environmental Defense Fund (EDF) (2020) California's cap and trade program step by step. https://
    www.edf.org/sites/default/files/californias-cap-and-trade-program-step-by-step.pdf
Harker J, Taylor P, Knight-Lenihan S (2016) Multi-level governance and climate change mitigation
    in New Zealand: lost opportunities. Climate Policy 17(4):485–500. https://doi.org/10.1080/146
    93062.2015.1122567
International Carbon Action Partnership (ICAP) (2020a) International carbon action partnership.
    https://icapcarbonaction.com/en/news-archive/484-rggi-plans-for-2030-stricter-cap-and-new-
    emissions-containment-reserve
International Carbon Action Partnership (ICAP) (2020b) USA—Regional Greenhouse Gas Initia-
    tive (RGGI). ETS latest information. https://icapcarbonaction.com/en/?option=com_etsmap&
    task=export&format=pdf&layout=list&systems%5b%5d=50
Isser SN (2016) A review of carbon markets: EU-ETS, RGGI, California, the clean power plan and
    the Paris agreement. SSRN Electron J 1–81. https://doi.org/10.2139/ssrn.2827620
Jiang J, Xie D, Ye B, Shen B, Chen Z (2016) Research on China's cap-and-trade carbon emission
    trading scheme: overview and outlook. Appl Energy 178:902–917. https://doi.org/10.1016/j.ape
    nergy.2016.06.100
Jordan A, Benson D, Wurzel R, Zito A (2011a) Policy instruments in practice. Oxford Handbooks
    Online 536–549. https://doi.org/10.1093/oxfordhb/9780199566600.003.0036
Jordan A, Benson D, Wurzel R, Zito A (2011b) Policy instruments in practice. Oxford Handbooks
    Online 1–17. https://doi.org/10.1093/oxfordhb/9780199566600.003.0036
Kern F, Rogge KS, Howlett M (2019) Policy mixes for sustainability transitions: new approaches
    and insights through bridging innovation and policy studies. Res Policy 48(10). https://doi.org/
    10.1016/j.respol.2019.103832
Mavrot C, Hadorn S, Sager F (2019) Mapping the mix: Linking instruments, settings and target
    groups in the study of policy mixes. Res Policy 48(10). https://doi.org/10.1016/j.respol.2018.
    06.012
Munnings C, Morgenstern RD, Wang Z, Liu X (2016) Assessing the design of three carbon trading
    pilot programs in China. Energy Policy 96:688–699. https://doi.org/10.1016/j.enpol.2016.06.015
Vedung E, Bemelmans-Videc ML, Rist RC (1998) Policy instruments: Typologies and theories. In:
    Vedung E, Bemelmans-Videc ML, Rist RC (eds) Carrots, sticks and sermons: policy instruments
    and their evaluation. Transaction Publishers, New Brunswick, pp 21–58
Wurzel RKW, Zito A, Jordan A (2013) From government towards governance? exploring the role
    of soft policy instruments. German Policy Studies 9(2):21–48

Chapter 3
The Political Economy of Regulation: An Analysis of the Mexican Emission Trading System

Daniela Stevens

Abstract The chapter argues that the design of carbon pricing policies takes place as a sequential, negotiated process whereby specific constituencies have privileged access to shape policy design because they have high stakes in regulations. These groups, identified ex ante based on the political economy of regulation and a stakeholder approach, exhibit two characteristics: first, they are high-interest actors, as a change in the status quo would impose concentrated costs on them; second, they are high-power actors, since their resources and participation in the national economy make them a critical sector. Using theory-guided process tracing and the policy stages heuristics framework, the empirical analysis explores the policymaking process of the Mexican pilot emission trading system and discusses key features of its design.

Keywords Carbon pricing · Political economy · Stakeholder analysis · Policy Stages Mexico · Emission trading system

Introduction

Middle-income countries are at a crossroads between meeting their emission-abatement commitments and their growing energy demands, in a context where the destructive consequences of climate change are increasingly evident. For Mexico, an industrializing middle-income nation, the price of mitigation could represent 15% of the GDP by 2040 (Veysey et al. 2016), yet the impacts of climate change could be more costly. This conundrum begs the question of whether countries that rely on fossil fuel or emission-intensive industries are planning to meet their emission-reduction pledges.

The chapter hypothesizes that the introduction of carbon pricing policies (CPPs) takes place as a sequential, negotiated process whereby constituencies that profit

D. Stevens (✉)
División de Estudios Internacionales, Centro de Investigación y Docencia Económicas, Carr. México-Toluca 3655, Colonia Santa Fe, Álvaro Obregón, 01210 Ciudad de México, CDMX, Mexico
e-mail: daniela.stevens@cide.edu

© The Author(s) 2022
S. Lucatello (ed.), *Towards an Emissions Trading System in Mexico: Rationale, Design and Connections With the Global Climate Agenda*, Springer Climate,
https://doi.org/10.1007/978-3-030-82759-5_3

from a fossil fuel-based economy influence policymakers to lower the stringency of CPPs, contingent on two characteristics. First, if a change in the status quo would impose concentrated costs on them (if they are high-interest actors), and second if their resources and participation in the national economy make them a critical sector (if they are high-power actors). The paper refers to this process as policy shaping, which, in a fossil fuel-based economy, tends to produce outcomes that lower the policy's mitigation potential. Using theory-guided process tracing (Falleti 2016), the empirical analysis explores the policymaking process of the Mexican Pilot Emission Trading System (pilot ETS) and proposes to use a stakeholder framework to identify ex ante the non-State actors with the interest and ability to *shape* the process.

Mexico was the first Latin American nation to tax carbon and to implement a cap-and-trade. This case resonates with other Latin American presidential and multi-partisan systems that currently price emissions and whose economies rely on emission-intensive sectors or fossil or mineral resources, like Colombia, Argentina, and Chile. The Mexican experience is also representative of the challenges and opportunities that nations seeking to implement ETSs—like Colombia, Chile, and Brazil—may find. The implications of tracing the patterns of public–private sector interaction matter, as all nations grapple with the concept of the cost of mitigation and the political and operative hurdles to reach efficient policy outputs that contribute to a fair de-carbonization.

This chapter contributes to the growing body of literature that explores the political economy determinants of national climate action (Fullerton 2011, Harrison 2015, Ervine 2017, Stevens 2021). By understanding climate change policies as political constructs embedded in structural dynamics, it provides criteria for an identification of stakeholders with the ability and resolve to shape CPPs. Further, it highlights the need to increase transparency to contribute to a transition to a low-carbon economy. The text is organized into four sections. Section "Theory and Hypothesis" discusses the theoretical framework and the hypothesis and section "Method" the methodology. The third section traces the pilot ETS, while the fourth draws lessons and closes with concluding remarks.

Theory and Hypothesis

Most environmental economists and political scientists agree that stakeholder engagement is desirable because it leads to legitimate policy outcomes and fosters private sector accountability (Starik 1995; Rodriguez-Melo and Mansouri 2011; Talley et al. 2016; Narassimhan et al. 2018). However, these analyses overlook the power dynamics of stakeholder intervention, which tend to produce outcomes significantly different from the optimal policy.

This chapter analyzes the policy process of the Mexican pilot ETS relying on the literature of the political economy of regulation (Stigler 1971; Posner 1974, 2013; Grossman and Helpman 1994, 2001). Since Stigler's pioneering work, *The Theory of Economic Regulation* (1971), academics have used terms like "regulatory capture" to evidence that policies are more than a tool to resolve market failures and that, in fact, the policymaking process does not necessarily yield optimal results due to the influence of business groups that have high stakes in regulation and seek to create or maintain competitive advantages.

Theoretical additions have refined the overall approach and the concept of capture, noting that influence is not binary but a matter of degree (Carrigan and Coglianese 2016), that elected officials and bureaucratic agents face different incentives (Laffont and Tirole 1991), and that different arenas of regulation involve specific public–private interactions (Sprengel and Busch 2010). Although the postulates of economic regulation have been criticized for overstating the power of business, they still constitute the "type of research needed to equip decision makers" to make better regulation (Carrigan and Coglianese 2016: 10).

Following this literature, the chapter proposes that constituencies will lobby throughout the policymaking process of CPPs to lower the stringency of the regulation if they are:

(a)   High-power stakeholders (HPS), or constituencies with more resources and larger output sizes and

(b)   High-interest stakeholders (HIS), or those with more costs to bear after a change in the status quo.

As Stigler notes, the political process is not akin to an ordinary market, but far more complex, uncertain, and embedded in power relation (Stigler 1971: 12). Similarly, the contention here is not that stakeholders buy ineffective policies, but that actors with power and interest are able to shape policy outcomes, and that governments are responsive to the extent that they rely on these sectors. In this sense, this is not a categorical capture but a policy shaping process, whereby strategic calculations lead to varying levels of policy stringency. With these considerations, the sections below define the criteria to identify the entities that engage in this policy shaping, both HPS and HIS.

(a)   *High-interest stakeholders*

Market-based mitigation policies have costly distributive implications because they entail resource reallocation and aim to change behaviors. A group has high interest in a policy if the instrument would generate concentrated costs or benefits, or if it changes a status quo of concentrated costs or benefits (Wilson 1974). The design of a CPP specifies who pays mitigation costs and how, but as a generality, emission-intensive sectors bear larger costs. When private actors estimate the costs of a potential CPP, they calculate significantly higher costs than government because the cost of capital is higher for private decision makers and they expect unpredictable variations in emissions (Mehling and Dimantchev 2017: 28). Furthermore, mitigation policies may decrease the competitiveness of emission-intensive sectors exposed to trading partners that do not price emissions. This phenomenon is called carbon leakage, whereby energy-intensive firms relocate to jurisdictions with laxer regulatory standards. The result is the same level of emissions but located in a different place.

Other sectors from the civil society, such as environmental organizations, are HIS as well, but they face more collective action issues because they advocate for dispersed benefits and lack the resources of industrial constituencies (Kraft 2017).

(b)   *High-power stakeholders*

The analysis considers two types of power: resource mobilization and structural control. Structural power refers to a sector's relevance in the national economy or its participation in the Gross Domestic Product (GDP). Groups might not exercise structural power as direct pressure, but policymakers are aware that imposing burdens on critical productive engines can harm the economy by lowering the overall economic output and growth rate or creating unemployment. If the regulated sectors are critical for economic growth, governments may attempt to minimize the costs of CPPs and propose lax policies. In turn, structurally powerful actors tend to have more economic capital and the ability to organize into representative bodies, that is, resource mobilization power.

The empirical analysis in section "The Pilot Emission Trading System in Mexico" suggests that only the stakeholders that displayed high power and high interest concurrently (HPS and HIS) had access to policy shaping in the design of the pilot ETS. The following section discusses the process tracing method by which the narrative demonstrates a correlation between the voiced opinions of stakeholders and the policymakers' choices, as well as a change in the initial policy proposals after stakeholders' recommendations.

Method

Jurisdictions have mixed reasons for deciding to adopt CPPs, which include addressing climatic or financial vulnerabilities, domestic and international commitments, or a combination of these factors (Rabe 2008; Krause 2013). However, this chapter does not seek to explain why countries enact CPPs, but to outline how and why they implement policies that deviate substantially from a more ambitious design. The chapter proposes that HPS and HIS mediate within the causal chain to influence the outcome, making policies less stringent than the original formulation.

Using a theory-guided process tracing method (Falleti 2016), this chapter reconstructs one intensive policy process—the Mexican pilot ETS—which had transformative effects on the outcome of interest, stringency. Intensive processes initiate after the triggering cause and end before the outcome (Falleti 2016: 457). The empirical observables include records of meetings, roll-call votes, official documents, newspapers, and half a dozen semi-structured interviews with Mexican officials and industry representatives conducted from 2017 to 2019.

The narrative shows "the how" of lower stringency, finding answers in the distributive effects of CPPs. Timing and order matter in this approach and can ultimately

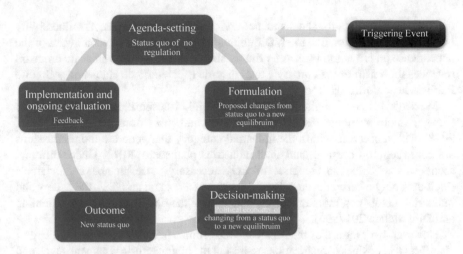

Fig. 3.1 Policy stages heuristic framework. *Source* Author elaboration based on deLeon (1999)

explain how power dynamics translate into lower stringency. Consequently, the narratives rely on the policy stages heuristic framework in Fig. 3.1 to assess each stage sequentially. The triggering factor may be endogenous or exogenous, yet the policy shaping process is inherently endogenous. The analysis considers the initial proposal as the most stringent design that a jurisdiction is willing to impose and traces the policy's origin from the agenda setting stage.

Although critics argue that the stages framework overly simplifies and idealizes a policymaking world laden with power relations and belief systems (see Sabatier 1991; Colebatch 2006), the claim here is not that it has explanatory power, but that it constitutes a useful tool to guide the process tracing and potentially makes systematic comparisons. The stages heuristic framework has been extensively used in the policy analysis world since its introduction by H. Lasswell in 1956, helping to process a policy's complexities more efficiently by assigning attention to each stage (Weible et al. 2012: 5). Although the terminology and number of stages vary widely across authors, the framework has helped create synergies with other fields like historical institutionalism (for example, in the use of the concepts of path dependency and feedback loops) (Howlett et al. 2014).

The Pilot Emission Trading System in Mexico

Projections estimate that Mexico can meet its emission-reduction targets using and profiting from well-designed carbon prices (Altamirano et al. 2016; McKinsey and Company 2013). Energy security needs, such as the increasing demand from industrialization processes and the growing population, have prompted Mexican governments to pursue energy reforms for decades. Additionally, governments have sought to lessen the country's fiscal dependence on fossil fuels and, in turn, its vulnerability in a volatile international market. Different administrations have been aware of the dependence on US natural gas as a primary source of energy, as well as on crude oil revenues for a fifth of the country's total income (Secretaría de Hacienda y Crédito Publico, henceforth SHCP or *Hacienda* 2017).

Mexico's primary emitters are the industrial, transportation, and electricity sectors, which contribute, respectively, 9%, 25%, and 26% to total emissions (INECC 2018). The subsectors within the industrial category that generate more emissions are construction, chemical, and steel industries (Canacero 2016). Under different scenarios that calculate the price of CO_2 necessary to reach abatement targets (including only a carbon tax or a hybrid system of a tax and an ETS), the energy and industrial sectors can more efficiently concentrate the costs of mitigation (Mehling and Dimantchev 2017: 36).

However, over a third of the Mexican GDP comes from industry, and the largest chambers and business organizations represent the manufacturing, cement, steel, and oil sectors. The following narrative, organized as policy stages, shows that these HPS were central in the policy shaping of the Mexican Pilot ETS.

(a) *Agenda-setting*

For over a decade, the international carbon bonds market has been an alternative for financing clean energy projects. Mexico is one of the main project recipients of projects within the United Nations' Clean Development Mechanism (CDM) and of other international projects. Outside the CDM, California invests in Mexican projects through certificates called Climate Reserve Tonnes (CRTs). Mexico also participated in cap-and-trade discussions in the context of NAFTA's Commission on Environmental Cooperation, but as a Non-Annex I country, its role was one of offset provider.

In 2016, the Secretariat of the Environment (SEMARNAT), the Mexican Stock Exchange (*Bolsa Mexicana de Valores* [BMV]), and its subsidiary MexiCO2 signed an agreement to develop an ETS simulation as a tool to comply with the commitments of the Paris Agreement. The platform MexiCO2 offers certificates of emission reductions as carbon credits in exchange of projects developed in Mexico, as well as a service by which companies can pay the carbon tax purchasing credits.

Prior to the announcement, the government took two fundamental steps that established the bases of cap-and-trade. First, the publication of the General Law on Climate Change (LGCC) in 2012, which included guidelines for developing mitigation instruments. Second, the creation of the National Greenhouse Gas Emissions Register (RENE) for facilities and companies in 2014. Furthermore, in 2015, SEMARNAT started a feasibility assessment of a mandatory system.

Different interviewees credit the deputy secretary of SEMARNAT, Rodolfo Lacy Tamayo, as the political entrepreneur who, as part of the "presidential inner circle," introduced a compliance carbon market into the agenda (Alarcón 2017; Escalona 2017; Muñoz Piña 2016). The idea, which arose from technical discussions on tools to mitigate emissions, found international traction at the Partnership for Market Readiness (PMR), an international organization built around the World Bank's governance that supports the formulation and implementation of mitigation policies. Given the relative novelty of an ETS and its technicalities, SEMARNAT and the PMR launched a "simulation market exercise" (2017–2018) to familiarize the Mexican private sector with cap-and-trade systems, before the compliance market started.

(b)   *Formulation and decision-making*

The experiences of formulating and approving the carbon tax in 2013 and the Energy Transition Act (ETA) in 2015 underscored the trade-off between stringency and feasibility, as well as the need to involve interested HPS in the decision-making process. However, unlike these pieces of legislation, the ETS was a policy program, so the executive would lead both the formulation and decision-making processes without debates in legislative committees or the floor, and without official records. Still, the Senate held fora where legislators discussed the future of cap-and-trade and voiced their desire to make the Environmental Commission permanent, since it only met intermittently (Comunicación del Senado, August 16, 2017).

Officials in different government bureaus usually decide to involve stakeholders from the outset based on strategic concerns related to feasibility given that the willingness of the entities that the ETS would regulate is essential for these systems to operate. Because in Mexico, firms can file a motion of *amparo*, preliminary negotiations with the high-interest and high-power stakeholders are fundamental to secure compliance. The amparo proceeding is a legal resource that protects citizens and entities against official mandates and procedures if they can demonstrate any violation of their rights, abuse of power, or unconstitutionality. The chances that a policy reaches enactment increase if key parties and the public sector reach a consensus in the formulation stage, so officials across the globe must choose between lowering the program's stringency or facing sabotage.

SEMARNAT created a Working Group that included representatives from the federal government and the private sector, without broader participation of academia or civil society. In their sessions, the group found "common ground on what are usually divergent positions" (International Carbon Action Partnership [ICAP] 2019a, b). In other words, although regulators had to make concessions, the stakeholder engagement was fundamental to make cap-and-trade a politically feasible alternative in Mexico.

Environmental officials anticipated legal obstacles by introducing a policy with high costs on stakeholders with structural power. Article 94 of the 2012 version of the General Law of Climate Change (LGCC) considered a voluntary carbon market as a possible strategy, but it did not warrant the establishment of a compliance market, a legal hurdle that concerned the agency. Interviewees affirmed that SEMARNAT could use a lax interpretation of the law's wording, arguing that while trading emissions was voluntary, the upper limit was mandatory (Interviewee 3, 2017), but that they favored amending the law in Congress. Finally, the amendments gained Senate approval in April 2018 and gave SEMARNAT the mandate to establish a compliance market.

In interviews during the formulation and decision-making stages, SEMARNAT representatives alluded to industrial security and the need to keep the discussions and the policy's design secret while affirming that the pilot would not operate under the same rules as the market (Nieto 2018). However, they acknowledged that the ETS would almost inevitably include a free allocation of permits to some industries including steel (Escalona 2017; Nieto 2018).

Moreover, SEMARNAT authorities admitted to a lack of communication with the SHCP, which presented potential challenges of interaction with the carbon tax. The transversal coordination of executive agencies is critical to avoid inefficiencies as jurisdictions add pricing mechanisms to the policy portfolio. In the Mexican context, coordination would involve at least three executive agencies, SEMARNAT, the Secretariat of Energy (SENER), and the fiscal SHCP. Their lack of communication formulating the ETS, evident to policymakers (Escalona 2017; Interviewee 1, 2016; Interviewee 3, 2017), raised serious concerns about policy interactions because it can result in weak price signals. SEMARNAT officials claimed that more rigorous coordination was not necessary because Secretariats know each other's position, "a commitment to reduce emissions at the lowest cost" (Escalona 2017). While Secretariat's relative independence prevents impasses within agencies, it might also result in a mitigation regime that inefficiently combines elements of price setting and quantity rationing.

Potential participants claimed that the costs of setting up the system were too high, that Mexico was not a large emitter globally, compared to the US and China, and that the ETS would lead to loss of competitiveness (Reboulen, November 14, 2017). Further, potential national and international participants of the market perceived that a new presidential administration could have a different take on climate change mitigation commitments from 2018 on. If indeed the personal closeness of a political entrepreneur to president Enrique Peña Nieto encouraged the development of the ETS, the replacement of this official could impair the costly efforts to set up a cap-and-trade with the president from an opposition party with an agenda that relies distinctly on fossil fuels, Andrés Manuel López Obrador.

The HPSs that were vocal about their discontent with the ETS were, predictably, energy-intensive industries. In general, Mexican industry has used the qualification of "revenue collection mechanism" as a criticism of carbon pricing policies in general, implying that the environmental objectives are an excuse to collect revenues. Another commonly reproduced criticism was that the final consumers would pay the price. For example, directives of one of the main steel consortiums, DeAcero, argued that an ETS would decrease productivity and result in overall market distortions, especially regarding salaries. The National Chamber of the Transformation Industries, CONCAMIN, echoed the "means to collect revenue" criticism and emphasized that consumers would pay the price (Reboulen, November 14, 2017). Industrial representatives could credibly threaten that the ETS would hurt the national economy, which, in turn, would damage the president's party, *Partido Revolucionario Institucional* (PRI) during 2018, an electoral year. Still, with the help of the German Agency for International Cooperation (GIZ) and the World Bank's PMR, SEMARNAT's expert bureaucracy kept conducting several studies that provided an analytical basis for the ETS's design.

Peña Nieto's administration did not enact the cap-and-trade during its six-year tenure (2012–2018) and rescheduled the launching to 2020 when López Obrador would be in power. In October 2018, less than two months before the end of the presidential term, SEMARNAT published the draft of the system's preliminary rules, alongside the details of a consultation process. However, the process was soon thereafter suspended—and the rules taken down from the official webpage—to allow the incoming presidency to conduct the procedure (ICAP 2018a, b).

The new administration started a public consultation process that took place between March 2018 and May 2019 and published the preliminary rules in October 2019 without significant changes. Besides the regulations outlined in the following section, the preliminary rules established a public–private consultative committee that would serve as a technical and advisory body for consultation and would be comprised by invitation of SEMARNAT's Under Secretariat for Planning and Environmental Policy. The Committee's composition differentiates between two groups of stakeholders: those with a voice and vote during the discussions, and consultation partners who only have a voice. Groups that were both HPSs and HISs in the process constitute the former, while academics and civil society organizations, that is, groups that were only HISs, the latter. The committee was inaugurated in June 2020.

(c)   *Outcome*

While legislators approved the carbon tax in two months, it took four years to enact the pilot ETS. In 2020, the system entered a pilot phase that allocates free credits and does not sanction non-compliance. The 2018 amendments to the LGCC approved the establishment of a compliance market under the condition that the pilot phase did not lead to "negative economic impacts for the participating sectors."

The trial program will last 36 months and will consist, in turn, of two periods: A pilot phase (January 2020 to December 2021), and a transition phase (January to December 2022). The ETS is expected to have a real price signal in the operational phase in 2023, but the rules have not been established.

Different international experiences demonstrate that opposition to cap-and-trade can be addressed if formulators introduce flexibility in the ETS design, which makes it feasible, yet economically inefficient. Arguably, the elements in the design of the policy most crucial to both feasibility and efficiency are: (a) the cap, (b) the mechanism of permit distribution, and (c) the participants covered.

According to the preliminary rules, the trial program will cover the energy and industry sectors and participants will be the companies whose annual emissions have been equal to or greater than 100 thousand tons of emissions of carbon dioxide (tCO_2) in 2016, 2017, 2018 or 2019 (Diario Oficial de la Federación [DOF] 2019). Compared to other jurisdictions, Mexico opted for a conservative inclusion threshold, unlike California, which set the threshold at 25 thousand tCO2, or Beijing's at 5 thousand tCO2. With this threshold, the pilot ETS will cover the 308 firms responsible for around 96% of sectoral emissions and 45% of national emissions. Whereas a larger number of entities is more conducive to market liquidity, most programs begin including only the largest emitters (such as industrial and power plants) since the inclusion of small emitters increases transaction costs (Butzengeiger et al. 2001).

Only direct carbon dioxide emissions will be covered during the trial program. By November 1st of each year, participants must submit a number of allowances to SEMARNAT equivalent to the emissions reported and verified from the immediately previous year. Participants must also present a report and a verification opinion regarding the emissions that they will report.

Regarding the level, a cap that is too high is problematic because it will not bring about the desired emission reduction. This leads, in turn, to an overallocation of permits. High caps and the subsequent overallocation of allowances can depress prices and further undermine the effectiveness of the program. A trajectory that is initially conservative can be consistently adjusted to achieve targets, which may imbue a sign of continuity irrespective of political and electoral calendars.

The political economy of permit allocation has been widely studied; while some argue that ETSs are part of the climate change mitigation toolbox only because they are subject to manipulation, others claim more categorically that permits are "constructed from political whole cloth" and distributed according to political, not technical criteria (Sagoff 2008).

Evidence supports the fact that most of the current cap-and-trade programs exist because their first phase entailed a free allocation of permits based on historical emissions, or grandparenting. For example, the overall cap of the first phase of the European Union ETS was set around 5% above their business as usual (BAU) levels, and sectors like cement were allocated 105% of their BAU emissions (McAllister 2009: 410; Weishaar 2014: 101; Ervine 2017: 9). Similarly, the ability to allocate free allowances "to address differential economic impacts" across industries, states, and Congressional districts was decisive for the implementation of the SO_2 cap-and-trade in the US (Schmalensee and Stavins 2013).

In most cases, there is some evidence that participants have advocated not only for free allocation, but also for a higher number of permits per firm. While interest groups have the right to access policymaking through different participatory mechanisms, the lack of transparency makes policy processes more prone to delays and blockage, and even raises suspicions regarding rent-seeking behaviors. Auctioning permits is considered the most efficient way to mitigate greenhouse gases, yet evidence also shows that most ETS participants oppose auctioning at least in the initial stages.

Mexico will allocate free permits at the beginning of the trial program based on the information reported to the National Registry of Emissions (RENE). However, SEMARNAT is in charge of implementing auctions from the second year of the pilot phase of the trial program, whereby regulated firms will be able to buy allowances. According to the preliminary rules, however, the inclusion of this mechanism is still "contingent on the behavior of the market." Moreover, the rules include the creation of a secondary market, that is, transferring permits between regulated companies.

A question remains regarding a case in which the Mexican government did not meet its commitments, namely, whether it would be able to purchase emission allowances from other markets. One of the ways in which Peña Nieto framed the ETS was to highlight its potential linkage to the California-Quebec ETS. In December of 2017, the government signed the "Declaration on Carbon Prices in the Americas" with Canada, Colombia, Chile, and the authorities of California, Washington State, Ontario, and Quebec, as an agreement to promote intra-regional carbon markets and a system of standardized prices in the framework of the One Planet summit in Paris. The signatories are committed to working together to strengthen the monitoring, reporting, and verification (MRV) systems of carbon emissions. The ultimate goal was a market linkage, taking advantage of the potential within the hemisphere. Even though the mention of linkage with this market was recurrent in the media, officials at SEMARNAT dismissed it even as a long-term goal (Escalona 2017; Interviewee 3, 2017), and assured that the cooperation with California has been strictly related to sharing technical expertise. A former SEMARNAT official affirmed that even if the Mexican market were ready, Californian market participants would be reticent to establish a linkage with a Mexican market that they fear will not endure political transitions (Interviewee 3, 2017).

(d)   *Implementation and evaluation*

Monitoring, reporting, and verification of emissions will be done in accordance with the monitoring plan issued by SEMARNAT, as well as with the technical provisions established in the LGCC eDuring the pilot phase, emissions must be reported through the Allowance Monitoring System, an electronic platform whereby issued, transacted, and canceled. SEMARNAT must still determine how to present the verification reports.

During the trial program's transition phase, SEMARNAT must establish the rules for the operational phase while considering the results of the program. However, if the agency does not issue the rules, the trial program's rules will remain in place for another 6 months after the end of the transition phase (DOF 2019).

The transition to a market that sends an authentic price signal depends on these key pending regulations. Particularly, congress will have to amend the LGCC again in order to include enforcement mechanisms and sanctions. Implementation during the pilot phase will be decisive to keep developing the program, as it will highlight the challenges and areas of opportunity for the ETS, and reveal a different phase of stakeholder engagement. As Mexico expands its climate policies portfolio, HISs keep gaining experience and building networks. New influential groups are emerging as well, alongside the incumbent HPSs, such as the private regulatory bodies needed to operate, administer, and oversee new markets.

Lessons Learned and Concluding Remarks

This chapter examined the policy process of making an emissions trading system in Mexico. It revealed the politicized nature of the formulation and decision-making stages and provided criteria to identify ex ante the actors likely to lower policy stringency based on potential distributive implications, namely, high-interest stakeholders with high power.

Stakeholder engagement is a normalized and desirable practice in democracies, and indeed necessary to make carbon pricing politically feasible. However, the case suggests that policymakers interact with stakeholders selectively, prioritizing the engagement of actors that have both power and interest. Although high-interest actors like academia are involved in informative, participative, and networking processes, their voices are secondary in the design of emissions trading. High-power stakeholders (HPS) tend to be among the top productive forces, and have the resources to mobilize. This selective engagement creates opportunities for an unbalanced policy shaping, which, in turn, may weaken the energy transition's perceived fairness. Not only does unequal access to policymaking stymie mitigation goals, but it also renders a policy's legitimacy questionable, and does not contribute to the achievement of a just transition.

The interested HPS opposed the ETS initially, arguing a variety of market distortions, such as loss of competitiveness, carbon leakage, depression of wages, and the low reliability of renewables. These constituencies could credibly threaten to pass on the costs to the society, or harm a country's economic output, which gave them an undeniable leverage vis-à-vis regulators. Paradoxically, whereas the influence of high-power prospective participants weakens a policy, it also increases probabilities of enactment. The participation of structurally powerful sectors is critical because regulators need their compliance, and because their inclusion in the policy process fosters trust among parties. Even though emission-intensive sectors generally undervalue the social cost of carbon, pricing emissions is unfeasible without their input.

Emerging actors such as renewable power producers have the potential to influence the design of mitigation policies as well, yet Mexican society still needs to experience the benefits of renewables firsthand. Although environmental NGOs have grown in institutional capacity, visibility, experience, and international connections, they still lack the resources and national economic participation of industrial constituencies.

The case warrants caution about the fossil fuel and power industries, absent in these processes because the state monopolized them until late 2013. Mexican energy reform by which State-owned Pemex lost its monopoly in the sector, began restructuring the oil and electricity sectors. Beginning in 2017, private entities could produce their own gasoline brand, and from 2018 onwards, gasoline imports became tax-free. These changes may increase the complexity of the policymaking scenario as stakeholders with conflicting interests emerge and grow in power.

Tracing this policy in Mexico bolsters our understanding of the phenomenon of designing ETSs in nations that share structural characteristics. The lessons are indicative of processes in other jurisdictions that share relevant macro-level traits like presidential multi-partisan political systems, and a reliance on emission-intensive sectors or fossil fuels like Colombia.

The findings have relevant policy implications pertaining to transparency and participatory processes. As the 2018 report of the IPCC acknowledges in the innovative Chaps. 4 ("Strengthening and Implementing the Global Response") and 5 ("Sustainable Development, Poverty Eradication and Reducing Inequalities") socio-cultural legitimacy is vital to increasing the ambitiousness and feasibility of mitigation targets, especially as they relate to industrial and private sector acceptance (IPCC 2018: 316, 389). Additionally, the report emphasizes the need for environmental justice via stringent policies, and fair share debates on responsibility, capability, and the right to development (IPCC 2018: 470). The key aspect is that policymakers hear all voices in the design and implementation of policies, so that the population perceives that moving toward a low-carbon economy is an inclusive, ethical, and fair process. Examples like the Yellow Vest Movement in France show that transitions cannot be imposed without transparency and social justice.

The chapter evidences that mitigation strategies depend both on structural factors and on interest-based strategic calculations and shows that selective stakeholder engagement does not necessarily lead to better policies. However, whereas transparent policymaking contributes to increased accountability and legitimacy, some degree of confidentiality may also be fundamental to imbue trust to negotiations with key participants.

Although the political economy perspective of the analysis highlights the costs and benefits of emissions commodification, outcomes are not the result of fixed and homogeneous dominant interests. Albeit slowly, actors adjust their strategies in time; firms that constantly seek ways to reduce costs may find approaches that may overlap with mitigation such as energy efficiency measures.

International agreements have paved the way to ratchet-up the ambitiousness of emission-reduction commitments, yet these pledges remain dependent on intricate domestic strategies. The alleged inability to "afford" climate policies expands across the globe, as developed and developing nations face the apparent but false dilemma of striving for economic growth or climate mitigation.

References

Altamirano JC et al (2016) Achieving Mexico's climate goals: an eight point action plan. World Resources Institute (WRI), Washington, DC
Butzengeiger S, Betz T, Bode S (2001) Making GHG emissions trading work—crucial issues in designing national and international emissions trading systems. In: EconStor, Hamburg Institute of International Economics
Canacero (2016) Suplemento de la Industria Siderúrgica, México. http://www.canacero.org.mx/en/assets/suplemento-siderurgia-2016.pdf. Accessed 12 July 2017
Cantala D, McKnight S, Sempere J (2013) Designing a greenhouse gas emission market for Mexico. Environ Ecol Res 1(3):135–141. https://doi.org/10.13189/eer.2013.010303.Vol.1(3),pp.135-141
Carrigan C, Coglianese C (2016) George Stigler's. 'The theory of economic regulation'. In: Lodge M, Page EC, Balla SJ (eds) The Oxford handbook of classics in public policy and administration. Oxford University Press

Colebatch HK (ed) (2006) Beyond the policy cycle: the policy process in Australia. Allen & Unwin, Crows Nest

deLeon P (1999) The stages approach in the policy process: what has it done? In: Theories of the policy process. Westview, Boulder

DOF (2019) Acuerdo por el que se establecen las bases preliminares del Programa de Prueba del Sistema de Comercio de Emisiones, 01/10/2019. http://www.dof.gob.mx/nota_detalle.php?cod igo=5573934&fecha=01/10/2019. Accessed 2020

EPA (2016) The social cost of carbon, 2016. https://19january2017snapshot.epa.gov/climatechange/ social-cost-carbon_.html. Accessed 13 March

Ervine K (2017) How low can it go? Analysing the political economy of carbon market design and low carbon prices. New Polit Econ 23(6):690–710. https://doi.org/10.1080/13563467.2018.138 4454

Falleti TG (2016) Process tracing of extensive and intensive processes. New Polit Econ 21(5):455–462. https://doi.org/10.1080/13563467.2015.1135550

Fullerton D (2011) Six distributional effects of environmental policy. National Bureau of Economic Research, Working Paper No. 16703

Grossman GM, Helpman E (1994) Protection for sale. Am Econ Rev 84:833–850

Grossman GM, Helpman E (2001) Special interest politics. Massachusetts Institute of Technology, Cambridge

Harrison K (2015) International carbon trade and domestic climate politics. Glob Environ Polit 15(3):27–48. https://doi.org/10.1162/glep_a_00310

Howlett M et al (2014) Streams and stages: reconciling Kingdon and policy process theory. Eur J Polit Res. https://doi.org/10.1111/1475-6765.12064

ICAP (2018a) UPDATED: Mexico releases draft regulations for pilot ETS system. https://icapca rbonaction.com/en/news-archive/586-mexico-releases-pilot-ets-regulations-for-system-starting-in-2019

ICAP (2018b) Emissions trading worldwide status report 2019. https://icapcarbonaction.com/en/ news-archive/586-mexico-releases-pilot-ets-regulations-for-system-starting-in-2019. Accessed 20 Aug 2020

ICAP (2019a) Emissions trading worldwide, status report 2019. ICAP, Berlin

ICAP (2019b) Emissions trading worldwide status report 2019. https://icapcarbonaction.com/en/ icap-status-report-2019. Accessed 20 Aug 2020

INECC (2018) Inventario Nacional de Emisiones de Gases y Compuestos de Efecto Invernadero, México, 2018. https://www.gob.mx/inecc/acciones-y-programas/inventario-nacional-de-emisio nes-de-gases-y-compuestos-de-efecto-invernadero. Accessed 30 April 2017

IPCC (2018) Special report. Global warming of 1.5 °C, WMO, UNEP, 2018. https://www.ipcc.ch/ sr15/. Accessed 20 Jan 2019

Kraft M (2017) Environmental policy and politics, 7th edn. Routledge, New York

Krause RM (2013) The motivations behind municipal climate engagement: an empirical assessment of how local objectives shape the production of a public good. Cityscape 15(1):125–141

Laffont J-J, Tirole J (1991) The politics of government decision-making: a theory of regulatory capture. Quart J Econ 106:1089–1127

Ley General de Cambio Climático (Mexico General Law on Climate Change) (2012). http://www. profepa.gob.mx/innovaportal/file/6583/1/ley_general_de_cambio_climatico.pdf. Consulted 28 April 2020

McAllister LK (2009) The overallocation problem in cap-and-trade: moving toward stringency. Columbia J Environ Law 34(2):395–445

McKinsey & Company (2013) Updated analysis of Mexico's GHG emissions baseline, the marginal abatement cost curve and project portfolios. United States Agency of International Development, Washington, DC

Mehling M, Dimantchev E (2017) Achieving the Mexican mitigation targets: options for an effective carbon pricing policy mix. MIT, Deutsche Gesellschaft für Internationale Zusammenarbeit (GIZ), SHCP

Narassimhan E et al (2018) Carbon pricing in practice: a review of existing emissions trading systems. Clim Policy 18:8
Posner R (1974) Theories of economic regulation. Bell J Econ 5(2):335–358
Posner R (2013) The concept of regulatory capture: a short, inglorious history. In: Carpenter, Moss (eds) Preventing regulatory capture: special interest influence and how to limit it. Cambridge University Press, New York, pp 49–56
Rabe BG (2008) States on steroids: the intergovernmental odyssey of American climate change policy. Rev Policy Res 25(2):105–128
Reboulen JA (2017) Siderurgia perdería competitividad con bonos de carbón. El Financiero, November 14th. http://www.elfinanciero.com.mx/monterrey/siderurgia-perderia-competiti vidad-con-bonos-de-carbon.html. Accessed 1 April 2018
Rodriguez-Melo A, Mansouri SA (2011) Stakeholder engagement: defining strategic advantage for sustainable construction. Bus Strat Env 20:539–552
Sabatier PA (1991) Toward better theories of the policy process. PS: Polit Sci Polit 24(2):144–156
Sagoff M (2008) The economy of earth: philosophy, law, and the environment, 2nd edn. Cambridge, CUP
Senado de la República (2017) Comunicación del Senado, Boletín de Prensa, August 16. http://com unicacion.senado.gob.mx/index.php/informacion/comision-permanente/boletines-permanente/ 46-grupos-parlamentarios/boletin-de-prensa/38074-analiza-senado-implementacion-de-un-sis tema-de-comercio-de-emisiones-de-carbono.html. Accessed 1 June 2018
SHCP (2017) Estadísticas Oportunas de Finanzas Públicas, México. http://www.shcp.gob.mx/POL ITICAFINANCIERA/FINANZASPUBLICAS/Estadisticas_Oportunas_Finanzas_Publicas/Pag inas/unica2.aspx. Accessed 10 June 2018
Sprengel DC, Busch T (2010) Stakeholder engagement and environmental strategy—the case of climate change. Bus Strateg Environ 20(2011):351–364. https://doi.org/10.1002/bse.684
Starik M (1995) Should trees have managerial standing? Toward stakeholder status for non-human nature. J Bus Ethics 14(3):207–217. https://doi.org/10.1007/bf00881435
Stigler GJ (1971) The theory of economic regulation. Bell J Econ Manag Sci 2(1):3–21. https://doi. org/10.2307/3003160
Talley JL, Schneider J, Lindquist E (2016) A simplified approach to stakeholder engagement in natural resource management: the five-feature framework. Ecol Soc 21(4). https://doi.org/10. 5751/es-08830-210438
Veysey J, Octaviano C, Calvin K, Martinez SH, Kitous A, McFarland J, Van der Zwaan B (2016) Pathways to Mexico's climate change mitigation targets: a multi-model analysis. Energy Econ 56:587–599. https://doi.org/10.1016/j.eneco.2015.04.011
Weible et al (2012) Understanding and influencing the policy process. Policy Sci 45(1):1–21. https:// doi.org/10.1007/s11077-011-9143-5
Weishaar SE (2014) Emissions trading design: a critical overview. Edward Elgar, Cheltenham
Wilson JQ (1974) Political organizations. Basic Books, New York

Interviews

Alarcón S (2017) Director of Carbon Trust Mexico, April 2017, Mexico City
Escalona V (2017) Subdirector of Mitigation Policy, SEMARNAT, July 2017, Mexico City
Interviewee 1, August 2016, Mexico City
Interviewee 2, April 2017, Mexico City
Interviewee 3, April 2017, Mexico City
Muñoz Piña C (2016) General Director of Revenue Policy, SHCP, August 2016, Mexico City
Nieto I (2018) Director of Sectoral Models of the National Institute for the Ecology and Climate Change, June 2018

Schmalensee R, Stavins RN (2013) The SO_2 allowance trading system: the ironic history of a grand policy experiment. J Econ Perspect 27(1):103–122

Secretaría de Gobernación, México (2019) Acuerdo por el que se establecen las bases preliminares del Programa de Prueba del Sistema de Comercio de Emisiones, DOF (Diario Oficial de la Federación). https://www.dof.gob.mx/nota_detalle.php? codigo=5573934&fecha=01/10/2019

Stevens D (2021) Institutions and agency in the making of carbon pricing policies: evidence from Mexico and directions for comparative analyses in Latin America. J Compare Policy Anal: Res Practice 23(4):485–504. https://doi.org/10.1080/13876988.2020.1794754

Chapter 4
Moving Towards an ETS in Mexico: The Case of International Cooperation

Neydi Cruz and Mireille Meneses

Abstract Mexico has participated in different international climate initiatives and has benefited from international collaboration. This cooperation, both at the political and technical levels, has been crucial for the design and implementation of the national carbon market. Through its climate diplomacy leadership, Mexico has played a key role in international carbon pricing initiatives, and in the technical sphere, the country has benefited from peer-to-peer international experiences and knowledge. This chapter analyzes those initiatives and their contribution to continue broadening collaboration towards a carbon market in the country. It explores how recent changes to the environmental agenda, adopted as of 2018 by the new federal administration, could hinder the implementation of the market mechanism.

Keywords International cooperation · Development assistance committee · Mitigation · Trade

Introduction

Since becoming a signatory to the United Nations Framework for Climate Change (UNFCCC) in 1994, Mexico has pursued the implementation of public policies to ensure sustainable development and to address environmental degradation. The exposure of the country to a climate agenda and the historic discussions around the decarbonization of the energy sector influenced the development of public policies and substantial reforms, particularly in the energy sector (Valenzuela and Studer 2016). This climate leadership has been materialized into domestic economic instruments to promote environmental protection, phase out fossil fuel subsidies and foster the use of alternative sources of energy. These policies in Mexico have sought to respond

Electronic supplementary material The online version of this chapter
(https://doi.org/10.1007/978-3-030-82759-5_4) contains supplementary material, which is available to authorized users.

N. Cruz (✉) · M. Meneses
Instituto Mora-CONACYT, Mexico City, Mexico

© The Author(s) 2022   65
S. Lucatello (ed.), *Towards an Emissions Trading System in Mexico: Rationale, Design and Connections With the Global Climate Agenda*, Springer Climate,
https://doi.org/10.1007/978-3-030-82759-5_4

to an incipient need to address the climate crisis and have evolved in parallel with the international conversations led by the UNFCCC.

Climate change discussions in the country gained relevance with the foundation and entry into force of the UNFCCC and when the negotiations of the Kyoto Protocol started back in 1995. Despite not being part of the countries with an emission reduction goal within the Kyoto Protocol (Annex I countries), Mexico was one of the first UNFCCC countries to ratify it in 2000 (United Nations 2020). Climate advocacy in the country has been variable across government administrations. Four presidential terms have passed since Mexico's adhesion to the UNFCCC, but only two terms stand out for the rate of policy development and bold commitments. This could have been partly influenced by personal beliefs or to keep pace with a constantly changing international agenda (Balderas Torres et al. 2020; Valenzuela and Studer 2016). Regardless of the political divisions in its government, Mexico's climate agenda advanced steadily (Meirovich 2014) and helped position itself as a pioneer in the Latin American region.

Mexico's advocacy for a strong environmental policy, and more recently for carbon pricing, positioned the country as a regional leader. However, as the entry into force of the Paris Agreement approached, some of the political work had been done, yet much of the technical problem remained. The implementation of carbon pricing was not clear to many countries and this indicated the lack of technical capacities for ensuring effective implementation.

Several countries and institutions that backed the implementation of carbon pricing also supported the founding of larger advocacy groups for promoting capacity-building through international cooperation. Thanks to the thought leadership demonstrated in the process, Mexico attracted the attention of international partners who supported the country in building its climate policy. Over the next sections, we will present the main political milestones that take Mexico to the adoption of one of the most ambitious efforts for climate change mitigation: a national carbon market.

The International and National Context of Market-Based Mechanisms

The Paris Agreement seeks to stop the division that existed in the Kyoto era between Annex I countries and the rest of the world. The Agreement recognizes the common but differentiated responsibilities of the members and calls for a coordinated response to climate change. It also acknowledges the importance of international cooperation and support among countries in order to achieve global mitigation and adaptation goals. The Paris Agreement itself explicitly refers to *international cooperation* on adaptation efforts and relating to the outcome of the global stocktake. However, it can be implicitly found in the cooperative approaches envisioned as part of Article 6.

Article 6 of the Paris Agreement promotes cooperation among countries in order to achieve their national mitigation goals. Article 6.2 refers to the use of voluntary cooperative approaches to facilitate emissions reductions that can be translated into internationally transferred mitigation outcomes (ITMOs). With this, the Paris Agreement recognizes the existence of market-based instruments as a means to catalyze climate finance, promote technology transfer and enable GHG emissions reduction. The rules on how this mechanism will work are still a work in progress, however, countries such as Canada, Japan, Switzerland and Sweden have started to negotiate bilateral agreements to test the Article ITMO transactions (Greiner et al. 2019).

Cooperative responses like these are required because most greenhouse gases accumulate over time and mix globally. Effective mitigation will not be achieved if individual agents advance their own interests independently (IPCC 2014). International cooperation can provide an enabling environment for the implementation of ambitious actions implied by limiting global warming to 1.5 °C. It could then be achieved in all countries and for all people, in the context of sustainable development. International cooperation is a critical enabler for developing countries and vulnerable regions (IPCC 2018).

Over the presidential period from 2006 to 2012, climate change and environment policies were considered a priority in Mexico in terms of political leadership and budget allocation to the Ministry of Environment and Natural Resources (SEMARNAT). During this six-year period, the country enhanced its climate leadership by hosting the Conference of the Parties (COP)-16 in Mexico, pursuing the establishment of the Green Climate Fund and publishing domestic planning instruments such as the National Strategy on Climate Change (NSCC) and the Special Program of Climate Change (SPCC). Other sectoral programmes to address issues such as deforestation and forest degradation were also established. All these efforts were materialized at the end of the administration with the enactment of the General Law on Climate Change (GLCC) in 2012 (Balderas Torres et al. 2020).

This climate leadership in Mexico was sustained over the subsequent presidential period (2012–2018), however, its importance decreased during the first couple of years when other social and economic issues were under the spotlight (Balderas Torres et al. 2020). Climate change in Mexico gained traction towards the COP-21 in 2015 when countries were expected to present their national emission reduction targets and agree on a new climate change accord. This presidential term showcased five major milestones in the domestic climate agenda: the establishment of a carbon tax applicable to fossil fuels (D.O.F. 2013); the submission of the Nationally Determined Contribution (NDC) to the UN; the approval of the Energy Transition Law (ETL) (D.O.F. 2015) to promote the development of renewable energy projects and the national strategy for reforestation and forest degradation projects (CICC 2017); and the creation of a national carbon market. Towards the end of this period and after the entry into force of the Paris Agreement, the GLCC was reformed to incorporate the provisions of the Agreement and the national emission reduction pledges. Besides streamlining the principles of the Agreement, the GLCC also established a mandatory Emissions Trading System (ETS) in the country to enter into force in 2019.

The experience of the country with market-based mechanisms as an instrument for climate change mitigation dates back to the entry into force of the Kyoto Protocol in 2005. The Protocol established flexibility mechanisms such as Joint Implementation (JI), the International Emissions Trading (IET) and the Clean Development Mechanism (CDM) to assist Annex I countries in the achievement of their emissions reduction targets. While the first two mechanisms focussed on the implementation of mitigation activities in developed countries, the CDM was the only project-based mechanism that opened the door for mitigation activities hosted in developing countries.

Upon the entry into force of the Protocol, the CDM was perceived as an opportunity for catalyzing investment in Mexico and this interest resulted in the creation of domestic instruments to facilitate understanding of the CDM. In 2006, SEMARNAT, the National Bank for Foreign Trade (BANCOMEXT) and the World Bank Group signed a Memorandum of Understanding that formalized the cooperation for the design of the Mexican Carbon Fund known as FOMECAR. This instrument operated as a technical assistance and financing mechanism to promote CDM projects in the country. FOMECAR responded to the need for a mechanism that would combine efforts to identify, promote and develop GHG mitigation projects under the CDM. In order to finance CDM projects in Mexico, other cooperation instruments in the matter were signed with countries such as Austria, Canada, Denmark, France, Germany, Italy, Japan, the Netherlands, Spain and Portugal (INE 2006).

These agreements enabled the development of CDM projects that increased climate finance to mitigation activities in the country, mainly driven by the private sector. The energy, waste and transport sectors benefitted largely from CDM transactions. The Climate Change Committee in the Energy Sector was instituted to identify potential mitigation opportunities under the CDM in Mexico. It was coordinated by the Ministry of Energy (SENER) and led by relevant energy institutions in Mexico, such as PEMEX. This Committee identified several renewable energy projects, as well as methane emissions reductions for the hydrocarbon sector. PEMEX identified nearly 44 potential CDM projects and therefore, signed agreements with the World Bank and the Inter-American Development Bank (IDB) to expand its opportunities under future carbon markets (INE 2006). Given the large contribution of the energy sector in Mexico to national GHG emissions, it could be argued that its involvement in the CDM, particularly the case of PEMEX, was one of the main drivers that triggered the interest of Mexico in pursuing market-based mechanisms as a cost-effective option for reducing GHG emissions.

To date, approximately 192 Mexican projects are registered under the CDM, however, only 72 projects have issued Certified Emission Reductions (CERs). The interest in CDM projects decreased between 2012 and 2013 as a result of the collapse in global market prices (Harvey 2012). Consequently, many of the projects registered during that period were unable to complete the validation and verification phases or were simply withdrawn in the process. Projects registered before 2013 continued to generate CERs in subsequent crediting periods, however, the demand for CDM credits decreased as well. The GLCC (D.O.F. 2012) highlighted the importance of channelling international resources to finance GHG mitigation projects through public policies. Mexico tried to reactivate the interest of CDM projects by including a compensation mechanism to meet obligations under the carbon tax. The rules for this compensation mechanism were approved until 2017 and so far, companies have not used CDM credits in lieu of their obligations under the carbon tax.

The Kyoto Protocol was an international instrument pioneering in the implementation of market instruments for meeting emission reduction goals. Different countries including Mexico, private actors, international institutions, and civil society organizations around the world recognized the need to include similar market mechanisms in the new global climate order. Such actors led the creation of various initiatives and alliances that sought to influence the use of carbon prices as a key policy instrument to combat climate change. These efforts materialized in the inclusion of market-based instruments to support the achievement of the Paris Agreement goal to limit global warming to below 2°.

Mexico has actively participated in several international initiatives, partnerships and readiness funds which aim to support not only the development of carbon markets, but also to enhance capacity building. These initiatives assisted the country in designing its national ETS from a political and technical level.

The international public advocacy work began in 2014 at the UN Climate Summit, where 73 subnational governments, 11 local governments and more than one thousand businessmen—representing 54% of the global GHG emissions, 52% of the world's GDP and almost half of the world's population of that time—subscribed the first global carbon pricing statement, giving a positive signal to the private sector and the need to invest in climate solutions (World Bank 2014). Mexico, at the head of the state level, was part of this launching.

The WB led this international call not only with the federal government, but also with congressmen. In June 2014, Mexico hosted the 2nd World Summit of the Global Legislators Organization for a Balanced Environment (GLOBE) where the WB Vice President and Special Envoy for Climate Change requested that legislators engage in carbon pricing instruments and ensure their governments support the declaration to be adopted later in September of that year.

In March of 2015, Mexico was the first developing country to submit its NDC to the UNFCCC. Climate international negotiators and practitioners referred to the 'Mexican model' when an NDC included considerations of adaptation, human rights and gender. It was also relevant that Mexico explicitly conditioned its emission reduction commitment to the inclusion of "international carbon pricing, carbon-sensitive levies, technical cooperation, access to low-cost financial resources and technology transfer" (México 2015) in the global agreement.

Global political support was needed for an ambitious and extensive agreement. Therefore, the French Presidency in charge of the climate talks innovated the format by organizing a segment of leaders and high-level representatives to initiate the COP-21. Over 150 heads of state and government gathered the first day, being "the largest group of leaders ever to attend a UN event in a single day" (UNFCCC 2015). In that context, high-level events and declarations took place on November 30th, 2015, sending political signals of what should be covered by the climate accord.

At that point, for example, it was not certain if parties were going to agree to include a reference of market-based instruments into the expected agreement. Then, climate diplomacy took place. The heads of state from Canada, Chile, Ethiopia, France, Germany and Mexico, as well as the Secretary General of the OECD joined the president of the WB[1] in a carbon pricing panel to urge other nations and companies to put a price on carbon. For Mexico, "carbon pricing is an effective means of reducing greenhouse gas emissions and promoting the use of cleaner fuels" (Gobierno de México 2015), said then the Mexican President[2] even when the country had a minimum carbon levy in place which was criticized also because the revenues were not allocated to environmental expenditure. To advise on what Mexico could proclaim, the Ministries of Environment, Finance and Foreign Affairs had to concur; before that point, there was not much joint work at the political level on this matter.

Also, during COP-21,[3] the Carbon Pricing Leadership Coalition (CPLC) was launched with the aim of promoting the use of carbon pricing as a policy instrument to reduce GHG emissions. Mexico supported the CPLC as a founding member and the journey began. From that moment, Mexico started to be a key political player within *carbon pricing momentum*. This international exposure triggered domestic discussions on the matter. To date, representatives from the public, private sector, NGOs and universities are part of the CPLC.[4]

In 2016, there were three important political accomplishments for carbon pricing. First, the heads of state of Mexico and Germany reaffirmed their cooperation on climate[5] with three specific commitments on carbon pricing—one being German assistance to Mexico and alliances with other partners (Germany-BMU 2016). The Second was the declaration *Setting a Transformational Vision for 2020 and beyond* from the carbon pricing panel[6] calling for a carbon pricing goal to double coverage of global emissions with explicit carbon prices to 25% by 2020 and to reach 50% by 2030. It was also recognized that such a goal could be achieved by broadening and

---

[1] Note from the authors: days before the media event, high-level representatives from China and the United States were confirmed to participate.

[2] In addition to the official high-level segment, the president of Mexico only participated in two events while at the COP-21: the launching of Mission Innovation and the carbon pricing panel.

[3] In the margins of the COP-21, Mexico joined a ministerial declaration on carbon markets promoted by New Zealand, committing to "work together to ensure the development of standards and guidelines for using market mechanisms that ensure environmental integrity and avoid any double-counting or double-claiming of emissions reduction units" (New Zealand 2018). However, this international cooperation did not occur on a bilateral basis.

[4] Through the work of regional working groups, the CPLC triggers the exchange of knowledge and best practices by facilitating dialogue between the public and private sectors on relevant issues, such as nature-based solutions, competitiveness, monitoring reporting and verification, among others. Mexico, as part of the Americas Working Group, has shared its experience in establishing a carbon tax and recently implementing the ETS.

[5] Joint Declaration between the Government of the Federal Republic of Germany and the Government of the United Mexican States on Climate Action and the Energy Transition and Biodiversity.

[6] This statement vision was released parallel to the signing ceremony of the Paris Agreement, which was politically relevant at the time.

deepening new and existing carbon prices and that international cooperation was key in such efforts. Mexico's contribution to this statement was robust. There were clear mitigation goals based on the market mechanisms already embedded in legislation: a carbon tax, clean energy certificates and green bonds (World Bank 2016).

Third, at the North American Leaders' Summit, the heads of state of Canada, Mexico and the United States (U.S.) embraced an unprecedented statement on climate, clean energy and environment, which led to the adoption of a partnership and an action plan. This was a historical commitment to 51 specific actions to ensure a more modern, secure and cleaner future in the region. Recognizing the role that carbon markets can play in helping achieve climate targets while driving innovation and support robust implementation of the Paris Agreement, the three countries were promising to share best practices and technical solutions to improve accounting effectiveness, as well as encouraging sub-national governments to share lessons learned about the design of effective carbon pricing systems and supportive policies and measures (Whitehouse 2016).

This North American signal was a breaking point and a spotlight to the world. However, five months later a new administration in the U.S. not only withdrew support to these vows, but also denied climate change. However, Canada, Mexico and the U.S. Climate Alliance[7] moved to a decentralized (national–subnational) cooperation to accelerate climate policy efforts across North America, including carbon pricing considerations. The new North American Climate Leadership Dialogue (NACLD) addressed among others, carbon pricing instruments. By September 2018, partners at the Global Climate Action Summit in San Francisco were still highly dedicated to incorporate the cost of carbon pollution into decision-making, but the 2018/19 progress update (Gobierno de México, Canada and US Climate Alliance 2019) submitted at the 2019 UN Secretary General's Climate Summit lacks an allusion to carbon pricing.

A year later, the first regional meeting[8] on carbon pricing took place in early 2017 in Mexico, organized by SEMARNAT and the World Bank. Again, the Ministries of Environment, Finance and Foreign Affairs convened to send a message on the importance of carbon pricing policies (Gobierno de México-SEMARNAT 2017a, b, c). The meeting also welcomed the beginning of crucial international collaboration among private, public and social organizations.[9] Later that year, Mexico hosted the Latin

---

[7] The U.S. Climate Alliance is a bipartisan coalition of then 15 U.S. Governors and now supported by 25 states.

[8] "The Advantage of International Cooperation in Achieving Regional Mitigation Goals in the Americas" was held in Mexico in January 2017, bringing together participants from governments, businesses, civil society and international organizations.

[9] "The spirit of international collaboration was reflected in among other actions, by the signing of a MoU between the Business Council for Sustainable Development of Mexico (CESPEDES) and the International Emissions Trading Association; as well as an MoU between the Mexican Ministry of Environment and Natural Resources, and Environmental Defense Fund." (CPLC 2017) A keystone was for local and global business associations which jointly developed a comparison of core policy elements of existing carbon markets, including private sector views on lessons learned and best practices (IETA 2017).

American Caribbean Carbon Forum combining efforts with the council meeting of the World Business Council for Sustainable Development (WBCSD) held in parallel, which included a carbon pricing position as part of its perspective on climate action and policies (WBCSD 2017).

The combination of political understanding and technical preparedness was successful for Mexico, both, within its institutions and with the international partners. Several stakeholders[10] looked at Mexico as a main ally and a regional leader on carbon pricing. This variety of institutions became relevant in the implementation of the carbon market in Mexico in the following years.

International Cooperation to Develop an ETS in Mexico

With the ratification of the Paris Agreement and the establishment of national emission reduction goals, the inclusion of market-based instruments to achieve mitigation objectives was a natural consequence of climate leadership in Mexico. The GLCC of 2012 offered a glimpse of a potential carbon market in the country. Consequently, Mexico began with the deployment of actions towards the establishment of a carbon market. This process started with the entry into force of the National Registry of Emissions (RENE). The ultimate objective of this instrument was to collect information on GHG emissions from facilities that emit more than 25,000 tCO_2e, which included entities in sectors such as energy, transportation, industry, among others. For the implementation of RENE, various international partners supported Mexico, such as the US Agency for International Development (USAID) and the German development agency (*Deutsche Gesellschaft für Internationale Zusammenarbeit*, GIZ) which developed training workshops to prepare private sector companies to comply with such reporting obligations and identify opportunities for reducing its carbon footprint (GIZ 2015).

Driven by the need to achieve a collective emissions reduction goal and by the opportunity to use market-based mechanisms to that end, developed countries assigned resources to support other nations in achieving such goals. In terms of ODA committed to market and financial mechanisms for fighting climate change, Germany provides almost half of the flows for this purpose among the five donors who support Mexico on the matter. The United States, Canada, Norway and the United Kingdom are the other donors, listed in descending order of ODA amounts.

This cooperation from the Development Assistance Committee (DAC) members to Mexico was dedicated to creating the necessary institutional preconditions and technical capacities of both public and private actors for establishing and implementing an ETS, as well as supporting the Mexican decision-making processes.

---

[10] Such as the governments of California, Canada, Chile, France and Germany, as well as colleagues from the World Economic Forum (WEF), Carbon Disclosure Project (CDP), the International Emissions Trading Association (IETA), the OECD, the International Monetary Fund (IMF) and the Environmental Defense Fund (EDF).

The institutions executing this cooperation are diverse: public (federal and local), private, academia, non-governmental organizations and international initiatives. Some specific collaboration (OECD 2020a, b) is to:

- Develop a national greenhouse gas emissions registry.
- Produce scientific analyses and political recommendations on sector coverage, emissions thresholds and economic impacts.
- Support ministries, companies and financial institutions through capacity building on their roles and responsibilities in the market.
- Improve Mexico's ability to manage carbon emissions by introducing carbon credits from different standards that will be transacted in public auctions.
- Support Mexico's participation in the Forest Carbon Partnership Facility's Readiness Fund.
- Establish international dialogues and exchanges with jurisdictions that dispose of an ETS.
- Disseminate lessons learned in the Mexican context.
- Enhance measurement, reporting and verification with the Pacific Alliance.
- Coordinate internationally on emissions-related policy initiatives.

German cooperation has been the major technical ally in setting the stage for the ETS in Mexico. The German Federal Ministry for the Environment (BMU) commissioned GIZ to work together with SEMARNAT in improving decision-making capacities for the design of the national ETS. This was implemented under the project called "Preparation of an Emissions Trading System in Mexico", known as SiCEM, which has been extended until 2023 (IKI n.d.).[11] By holding technical workshops and study tours for those responsible for ETS implementation, SiCEM has contributed to increasing preparedness in the country. This benefited not only government officials, but also companies. SiCEM seeks to promote dialogue between market actors and to strengthen technical capacities by improving understanding of the roles and responsibilities of market participants.

According to statistics from the Organization of Economic and Cooperation Development (OECD), from 2010 to 2018, Mexico received a total commitment of USD $4.64718 billion, constant prices, 2017 (OECD 2020a, b) of bilateral aid—both, grants and loans—from DAC members, for activities targeting global environmental objectives—known as Rio Markers (see Annex 1). Around 50% of that bilateral aid was for climate change mitigation. (See Table 4.1.)

Although this USD $4.6 billion represented barely 2% of the total bilateral aid to all developing countries on the same issues and time, during these nine years, 14 developed countries committed to support Mexico with 265 climate projects for mitigation and 96 projects for adaptation.

Just to have an idea of the increased international support for environmental purposes, during the eight previous years (2002–2009) DAC donors committed to Mexico only 6% of what they committed the following nine years. (See Fig. 4.1.)

---

[11] IKI (n.d) SiCEM-Preparación de un Sistema de Comercio de Emisiones en México. Retrieved from http://iki-alliance.mx/portafolio/preparation-of-an-emissions-trading-system-ets-in-mexico/.

Table 4.1 Bilateral ODA commitment to Mexico from DAC members

For Rio Markers (biodiversity, climate change and desertification) USD millions, constant prices, 2017				For all sectors and purposes
Years	2002–2009	$285.48		$2,656.49
	2010–2018	$4,647.18		$6,927.36
Rio Markers (2010–2018)	Biodiversity	$488.42	*11%*	
	Climate change mitigation	$2,344.72	*50%*	
	Climate change adaptation	$276.19	*6%*	
	Desertification	$0.75	*0%*	
	Environment	$1,537.09	*33%*	
	Total	*$4,647.18*		

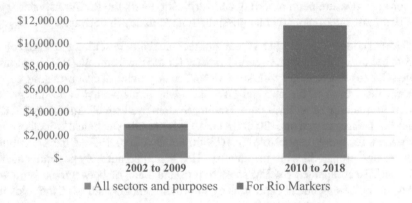

Fig. 4.1 Bilateral ODA commitment to Mexico from DAC members

In regards to multilateral support, in 2007, the Mexican government acquired the first debt (loan accountable as official development assistance, ODA) to achieve climate goals. In 2010, Mexico established debt instruments with the Inter-American Development Bank (IADB) for climate purposes. Since then, Mexico has taken part in climate investment funds (CIF), projects with the International Finance Corporation (IFC), the Global Environment Facility (GEF), the World Bank (WB), the International Fund for Agricultural Development (IFAD) and the European Development Bank (EIB); listed in order of length of the partnership.

The debt instruments have evolved from bilateral to multilateral commitments. According to the *OECD Climate-related development finance data visualization portal*, in 2017, 71% of the support to Mexico comes from multilateral institutions—mainly the IADB, and only 29% is bilateral—including some grants (OECD 2020a, b).

With the information available in the OECD database, as well as official reports from Mexico and its international partners, we can identify the main environmental topics where Mexico has received international cooperation and highlight those related to climate mitigation (see Annex 1), so as to finally focus on the alliances to develop the Mexican ETS.

It is important to highlight that a third of bilateral and multilateral mitigation aid is for energy-related projects (promotion of energy efficiency and renewable energies, generation of electrical energy and support to the energy transition in the country, especially energy auctions). Hence, confirming that the majority of the cooperation received is earmarked for mitigation projects.

According to the Gupta et al. (2014), investment in CDM activities around the world amounted to over USD$ 400 billion between 2004 and 2012. Investment in the energy sector is estimated at almost USD$25 billion in 2011. Moreover, 100 carbon funds were created to finance carbon projects. These reported a capital of nearly USD$ 14.2 billion, of which 48% corresponds to private capital, 29% to government funding and 23% to hybrid sources (public and private). Nevertheless, the amount of private resources provided to developing countries remains unknown (IPCC 2014).

Besides leveraging the political advocacy for carbon pricing instruments, the financial and technical support provided by the World Bank has been instrumental in the development of carbon pricing policies in Mexico. The Partnership for Market Readiness (PMR) is an initiative managed by the World Bank to assist developing countries in the preparation and implementation of policies to mitigate climate change. It is mainly focussed on carbon pricing instruments. Countries forming part of this initiative benefit from grant funding to develop market-based instruments for GHG emission reduction and to facilitate NDC implementation. The PMR also builds a knowledge-sharing platform that allows countries to exchange lessons learned and best practices to trigger collective and innovative climate action.

Since 2012, the PMR has been supporting Mexico in the implementation of Nationally Appropriate Mitigation Actions (NAMAs) across different sectors (i.e. transport and housing) and recently, became a cornerstone for the design and establishment of the ETS in the country. The initial PMR working programme approved for Mexico was adapted upon the country's request to reflect the need for developing a national carbon market (PMR 2018).

In October 2017, SEMARNAT, together with the Mexican Stock Exchange and MexiCO2 announced the launching of a carbon market simulation exercise. This exercise was implemented online over ten months and was carried out with the support of the World Bank through the PMR (SEMARNAT 2018). More than 90 private sector representatives took part, seeking to enhance their capacities and improve understanding of the dynamics of a compliance carbon market. This simulation was the beginning of a series of technical workshops designed for non-state actors ahead of the entry into force of the mandatory carbon market in Mexico.

The World Bank and GIZ enabled the dialogue between the private and public sectors and fostered capacity-building through technical studies and workshops. The knowledge acquired by government officials will be reflected in the development of policy and regulatory instruments for the formal establishment of the ETS.

The support received was instrumental for the alignment of the regulatory framework and policy instruments with the emission reduction commitments assumed internationally.

Mexico's Regional Pivotal Role

Representatives of Colombia, Chile, Mexico and Peru from the Working Group on Environment and Green Growth of the Pacific Alliance identified a major gap among these countries for the implementation of market-based policies. The Pacific Alliance countries determined that monitoring, reporting and verification (MRV) systems for emissions accounting was an area for further improvement and was, therefore, included in the Cali Declaration of 2017 (Alianza del Pacífico 2017).

Pacific Alliance countries are committed to collaborating on an analysis of the scope of MRV systems in the region. Although this statement did not explicitly reference market mechanisms for driving GHG emissions reductions, it attracted the attention of other countries willing to support such an endeavour and was used as an indication of the regional interest in pursuing carbon pricing. Thus, Chile, Colombia and Mexico, supported by other national and sub-national governments in the Americas, as well as international institutions, adopted the Declaration on Carbon Pricing in the Americas (IETA 2018; Cruz et al. 2018). As part of this declaration, national and subnational country members established a platform for cooperating to improve technical capacities for the design and implementation of carbon prices, to explore synergies between public and private actors and to set comparable MRV criteria, among others. A decisive component during the negotiation was to ensure financial mechanisms to support this platform. From the beginning, the two main donors to the initiative were the World Bank through the PMR, and the UN Economic Commission for Latin America and the Caribbean (ECLAC), with the funding support of the European Union, via EUROCLIMA. SEMARNAT was a key facilitator, especially in terms of inclusion and follow up with ECLAC and EUROCLIMA.

The strong collaboration with the government of California to develop and implement carbon pricing systems and other market-based instruments deserves special mention. Work initiated in 2014[12] towards the development of, inter alia, rigorous monitoring, reporting and verification to support carbon pricing or regulatory mechanisms, including potential linkage of both carbon markets (CalEPA 2018). It showed early impacts due to the financial resources available and specialized advisors based in Mexico.

California and Quebec were among the first cap and trade mechanisms to become fully linked. SEMARNAT and Quebec also agreed to collaborate,[13] but it was never

---

[12] In 2014, a Memorandum of Understanding to enhance cooperation on Climate Change and the Environment was signed between the State of California and SEMARNAT and Mexico's National Forestry Commission.

[13] A Memorandum of Understanding was signed in October 2015 with the purpose of promoting and carrying out cooperative activities related to environmental issues including, inter alia, climate

operationalized. Another less successful effort was working with Ontario, also on carbon pricing,[14] but it never went beyond meetings. The political partnership, especially with both California and Quebec, continued through the CPLC and the NACLD.

Political agreements are not the only source of cooperation. In recent years, SEMARNAT has participated in the negotiation of environment or sustainable development chapters within trade agreements. A new chapter on the environment was incorporated in the modernization of the North American Free Trade Agreement that derived in the recent United States-Mexico-Canada [Free Trade] Agreement (USMCA), and a new parallel Agreement on Environmental Cooperation (AEC) was adopted. Although it was impossible to mention climate change due to the US position on the matter, there are climate-related areas of cooperation including market mechanisms and other flexible and voluntary mechanisms under the "supporting green growth and sustainable development" part of the AEC Work Program (US-EPA 2018). What is remarkable about this allusion to trilateral cooperation is the long-term opportunity to develop technical cooperation, develop studies, compare MRV systems and other carbon market issues, among Mexico's most important economic partners.

At the same time, the trade portion of the new European Union–Mexico Global Agreement considers the joint work in preparing and adopting carbon pricing actions including emission trading systems (European Commission 2018). See Annex 2 for Mexico's international advocacy for carbon pricing and international cooperation for the ETS.

Moving Towards Implementation

According to the rules for the trial phase of the ETS, SEMARNAT must define the market compensation scheme and must decide on the national or international protocols to allow for offsetting emissions within the national ETS. The PMR is also supporting Mexico in the development of a registry for emission reductions generated as part of the ETS and the development of such compensation protocols (PMR 2018). There is no public information on when it will happen, however, these should be ready by the end of 2022 during the transition phase of the ETS towards its formal implementation.

It should be noted that since 2013, Mexico has had a voluntary carbon credit market in which various companies have participated in order to offset their emissions on a voluntary basis (CAR 2015). However, it was not until the reform of the GLCC that

---

change and emissions trading, according to their respective competencies, budgetary availability, and based on principles of equality, reciprocity, information exchange and mutual benefit (Quebec-MRIF 2015).

[14] Joint declaration between Mexico, Ontario and Quebec, committing to working together to fight climate change and seeking progress on the common commitment to pricing carbon (Gobierno de México 2016) (Ontario 2016).

the mandatory carbon market emerged with specific emission reduction obligations. According to the GLCC reform, the market was to be implemented gradually in two phases: a pilot phase starting in 2019 and a formal phase, starting in 2022. Due to delays in the approval of the rules of operation of the pilot phase, the entry into force did not begin until January 2020. Consequently, the operation of the formal phase of the market was deferred to 2023 (D.O.F 2019).

The Partnership for Market Implementation (PMI), another initiative hosted by the WBG, will be the successor of the PMR. The PMI will assist participant countries in the implementation of domestic carbon pricing instruments and other related activities such as capacity development for the creation of institutional structures, enhancing GHG data collection and MRV frameworks, as well as NDC alignment (World Bank 2020a, b). The partnership between Mexico and the PMR concludes in July 2020. However, it has not been disclosed if Mexico will become one of the 30 jurisdictions to benefit from this initiative. Nevertheless, it is natural for the country to pursue opportunities under the PMI as a continuation of previous efforts with the PMR.

Conclusion

Over the last decade, Mexico has contributed to forging current climate governance. Due to its legislative advances, increased ambition, concrete climate actions and opportunities for investment, Mexico attracted attention from various partners. This international spotlight was a long-term consequence of a demanding civil society, a vigilant private sector, an interested congress, and of course, committed officials. Note that this document did not cover other environmental subjects related to climate, such as forests, oceans, biodiversity, education, green growth, air quality and climate pollutants, etc. in which Mexico also showed leadership. Taken altogether, these actions have made for a robust scenario.

Especially around the adoption of the Paris Agreement, Mexico was part of many climate initiatives, meetings, declarations and working groups. High-level declarations set the political, and sometimes legal framework, to land international support—either financial or technical. Top down signals mean mandates and opportunities for implementation.

International cooperation in Mexico largely determined the pace at which the national ETS evolved. Mexico's climate leadership and the political advocacy it showed in the international arena sent strong signals that attracted the attention of international donors. In less than three years of placing carbon pricing at any possible international pact, and participating at the most strategic and highest boards, Mexico gained visibility, trust and created opportunities.

Thanks to this international support, mainly from Germany and the World Bank, it was possible to develop key regulatory instruments that would have not happened in the short term. This also helped to prepare the country for participation in international carbon markets. The development of national capacities for the implementation of the ETS in Mexico continues, thanks to the support of international partners. However, it must be kept in mind that there are still questions pending regarding the implementation of a future global carbon market within the Paris Agreement.

The resources—monetary or in-kind—from international partners, made it possible to complete one of the most ambitious environmental policy projects in Mexico during a time of significant budget cuts. Furthermore, the strengthening of capacities and knowledge transfer was promoted beyond the public sector and reached the private sector and civil society. Thanks to such support, non-state actors are now in a better position to assume their roles and facilitate compliance with national climate policies.

Some of the bilateral agreements signed with other international partners resulted only in a press release without ever being implemented. This could be attributed to many factors, such as a sudden change in national priorities, politics or the lack of enough specialized human resources by the Mexican counterpart. However, these helped to build *momentum* and led to other initiatives that were effectively materialized in action plans.

Throughout this process, the Ministries of Environment, Finance and Foreign Affairs benefited not only from the transfer of knowledge promoted by its international partners, but also from the support of external experts who collaborated on projects led by the World Bank and GIZ. International cooperation is deemed effective when this transfer of knowledge satisfies specific needs of the country and remains beyond a specific project. In the case of Mexico, the technical capabilities developed as a result of international support will only thrive if there is political willingness to maintain the policy and regulatory instruments that enabled such capacity building to take place. The knowledge acquired will also persist conditioned to satisfying primary needs of the country. If there is a substantial change in the national goals, there is a risk that the capacities generated will remain stranded, leaving regulatory instruments in disuse.

Mexico's experience through diverse international initiatives advocating for carbon pricing served as a benchmark for other countries in the Latin American and Caribbean region, such as Chile, Colombia, Costa Rica and Peru. The participation of the region in these technical and political initiatives in parallel to the discussions under Article 6 of the Paris Agreement, triggered interest on the part of these countries for enhancing their preparedness for the imminent entry into force of a potential global carbon market.

Unfortunately, the new administration that took place in late 2018 adopted a different environmental agenda. This administration is focussing on a new oil refinery, has cancelled renewable energy auctions, has significantly reduced the budget for the environment—and for international offices across all Ministries—, and has restricted international participation of government officials and high-level representatives. In less than six months of running the office, **the new government sent completely opposite signals on climate action**. Data is not yet available to estimate how this affects the flows of ODA and multilateral support. However, there is no doubt that the country is facing a slowdown when it comes to achieving the international climate agenda.

This is relevant to the implementation of the ETS as the mechanism already went through an initial delay. The full establishment of the ETS is planned to be completed in 2023, one year before the termination of the current presidential term. In 2025, countries should present the second update of their NDCs. The adequate and sustained implementation of the ETS in Mexico could allow the country to increase its climate ambition and hence, mitigation commitments. The ETS in Mexico could be a turning point in the country's climate agenda and, if used wisely, international cooperation can help accelerate the achievement of national mitigation and adaptation goals.

The climate change agenda in Mexico is uncertain. Given the lack of interest and mixed political signals that have characterized the first couple of years of the current administration, we foresee three possible scenarios for international cooperation in the country. In the first scenario, international cooperation continues but focuses on strengthening the capacities of non-state actors. This is particularly relevant for the ETS, as the private sector could be the one advocating for full implementation of the national carbon market if concrete business opportunities are identified.

In a second scenario, international stakeholders could demand that Mexico's federal government meet previous commitments. However, an excess of international pressure could have secondary effects on a government with a weak agenda and low human resources. As previously mentioned, there were some bilateral agreements that the Mexican government was unable to materialize.

In a final scenario, the climate agenda in Mexico is delayed, including the implementation of the ETS. The lack of certainty, rules and institutional infrastructure for the implementation of the ETS could put the initial efforts of the private sector and international partners in developing mitigation activities at stake, therefore, hindering climate finance opportunities for the country. There is no doubt that international alliances have been key to the carbon market in Mexico. However, it is important to bear in mind that even if the ETS is successful, any carbon market and non-market mechanism is a transitory process to a decarbonized world.

Annex 1. International Climate Cooperation for Mexico, 2010–2018

Purpose of the bilateral cooperation in mitigation/Keyword		International partner													
		Canada	Denmark	Finland	France	Germany	Ireland	Italy	Japan	Korea	Norway	Spain	UK	USA	California
Agriculture	Agriculture	x									x		x		
Air pollution	Air	x										x			
Banking and financial services	Banking												x	x	
Education	Education	x						x				x	x		
Renewable and transition/Oil and gas	Energy	x	x	x	x	x		x		x		x	x	x	
Carbon market/ETS*	ETS	x				x					x		x		x
Forestry	Forestry	x											x	x	
Green growth	Green growth									x			x		
Design of the intended national determined contributions	iNDC		x												
Climate policies	Policy				x	x				x		x	x	x	
Private sector	PPP	x			x	x						x	x	x	
Reduce emissions from deforestation and forest degradation	REDD										x			x	

(continued)

(continued)

Purpose of the bilateral cooperation in mitigation/Keyword		International partner													
		Canada	Denmark	Finland	France	Germany	Ireland	Italy	Japan	Korea	Norway	Spain	UK	USA	California
Sustainable transport and storage	Transport	x											x	x	
Urban dev./Housing/cities	Urban				x	x	x						x		
Supply and sanitation	Water				x	x						x			

*Australia is not a bilateral partner. However, it funds multilateral organizations to conduct climate change mitigation activities, specifically: Global Green Growth Institute; Partnership for Market Readiness; Scaling-Up Renewable Energy Program; and UNDP Low-Emission Capacity Building Program from which Mexico has benefited

Annex 2. Mexico's international advocacy for carbon pricing and international cooperation for the ETS

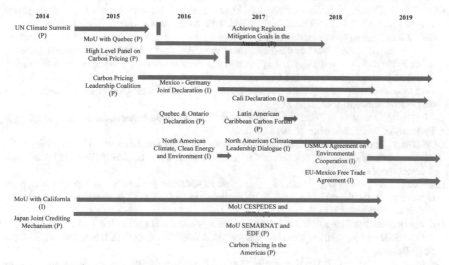

(P) Carbon pricing is the Principal element of the initiative or meeting.
(I) Carbon market reference is Included, among other elements of cooperation.

References

Alianza del Pacífico (2017) Declaración de Cali. https://alianzapacifico.net/
Balderas Torres AB, Vargas PL, Paavola J (2020) The systemic and governmental agendas in presidential attention to climate change in Mexico 1994–2018. Nat Commun 11(1):1–11. https://doi.org/10.1038/s41467-019-14048-7
BMU (Federal Ministry for Environment, Nature Conservation and Nuclear Safety, Germany) (2016) Report Amerikas. German-Mexican Climate Action Declaration. 12 April 2016. https://www.bmu.de/en/report/german-mexican-climate-action-declaration/
CALEPA (California Environmental Protection Agency) and Ministry of Environment and Natural Resources of Mexico (2018) California-Mexico Memorandum of understanding on climate change & the environment 2014–2018 summary report. https://calepa.ca.gov/wp-content/uploads/sites/6/2019/03/California-Mexico_MOU_Summary_Report_2014-2018.pdf
CAR (Climate Action Reserve) (2015) Introduction to carbon markets in Mexico. Climate Action Reserve. https://www.climateactionreserve.org/wp-content/uploads/2015/12/Climate-Action-Reserve_Mexico-Carbon-Markets-Memo-ENGLISH.pdf
CICC (Comisión Intersecretarial de Cambio Climático) (2017) Estrategia Nacional REDD+ 2017–2030. Repositorio Digital Especializado. México. http://www.monitoreoforestal.gob.mx/repositoriodigital/items/show/546.
CPLC (Carbon Pricing Leadership Coalition) (2017) Blog. The advantage of International Cooperation in achieving regional mitigation goals in the Americas. January 26, 2017. https://www.carbonpricingleadership.org/blogs/2017/1/25/the-advantage-of-international-cooperation-in-achieving-regional-mitigation-goals-in-the-americas

Cruz N, Churie-Kallhauge A, Forrister D, Keohane N (2018) Declaration on carbon pricing in the Americas: building momentum among continents. IETA Insights: Greenhouse Gas Market Report. No. 3. https://www.ieta.org/2018-IETA-Insights-No-3/

DOF (Diario Oficial de la Federación) (2012) Decreto por el que se expide la Ley General de Cambio Climático, SEMARNAT, Diario Oficial de la Federación, 6 June 2012, México

DOF (Diario Oficial de la Federación) (2013) Decreto por el que se reforman, adicionan y derogan diversas disposiciones de la Ley del Impuesto al Valor Agregado; de la Ley del Impuesto Especial sobre Producción y Servicios; de la Ley Federal de Derechos, se expide la Ley del Impuesto sobre la Renta, y se abrogan la Ley del Impuesto Empresarial a Tasa Única, y la Ley del Impuesto a los Depósitos en Efectivo (Continúa en la Tercera Sección). Diario Oficial de la Federación. 11 de Diciembre de 2013. México, D.F.

DOF (Diario Oficial de la Federación) (2015) Ley de Transición Energética. Diario Oficial de la Federación. 24 de diciembre de 2015. México

DOF (Diario Oficial de la Federación) (2019) Acuerdo por el que se establecen las bases preliminares del Programa de Prueba del Sistema de Comercio de Emisiones. Diario Oficial de la Federación. 1 de octubre de 2019. México

European Commission. Trade. News archive. New EU-Mexico agreement: the agreement in principle and its texts. Brussels, 26 April 2018—updated in May 2020. https://trade.ec.europa.eu/doc lib/press/index.cfm?id=1833

GIZ (Deutsche Gesellschaft für Internationale Zusammenarbeit) (2015) Integrated solutions that work. Integrated company report. https://www.giz.de/en/downloads/giz2015-en-Integrated-Com pany-Report.pdf

Gobierno de México (2015) Presidencia. Prensa. Participación del Presidente de los Estados Unidos Mexicanos, Enrique Peña Nieto, durante la presentación de la Iniciativa Carbon Pricing. 1 December 2015. http://www.gob.mx/presidencia/prensa/participacion-del-presidente-de-los-estados-unidos-mexicanos-enrique-pena-nieto-durante-la-presentacion-de-la-iniciativa-carbon-pricing

Gobierno de México (2016) Secretaría del Medio Ambiente y Recursos Naturales (SEMARNAT). Prensa. Pacchiano Alamán refrenda compromiso de México en II Cumbre de Cambio Climático de las Américas. 31 de agosto de 2016. https://www.gob.mx/semarnat/prensa/pacchiano-alaman-refrenda-compromiso-de-mexico-en-ii-cumbre-de-cambio-climatico-de-las-americas?idiom=es-MX

Gobierno de México (2017a) Secretaría del Medio Ambiente y Recursos Naturales (SEMARNAT). Prensa. Inicia foro internacional para fijar precio al carbono. 25 January 2017. https://www.gob.mx/semarnat/prensa/inicia-foro-internacional-para-fijar-precio-al-carbono?idiom=es

Gobierno de México (2017b) Secretaría del Medio Ambiente y Recursos Naturales (SEMARNAT). Prensa. SEMARNAT y Gobierno de Quebec firman acuerdo para fortalecer cooperación en cambio climático. 12 de octubre de 2015. https://www.gob.mx/semarnat/prensa/semarnat-y-gob ierno-de-quebec-firman-acuerdo-para-fortalecer-cooperacion-en-cambio-climatico

Gobierno de México (2017c) Secretaría del Medio Ambiente y Recursos Naturales (SEMARNAT). Blog. Acciones de México para establecer un mercado de carbono. 12 December 2017. https://www.gob.mx/semarnat/articulos/acciones-de-mexico-para-establecer-un-mercado-de-carbono?idiom=es

Gobierno de México. Canada. United States Climate Alliance (2019) North American climate leadership dialogue update on progress 2018/19. December 2019. http://www.usclimatealliance.org/international-cooperation#:~:text=Canada%2C%20Mexico%20and%20the%20U.S.,pol icy%20efforts%20across%20North%20America

Greiner S, Chagas T, Krämer N, Michaelowa A, Brescia D, Hoch S, Climate Focus CF (2019) Moving towards next generation carbon markets: observations from Article 6 pilots. https://www.climatefocus.com/sites/default/files/Moving-toward-next-generation-car bon-markets_update-june-2019-1.pdf

Gupta S, Tirpak DA, Burger N, Gupta J, Höhne N, Boncheva AI, Kanoan GM, Kolstad C, Kruger JA, Michaelowa A, Murase S, Pershing J, Saijo T, Sari A (2007) Policies, instruments and co-operative arrangements. In: Metz B, Davidson OR, Bosch PR, Dave R, Meyer LA (eds) Climate change 2007: mitigation. Contribution of Working Group III to the Fourth Assessment Report of the Intergovernmental Panel on Climate Change. Cambridge University Press, Cambridge, United Kingdom and New York, NY, USA, pp 13–12

Gupta S, Harnisch J, Barua DC, Chingambo L, Frankel P, Garrido Vázquez RJ, Gómez-Echeverri L, Haites E, Huang Y, Kopp R, Lefèvre B, Machado-Filho H, Massetti E (2014) Cross-cutting investment and finance issues. In: Edenhofer O, Pichs-Madruga R, Sokona Y, Farahani E, Kadner S, Seyboth K, Adler A, Baum I, Brunner S, Eickemeier P, Kriemann B, Savolainen J, Schlömer S, von Stechow C, Zwickel T, Minx JC (eds) Climate change 2014: mitigation of climate change. Contribution of Working Group III to the Fifth Assessment Report of the Inter-governmental Panel on Climate Change. Cambridge University Press, Cambridge, United Kingdom and New York, NY, USA

Harvey F (2012) Global carbon trading system has 'essentially collapsed.,' The Guardian 10

ICAP (International Carbon Action Partnership) (2017) Emissions trading worldwide: status report 2017. ICAP, Berlin. https://icapcarbonaction.com/en/?option=com_attach&task=download&id=447

IETA (International Emissions Trading Association) (2017) News. Press releases. IETA and CESPEDES sign MOU on Mexican Carbon Market Development and Cooperation. Mexico City, 25 January. https://www.ieta.org/page-18192/4568654

IETA (International Emissions Trading Association) (2018) IETA Insights: Greenhouse Gas Market Report. No. 3. https://www.ieta.org/2018-IETA-Insights-No-3/

IKI (International Climate Initiative) (n.d) SiCEM-Preparación de un Sistema de Comercio de Emisiones en México. http://iki-alliance.mx/portafolio/preparation-of-an-emissions-trading-system-ets-in-mexico/

INE (Instituto Nacional de Ecología) (2006) México tercera comunicación nacional ante la Convención Marco de las Naciones Unidas sobre el cambio climático. México

IPCC (Intergovernmental Panel on Climate Change) (2014) Approved summary for policymakers. IPCC Fifth Assessment Synthesis Report. 1 November 2014

IPCC (Intergovernmental Panel on Climate Change) (2018) Summary for policymakers. In: Masson-Delmotte V, Zhai P, Pörtner H-O, Roberts D, Skea J, Shukla PR, Pirani A, Moufouma-Okia W, Péan C, Pidcock R, Connors S, Matthews JBR, Chen Y, Zhou X, Gomis MI, Lonnoy E, Maycock T, Tignor M, Waterfield T (eds) Global warming of 1.5 °C. An IPCC Special Report on the impacts of global warming of 1.5 °C above pre-industrial levels and related global greenhouse gas emission pathways, in the context of strengthening the global response to the threat of climate change, sustainable development, and efforts to eradicate poverty. In Press

Japan. Ministry of the Environment. Joint Crediting Mechanism (JCM) (n.d.) Information on JCM partner countries. Mexico. July 25th, 2014. https://www.carbon-markets.go.jp/eng/jcm/initiatives/mexico.html

Meirovich HG (2014) The politics of climate in developing countries: the case of Mexico. Doctoral dissertation, Georgetown University

México. Gobierno de la República (2015) Intended nationally determined contribution. March 30, 2015

New Zealand. Ministry for the Environment (2018) Ministerial declaration on carbon markets. 19 April 2018. https://www.mfe.govt.nz/climate-change/why-climate-change-matters/global-response/new-zealand-and-international-carbon

OECD (Organization of Economic and Cooperation Development) (2020a) Climate-related development finance data visualisation portal. https://public.tableau.com/views/Climate-relateddevelopmentfinance-RP/CRDF-Recipient?:embed=y&:display_count=no&%3AshowVizHome=no%20#3. Accessed 1 June 2020

OECD (Organization of Economic and Cooperation Development) (2020b) Dataset: aid activities targeting global environmental objectives. https://stats.oecd.org/. Accessed 16 May 2020

Ontario. Office of the Premier (2016) Newsroom. Ontario Working with Québec and Mexico to Advance Carbon Markets. 31 August 2016. https://news.ontario.ca/opo/en/2016/08/ontario-wor king-with-quebec-and-mexico-to-advance-carbon-markets.html

Paris Agreement, Paris, 12 December 2015, United Nations Treaty Series, No. 54113. https://tre aties.un.org/Pages/showDetails.aspx?objid=0800000280458f37&clang=_en

PMR (Partnership for Market Readiness) (2018) Mexico PMR Project implementation status report (ISR). https://www.thepmr.org/country/mexico-0

Quebec. MRIF (ministère des Relations internationales et de la Francophonie) (2015) Partenariats. http://www.mrif.gouv.qc.ca/content/documents/fr/ententes/2015-12.pdf

SEMARNAT (Secretaría de Medio Ambiente y Recursos Naturales) (2018) Sexta Comunicación Nacional y Segundo Informe Bienal de Actualización ante la Convención Marco de las Naciones Unidas sobre Cambio Climático. México. http://cambioclimatico.gob.mx:8080/xmlui/handle/publicaciones/117

The Whitehouse. Office of the Press Secretary (2016) North American climate, clean energy, and environment partnership action plan. June 29, 2016. https://obamawhitehouse.archives.gov/the-press-office/2016/06/29/north-american-climate-clean-energy-and-environment-partnership-action

UNFCCC (United Nations Climate Change). News. Leaders and high-level segment. December 2015. https://unfccc.int/news/cop-21cmp-11-information-hub-leaders-and-high-level-segment

UNFCCC (United Nations Climate Change) Process and Meetings (2020) The Paris Agreement. https://unfccc.int/process-and-meetings#:a0659cbd-3b30-4c05-a4f9-268f16e5dd6b

United Nations—General Assembly resolution 70/1 Transforming our world: the 2030 agenda for sustainable development, A/RES/70/1 (21 October 2015)

United Nations (2020) United Nations treaty collection. UN

United Nations Climate Change (UNFCCC) News. Leaders and high-level segment. December 2015. https://unfccc.int/news/cop-21cmp-11-information-hub-leaders-and-high-level-segment

United States Environmental Protection Agency (2018) 2018 Agreement on Environmental Cooperation among the Governments of the United States of America, the United Mexican States, and Canada. https://www.epa.gov/international-cooperation/2018-agreement-environmental-coo peration-among-governments-united-states

U.S. Climate Alliance (2018) Publications. Joint Statement on North American Climate Leadership. 13 September 2018. https://www.usclimatealliance.org/publications/2018/9/26/joint-sta tement-on-north-american-climate-leadership

Valenzuela JM, Studer I (2016) Climate change policy and power sector reform in Mexico under the golden age of gas. Polit Econ Clean Energy Transit 410

WBCSD (World Business Council for Sustainable Development) (2017) WBCSD Council meeting in Mexico. Business perspective on climate action & policies. 18 October 2017. https://www.cbcsd.cz/wp-content/uploads/2017/11/Climate-action-policies-for-CM.pdf

World Bank (2014) Press release. 73 countries and more than 1,000 companies and investors support a price on carbon. September 22, 2014. https://www.worldbank.org/en/news/press-rel ease/2014/09/22/73-countries-1000-companies-investors-support-price-carbon?wptouch_prev iew_theme=enabled

World Bank (2016) Speeches & transcripts. Carbon pricing panel—setting a transformational vision for 2020 and beyond. April 21, 2016. https://www.worldbank.org/en/news/speech/2016/04/21/carbon-pricing-panel---setting-a-transformational-vision-for-2020-and-beyond

World Bank (2020a) State and trends of carbon pricing 2020, May. World Bank, Washington, DC. https://openknowledge.worldbank.org/bitstream/handle/10986/33809/978146481 5867.pdf?sequence=4&isAllowed=y

World Bank (2020b) Partnership for market implementation. [online]. https://www.worldbank.org/en/topic/climatechange/brief/partnership-for-market-implementation

Part II
Legal Frameworks and Design Perspectives for a Mexican ETS. Building the Blocks

Chapter 5
The International Influence of the Emissions Trading System in Mexico

Alicia Gutierrez González

Abstract This article aims to give an overview of the international influence of the Emissions Trading System (ETS) in Mexico. It is divided into three parts. First, it briefly examines both the international Climate Change regime through the description of such instruments as the 1997 Kyoto Protocol and the 2015 Paris Agreement, and the national regime by reviewing as the 2012 General Law on Climate Change (*LGCC*), the National Emissions Registry (*RENE*) and its Regulations, as well as other instruments regarding mitigation from carbon tax and clean energy. Second, it analyzes the legal framework of the pilot phase of the ETS in Mexico (under the cap and trade principle) which seeks to reduce carbon dioxide emissions (CO_2) only in the energy and industry sectors whose emissions are greater than 100 thousand direct tonnes of CO_2. In doing so, it also explains the relevance of implementing an ETS as a cost-effective mitigation measure to achieve the Nationally Determined Contributions (NDCs) in order to reduce 22% greenhouse gas (GHG) emissions by 2030 (increasing to 36% if there is international support and financing) and 50% by 2050 as a developing country. Third, it focuses on the European Union Emissions Trading System (EU ETS) experience and shows that all its phases must be done gradually by adopting the learning-by-doing approach.

Keywords Emissions trading system · Mexico · Paris agreement · European unión · Climate change

International Instruments

Environmental problems are currently the subject of international concerns (Malcolm 2014, p. 613). The question of the relationship between the protection of the environment and the need for economic development is a problem that developing countries have been dealing with due to the fact that it is very expensive for them to find a response to this in an environmentally safe way (Malcolm 2014, p. 617).

A. Gutierrez González (✉)
Facultad de Estudios Globales de la Universidad Anáhuac México, Estado de México, México
e-mail: alicia.gutierrez@anahuac.mx

© The Author(s) 2022
S. Lucatello (ed.), *Towards an Emissions Trading System in Mexico: Rationale, Design and Connections With the Global Climate Agenda*, Springer Climate,
https://doi.org/10.1007/978-3-030-82759-5_5

Currently, the international community is facing the main challenge of development with environmental protection. On the one hand, some nations invoke state sovereignty, and on the other, several countries reiterate the need for international cooperation (Malcolm 2014, p. 631). In this regard, Antonio Cassese explains that "the environment has come to be regarded as a common amenity; as an asset in the safeguarding of which all should be interested, regardless of where the environment is or may be harmed" (Cassese 2005, p. 487).

The protection of a clean and healthy environment requires international cooperation, because national actions by themselves may be insufficient (Dixon et al. 2011, p. 441), therefore, the importance of adopting and implementing international principles and rules in treaties, agreements, protocols, and so on.

At the international level, the United Nations has played an important role regarding climate change. The 1992 United Nations Framework Convention on Climate Change[1] (UNFCCC) establishes under Article 2 that: "The ultimate objective of this Convention and any related legal instruments …… is to achieve …. stabilization of greenhouse gas concentrations in the atmosphere at a level that would prevent dangerous anthropogenic interference with the climate system".

This objective shows that anthropogenic emissions are the principal danger in climate change, and therefore, the greenhouse gas concentrations must be limited, so that they can be stabilized (Dupuy and Viñuales 2018, p. 177). Regarding the UNFCCC, it sets out in Article 3 the principle of common but differentiated responsibilities (among other principles), which states that:

> The Parties should protect the climate system for the benefit of present and future generations of humankind, on the basis of equity and in accordance with their common but differentiated responsibilities and respective capabilities. Accordingly, as parties, developed countries should take the lead in combating climate change and the adverse effects thereof (Convention of Climate Change 1992).

This principle is twofold because it states the responsibility of developed countries to take on obligations for the protection and preservation of the environment and it establishes different environmental obligations for developing countries. The ultimate objective of this Convention is to stabilize greenhouse gas emissions at a level that would prevent dangerous anthropogenic interference with the climate system (Birnie et al. 2009, p. 358). The UNFCCC entered into force on 21 March 1994. Since then, there have been other global agreements regarding protection of the atmosphere and the environment. Unfortunately, the emissions of greenhouse gases continue to rise and there is a climate emergency, which includes:

(i)   frequent droughts;
(ii)   floods;
(iii)   storms;
(iv)   earthquakes;
(v)   loss of biodiversity; and so on.

---

[1] The 1992 Convention on Climate Change has been ratified by 194 States.

It is worth mentioning that the principle of common but differentiated responsibilities was set out in Principle 7 of the 1992 Rio Declaration on Environment and Development adopted in the United Nations Conference on Environment and Development (Rio Declaration on Environment and Development 1992). This principle states that the countries "shall cooperate to conserve, protect and restore the health and integrity of the Earth's ecosystem". For this reason, the Rio Earth Summit had the goal of preventing dangerous human interference in the climate system. Thus, the Rio Declaration and the UNFCC show that protecting the environment requires international cooperation and that the responsibility to undertake obligations is different regarding developed and developing countries.

The international influence of the first emission trading started with the 1997 Kyoto Protocol to the United Nations Framework Convention on Climate Change (Kyoto Protocol 1997)[2] which establishes under Article 3 that:

> 1. The parties included in Annex 1 shall, individually or jointly, ensure that their aggregate anthropogenic carbon dioxide equivalent emissions of the greenhouse gases listed in Appendix A do not exceed their assigned amounts…with a view to reducing their overall emissions of such gases by at least 5 per cent below 1990 levels in the commitment period 2008 to 2012.
>
> 2. Each party included in Annex 1 shall, by 2005, have made demonstrable progress in achieving its commitments under this Protocol…

An achievement of this Protocol was the establishment of quantitative restrictions on emissions from developed countries. Moreover, it sets up three market mechanisms:

(i)   Joint Implementation (technology development and transfer, Article 6);
(ii)   The Clean Development Mechanism (implementation of measures in developing countries, Article 12) and;
(iii) The Emissions Trading (global trading in emissions rights, Article 17).

Thus, Article 6 states that:

> 1. For the purpose of meeting its requirements under Article 3, any Party included in Annex 1 may transfer to, or acquire from, any other such Party emission reductions units resulting from projects aimed at reducing anthropogenic emissions by sources or enhancing anthropogenic removals by greenhouse gas sinks in any sector of the economy, provided that:
>
> (a) Any such Project has the approval of the Parties involved…
>
> (d) The acquisition of emission reduction units shall be supplemental to domestic actions for the purposes of meeting commitments under Article 3…
>
> 3. A Party included in Annex 1 may authorize legal entities to participate, under its responsibility, in actions leading to the generation, transfer or acquisition under this Article of emission reduction units…

In the first market mechanism, i.e., the Joint Implementation, the name of the units is called the Emission Reduction Units (ERUs). The objective of the Article is to enable developed countries to undertake joint implementation projects, whereby a country can earn emission credits by investing in projects in other countries.

---

[2] The 1997 Kyoto Protocol has been ratified by 193 Parties.

In the second market mechanism as mentioned above, i.e., the Clean Development Mechanism (CDM), the units are called Certified Emission Reductions (CERs). Article 12 of the 1997 Kyoto Protocol states that:

3. Under the clean development mechanism

(a) Parties not included in Annex 1 will benefit from Project activities resulting in certified emission reductions; and

(b) Parties included in Annex 1 may use the certified emissions reductions accruing from such Project activities to contribute to compliance with part of their quantified emission limitation and reduction commitments under Article 3."

The CDM promotes sustainable development and allows developed countries some flexibility due to the fact that they can trade the CERs to meet a part of their emission reduction goals under the 1997 Kyoto Protocol.

The third mechanism is the Emissions Trading, i.e., it has its foundation under Article 17, which authorizes developed countries to have environmental projects designed to reduce emissions.

Titenberg explains that "Emissions trading allows more flexibility in the timing of control investments. Under emissions trading, facilities have the ability to time their expenditures so that they coincide with optimal capital replacement schedules and prevailing market conditions" (Titenberg 2006, pp. 5–6). It can be said that the European Union adopted this mechanism to create the European Union Emissions Trading System (EU ETS).

The central features of the 1997 Kyoto Protocol were the market-based approaches such as the emissions trading and the clean development mechanism. The CDM allows emission reduction projects in developing countries to earn certified emission reduction credits, each equivalent to one tonne of CO_2. It is worth mentioning that the European Union (EU) favoured strong targets and limited flexibility, while the United States (US) and other non-EU developed states generally favoured weaker targets and greater flexibility (Bodansky et al. 2017, p. 160).

Currently, Mexico has three different GHG emission reductions mechanisms. The first is the 1997 Kyoto Protocol CDMs regarding CERs in areas such as: biomass energy, coal bed/mine methane, EE industry, EE own generation, fugitive, geothermal, HFCS, hydro, landfill gas, methane avoidance, n20, solar, transport and wind. At the start of CDM implementation, Mexico thus far is hosting 225 projects. Within Latin America, this corresponds to 18% in CDM projects and 15% in projects producing CERs (UNEP/CDM, 2020). The second mechanism is a voluntary carbon market in Mexico, where companies can buy Verified Emissions Reductions (VER) or carbon credits. With the 2020 ETS pilot programme, Mexico has the third mechanism for reducing GHG emissions.

The atmosphere as a shared resource is an area of common concern (Birnie et al. 2011, p. 337). Hence, it is crucial to take measures that include all countries worldwide. As widely known, GHG emissions from fossil energy sources contribute to global climate change. This is one of the reasons why emission trading in the 1997 Kyoto Protocol has become an important issue in terms of a global policy to reduce GHG.

The 2015 Paris Agreement is the most important international instrument world-wide regarding the mitigation of GHG. It was adopted by COP 21 on 12 December 2015 and entered into force on 4 November 2016 (Paris Agreement 2015). This Agreement makes no mention of carbon markets under Articles 6.1 and 6.2, therefore, it allows parties to take voluntary mechanisms in order to implement their nationally determined contributions (NDC) and to use internationally transferred mitigation outcomes to meet its reductions of GHG emissions (Sands et al. 2018, p.324).

Article 6 (4) states that:

"4. A mechanism to contribute to the mitigation of greenhouse gas emissions and support sustainable development is hereby established under the authority and guidance of the Conference of the Parties serving as the meeting of the Parties to this Agreement for use by Parties on a voluntary basis. It shall be supervised by a body designated by the Conference of the Parties serving as the meeting of the Parties to this Agreement, and shall aim:

(a)   To promote the mitigation of greenhouse gas emissions while fostering sustainable development;

(b)   To incentivize and facilitate participation in the mitigation of greenhouse emissions by public and private entities authorized by a Party;

(c)   To contribute to the reduction of emission levels in the host Party, which will benefit from mitigation activities resulting in emission reductions that can also be used by another Party to fulfil its nationally determined contribution; and

(d)   To deliver an overall mitigation in global emissions.

This Article promotes the mitigation of GHG through an ETS, supports sustainable development, and involves parties in cooperation. The ultimate purpose of the 2015 Paris Agreement is "to strengthen the global response to the threat of climate change". Regarding long-term mitigation goals, they can be formulated in terms of limiting temperature increase (2 °C or 1.5 °C above preindustrial levels); and a GHG emissions reduction goal (50% by 2050). (Bodansky et al. 2017, p. 228).

According to the Intergovernmental Panel on Climate Change (IPCC) greenhouse gases are:

Those gaseous constituents of the atmosphere, both natural and anthropogenic, which absorb and emit radiation at specific wavelengths within the spectrum of thermal infrared radiation emitted by the Earth's surface, by the atmosphere itself, and by clouds. This property causes the greenhouse effect. Water vapour (H_2O), carbon dioxide (CO_2), nitrous oxide (N_2O), methane (CH_4), and ozone (O_3) are the primary greenhouse gases in the Earth's atmosphere. Moreover, there are a number of entirely human-made greenhouse gases in the atmosphere, such as the halocarbons and other chlorine- and bromine-containing substances, dealt with under the Montreal Protocol. Besides CO_2, N_2O, and CH_4, the Kyoto Protocol deals with the greenhouse gases sulphur hexafluoride (SF_6), hydrofluorocarbons (HFCs), and perfluorocarbons (PFCs) (IPCC 2012).

The main GHGs include carbon dioxide (CO_2), methane (CH_4), and nitrous oxide (N_2O). Some of the GHGs exist in nature and they include water vapour, carbon dioxide, methane, and nitrous oxide; others are exclusively human-made such as fluorinated gases. This is referred to as the enhanced greenhouse effect or the anthropogenic greenhouse effect as it is primarily due to human activities (Tan 2014, p. 349).

The international documents mentioned above reaffirm that protecting the environment requires international cooperation. Nevertheless, on 4 November 2019, the United States of America notified the Secretary-General of its decision to withdraw from the 2015 Paris Agreement which shall take effect on 4 November 2020 in accordance with article 28 (1) and (2) of the agreement. This unfortunate decision shows the lack of international cooperation from that country, which is urgently required in order to combat climate change. To this date, 189 Parties of 197 Parties to the Convention have ratified (United Nations, September 2020).

Mexico's NDCs and Other GHG Mitigation Mechanisms in Line with the 2015 Paris Agreement

Mexico signed the Paris Agreement on 22 April 2016. It was approved by the senate on 14 September 2016 and published on 21 September 2016 in the Official Journal of the Federation (D.O.F.). As a Party of this international instrument, Mexico has amended its 2012 General Law on Climate Change (LGCC) with the aim of harmonizing the law with the objectives established in the Article Two of the Paris Agreement.

Additionally, Mexico has undertaken reforms regarding GHG mitigation with the 2015 Paris Agreement. The 2015 Energy Transition Law (*Ley de Transición Energética*) was published in D.O.F on 24 December. This law promotes the sustainable and efficient use of energy. According to the law, the Secretariat of Energy (SENER) should promote the generation of clean energy to reach the levels set forth in the LGCC for the electric power industry, including a minimum share of clean energies in electricity generation of 25% by 2018, 30% by 2021, and 35% by 2024 (Article 3 transitory of the 2015 Energy Transition Law). Clean electricity has been set at 5% for 2018 and the quota will be gradually increased to meet the target of 35% by 2024. This energy policy permits the issuance of clean energy certificates (*certificados de energia limpia*) which aim to promote the generation of clean energy as well as reinforce GHG mitigation. In addition, SENER has recently included the clean energy target of 39.9% by 2033 and 50% by 2050 (SENER 2020).

It is worth mentioning that the 1988 General Law on Ecological Balance and Protection of the Environment (LGEEPA) was amended in 2016, 2017, and 2018 with the objective of including the climate change agenda. Following this, the economic instruments to mitigate climate change in Mexico are: (i) the carbon tax, the Climate Change Fund, and the ETS pilot programme.

Furthermore, the 2015 Energy Transition Law also promotes clean energy production. Together with the LGCC, this law could help Mexico to achieve its NDC goals, mitigate the effects of climate change, and produce less CO_2 using green energies.

The Law on Special Tax on Production and Services (*Ley del Impuesto especial sobre producción y servicios*) was amended in 2012 with the aim of putting a carbon tax applying to CO_2 emissions to all sectors covering all fossil fuels, except natural gas (Article 2, I). It is an excise tax, which is capped at 3% of the fuel sales price. Companies may choose to pay the tax with credits from CDM projects conducted in Mexico or CERs that are also eligible for compliance in the EU ETS, equivalent to the market value of the credits at the time of paying the tax. This carbon price must be paid monthly and the carbon tax is updated yearly (World Bank 2020).

In October 2013, the carbon tax on fossil fuel production was introduced as part of the tax reform package and, in November of the same year, a voluntary carbon exchange, MEXICO2 was established to trade carbon credits. The carbon tax began in January 2014 and the certified emissions reductions can be used to meet 20% of the carbon tax obligation (IETA 2018).

Climate mitigation goals in Mexico are fivefold: clean energy transition, energy efficiency and sustainable consumption, sustainable cities, reduction of short-lived climate pollutants, sustainable agriculture, and protection of natural carbon sinks.

The 2015 Paris Agreement calls on parties to define an intended nationally determined contribution (INDC). This is a voluntary mechanism, but the key to meeting the Agreement's objectives. This Agreement does not establish specific goals for emissions reductions by parties, which differs from the 1997 Kyoto Protocol (Sands et al. 2018, p. 321). In 2015, Mexico published and submitted to the UNFCCC its commitment to unconditionally reduce "25% of greenhouse gases and short-lived climate pollutants emissions (below a Business as usual Baseline, (BAU)) for the year 2030, i.e., a reduction of 22% of GHG and a reduction of 51% of Black Carbon" (INDC 2015).

Mexico's INDC is divided into two parts: (i) mitigation and (ii) adaptation. As for mitigation, Mexico includes two types of measures: unconditional and conditional. The unconditional set of measures are those that Mexico will implement with its own resources, while the conditional actions are those that Mexico could develop if a new multilateral climate regime is adopted and if additional resources and transfer of technology are available through international cooperation.

It is important to mention that Mexico assumes an unconditional international commitment to carry out certain mitigation actions and reaffirms its commitment to combat climate change, to the multilateral rules-based climate regime that requires the participation of all countries, and to sustainable development, as well as its solidarity with the most vulnerable countries.

The Mexican government explained in its INDC that the conditional reductions of 25% could increase up to 40% conditionally, subject to a global agreement addressing major topics including international carbon pricing, carbon border adjustments, and technical cooperation. It also includes access to low-cost financial resources and technology transfer, all at a scale commensurate to the challenge of global climate change. Within the same conditions, GHG reductions could increase up to 36% and Black Carbon reductions to 70% in 2030 (INDC 2015).

As for unconditional reduction, Mexico is committed to unconditionally reducing 25% of its GHGs and short-lived climate pollutants emissions (below BAU) for the year 2030. This commitment implies a reduction of 22% of GHG and a reduction of 51% of Black Carbon" (INDC 2015). "This commitment implies a net emissions peak starting from 2026, decoupling GHG emissions from economic growth: emissions intensity per unit of GDP will reduce by around 40% from 2013 to 2030. Within the same conditions, GHG reductions could increase up to 36%, and Black Carbon reductions to 70% in 2030" (INDC 2015).

An achievement for Mexico as of 2015, is the legal mandate for all entities emitting more than 25,000 tCO$_2$ per year to report their emissions of (i) carbon dioxide (CO$_2$); (ii) methane (CH4); (iii) nitrous oxide (N$_2$O); (iv) hydrofluorocarbons (HFCs); (v) perfluorocarbons (PFCs); (vi) sulphur hexafluoride (SF6) and; (vii) black carbon in the National Emissions Registry (RENE).

It is worth mentioning that Mexico is a developing country, highly vulnerable to the effects of climate change. National emissions of GHG represent only 1.4% of global emissions and the net per capita emissions, inclusive of all sectors, are 5.9 tCO$_2$e (INDC 2015).

According to the 2018 Global Carbon Atlas, Mexico is in 12th place with 477 MtCOs (Global Carbon Atlas 2018). It is expected that Mexico will be the world's seventh largest economy in 2050 (PWC 2017).

The GHG reduction target for 2020 for Mexico is: (i) 30% below BAU GHG emission baseline (ambitious); (ii) by 2030: 22% below BAU GHG emissions baseline, this includes the NDC; and by 2050 below 2000 GHG levels; 50% included in the LGCC (ambitious).

The Legal Framework of the Pilot Phase of the ETS in Mexico Under the Cap and Trade Principle

The Mexican 2012 Climate Change Law was amended on 13 July 2018 with the aim of complying with the 2015 Paris Agreement and implementing a national emissions trading system that could help the country meet its mitigation goals. According to Articles 87, 88, 94, and 95 of the LGCC, an Emissions Trade System and a National Emissions Registry will be established with the aim to promote emission reductions in the most cost-efficient way, which is also measurable, reportable, and verifiable.

A regulation to the LGCC regarding the National Emissions Registry (RENE) was published on 28 October 2014, establishing the requirements for reporting GHG emissions (Regulation of the LGCC 28 October 2014). According to Articles 3, 4, 6, and 9 of the Regulation of the LGCC, the energy, transport, industrial, agricultural, residues, commerce, and services sectors are required to report their GHG emissions in the Annual Operational Certificate (Cédula de Operación Anual, COA). The COA is the main bottom-up environmental reporting instrument used by industries to report substances such as air, soil, wastes, pollutants, and GHG emissions.

The adoption of ETS in Mexico in the short and long term is built on the existing monitoring regulations in the National Emissions Registry. It can be said that the climate policies in Mexico are directly affected by international regulations and experience due to the fact that the European Union Emission Trading System has been taken into account in the pilot phase of the ETS in Mexico, as established in the second transitory Article of the LGCC, on 13 July 2018. (LGCC, 13 July 2018).

Mexico is currently in the pilot phase (2020–2023) of an Emissions Trading System under the cap and trade principle, with the purpose of reducing carbon dioxide emissions in the energy and industry sectors this phase, whose emissions are greater than 100 thousand direct tonnes of CO_2 per year. Around 300 entities are covered representing ~37% of national emissions (International Carbon Action Partnership, (ICAP 2020a).

It is important to mention that during the pilot phase there is no economic impact on regulated entities. However, in case of non-compliance, the entities will lose the chance to bank unused allowances into the next compliance periods within the pilot phase of the ETS (ICAP 2020a). On 1 October 2019, the rules and terms for the implementation of the ETS in Mexico were published in the Official Journal of the Federation (D.O.F., 1 October 2019).

Regarding the caps for 2020 and 2021, they were published on 27 November 2019 by the Ministry of Environment and Natural Resources (SEMARNAT). Thus, the caps for 2020 are 271.3 $MtCO_2$ and for 2021 are 273.1 $MtCO_2$ (1) (SEMARNAT, November 27, 2019a). The free allowances allocations for the different sectors were also published by SEMARNAT (SEMARNAT, 27 November 2019b).

As mentioned above, the pilot phase of the ETS is under the cap and trade principle with the purpose of reducing carbon dioxide emissions in the energy and industry sectors, whose emissions are greater than 100 thousand direct tonnes of CO_2. This was possible with the support of GIZ (Deutsche Gesellschaft fur internationale Zusammenarbeit GmbH).[3]

It should be noted that an ETS is a market-based instrument, which is used to reduce GHG related to the cap and trade principle. The cap establishes the global limit of GHG, which may be emitted and will be broken down into a number of allowances. These allowances may be traded under the companies regulated by an ETS. The final objective of this is to allow enterprises to comply with their obligations in a cost-efficient manner.

Following this, the Mexican ETS pilot programme started on 1 January 2020 and the scope of the implementation of the ETS includes only CO_2 emissions from the energy and industrial sector.

The energy sector includes:

(i)   Exploitation;
(ii)   Production;
(iii)   Transportation and distribution of hydrocarbons; and
(iv)   Generation, transmission, and distribution of electricity.

---

[3] The German Federal Ministry for the Environment, Nature Conservation and Nuclear Safety (BMU) has supported the project preparation of an Emissions Trading System in Mexico as a part of the International Climate Initiative.

And from the industrial sector are included:

(i)   Automotive;
(ii)   Cement;
(iii)   Chemical;
(iv)   Foods and beverages;
(v)   Glass;
(vi)   Steel;
(vii)   Metallurgical;
(viii) Petrochemical;
(ix)   Paper, and so on.

The participants of these sectors will be subject to:

(i)   Emissions reporting and verification obligations (taking into account their emissions during the previous year); and
(ii)   Obligations related to the delivery of emission rights or allowances (taking into account their emissions during the previous year).

The pilot phase runs from 2020 to 2021 and the transition phase in 2022 constitutes the ETS pilot programme in Mexico. After that, the operational phase will be in place. It covers direct CO_2 emissions from entities in the energy and industry sectors, which generate at least 100,000 tCO_2 per year. Around 300 entities are covered (37% of national emissions).

It is important to mention that during the pilot phase, allowances will be free of charge for all the participants and the amount of allowances issued by SEMARNAT and allocated to all of them will be determined based on historical information reported by the National Emissions Registry.

Table 5.1 shows the reserves and the sectoral allocation of allowances for 2020 and 2021 published by SEMARNAT.

As for auctions reserve (5%), new entrants reserve (10%), and the general reserve (5%), they have their foundation under Articles 13, 14, and 15 of the notice of the rules for the establishment of an ETS in Mexico (D.O.F., 1 October 2019).

The implementation of an Emissions Trading System in Mexico is to reduce a cost-effective mitigation measure to achieve its Nationally Determined Contributions of reducing 22% GHG emissions by 2030 (increasing to 36%, in case of international support and financing) and 50%, by 2050.

The verified annual CO_2 emissions report is made to the National Emissions Register and to the ETS registry. Nevertheless, the 2012 General Law on Climate Change envisages a possible linkage between Mexican ETS and ETS in other countries (LGGC, 13 July 2018, Transitory Article 2). SEMARNAT and the National Institute for Ecology and Climate Change (INECC) are the institutions involved in the establishment and implementation of an ETS. SEMARNAT will review the pilot phase each year and will publish the emission reduction reports supported by INECC.

Table 5.1 Reserves and sectoral allocation of emissions rights for 2020 and 2021

	2020 MtCO$_2$	2021 MtCO$_2$ (1)
CAP	271.3	273.1
Auctions reserve (5%)	13.6	13.7
New entrants reserve (10%)	27.1	27.3
General reserve (5%)	13.6	13.7
Electricity generation	138.1	138.1
Cement	30.2	30.2
Chemical industry	7	7
Glass	2.7	2.7
Iron and steel	14.7	14.7
Lime	0.6	0.6
Mining	2.1	2.1
Oil and gas	35.5	35.5
Refinement	17.8	17.8
Petrochemicals	5.7	5.7
Paper	2.3	2.3
Foods and beverages	7.7	7.7
Others	7	8.8

Source Compiled by author (SEMARNAT, 27 November 2019b)

The European Union Experience in the Emissions Trading System (EU ETS)

The EU ETS is "a cornerstone of the EU's policy to combat climate change and its key tool for cost-effective reduction of greenhouse gas emissions. It is the world's first major carbon market and remains the biggest" (European Union 2020). It works on the cap and trade principle. It covers 45% of the EU's greenhouse gas emissions and the main objective is to reduce the cap so that the total emissions fall. The other 55% comes from emitters such as household, transport users, and agriculture. Thus, the EU ETS is a carbon market operating across 31 countries in the European Union and in the European Economic Area, covering more than 11,000 greenhouse gas emitters including power stations and manufacturing plants, plus some EU flights (CISL 2015).

A cap is set on the total amount of certain GHG that can be emitted by installations covered by the system. Within the cap, companies receive or buy emission allowances, which they can trade with one another as needed. They can also buy limited amounts of international credits from emission-saving projects around the world. The limit on the total number of allowances available ensures that they have a value. After each year, a company must surrender enough allowances to cover all its emissions, otherwise, heavy fines are imposed. If a company reduces its emissions, it can keep the spare allowances to cover its future needs or else sell them to another company that is short of allowances. Trading injects flexibility that ensures emissions are cut, where it costs least to do so (ICAP 2020b).

Although the EU ETS created an explicit price on carbon emissions, which reduced emissions by 26% between 2005 and 2017, this is not enough to combat climate change (Delbeke and Vis 2019, pp. 15–16). Nonetheless, the EU ETS includes a symbolic value, which goes beyond economic consideration and demonstrates the will of the EU to stand as a leader in the international environmental policy context (Borghesi and Montini 2016, p. 3).

Based on the EU ETS experience, Mexico started with monitoring and reporting for the ETS pilot programme with the CO_2 emissions only in the energy and industrial sectors. It is worth mentioning that emissions from fixed and mobile sources are also considered.

In the EU ETS, most activities are defined based on a capacity threshold to decide which facility will take part in the ETS. The capacity threshold for combustion activity is expressed by the total rated thermal input and the capacity threshold for industry activities is expressed by production capacity. In the current situation in Mexico, an emission permit is not compulsory. An environmental license called the unique environmental license for the prevention and control of atmospheric pollution is given to the industries.

According to the 2015 report called "10 years of carbon pricing in Europe: a business perspective", published by the Institute for Sustainability leadership of the University of Cambridge, diverse enterprises explained that they made notable progress in reducing their carbon emissions during its lifetime (CISL 2015). Companies saved money through greater efficiency and then got a carbon bonus over their competitors, either by being able to sell allowances or not needing to buy them. The report also shows the flexibility of the EU ETS due to the fact that the benefit of reducing emissions for some is to reduce the amount of allowances they buy, and for others, to leave them with allowances to sell. Mexico has to learn from this flexibility and has to monitor, report, and verify its CO_2 emissions every year. The report concluded that the existence of the EU ETS and its reporting requirements has helped companies to focus on carbon. However, companies argue that the carbon price is too low to drive technical innovation.

It is important to mention that there are four trading periods within the EU ETS. The first and the second have concluded, the third is ongoing until 2020, and the fourth has been set up for the years 2021–2030. All of them are governed by the EU ETS Directive 2003/87/EC of the European Parliament and of the Council of 13 October 2003 establishing a scheme for greenhouse gas emission allowance trading within the Community and amending Council Directive 96/61/EC.

The phases of the EU ETS are as follows:

(i)   The first phase
      The first trading period or phase 1 lasted from the launching of the EU ETS in 2005 until the end of 2007, i.e., 3 years (pilot phase or the pre-Kyoto Period).
(ii) The second phase

The second trading period or phase 2 began in 2008 and ended in 2012, i.e., 5 years (coinciding with the first commitment period under the 1997 Kyoto Protocol).

Thus, the first and second phase had a free allocation to industry, which was decided on the national level.

(iii)   The third phase
         In the current third phase or trading period (2013–2020), 8 years, there is no free allocation for electricity production and the free allocation to industry is based on EU harmonized rules outlines in the Benchmarking Decision or Commission Decision of 27 April 2011 determining transitional Union-wide rules for harmonized free allocation of emission allowances pursuant to Article 10a of Directive 2003/87/EC of the European Parliament and of the Council (notified under document C(2011) 2772) (2011/278/EU).

In this third phase, the EU ETS established the centralized registry system (the single European Union Registry).

(iv)   The fourth phase

The fourth phase or trading period will start in 2021 and will run until 2030, i.e., 10 years. The aims of the EU ETS are:

(i)   to increase the pace of emissions cuts;
(ii)   to establish better-targeted carbon leakage framework; and
(iii)   to provide funds for low-carbon innovation and energy sector modernization

It is worth mentioning that the EU ETS is the central pillar of the EU Climate Change Policy. Due to the fact that the European Union aims to be climate-neutral by 2050, the Commission will review and propose a revision of climate policy instruments by June 2021 as part of the EU's Green Deal. This includes the EU ETS and a possible extension of emissions trading to new sectors.

The EU ETS applies to CO_2 emissions from industry, power, and aviation, and includes industrial processes emissions. It also covers N20 emissions from the certain chemical sector and PFC emissions from primary aluminium production. Mexico applies only to CO_2 emissions from the industry and electricity sectors. (ICAP 2020b).

As mentioned above, for the first two trading periods the EU ETS had focussed on CO_2 only for the sake of simplicity because it is the main anthropogenic GHG and because of the large monitoring uncertainty in different sectors, as in the waste sector (European Commission 2006). For the third trading period, nitrous oxide (N_2O) emissions from fertilizer manufacturing, perfluorocarbon (PFC) emissions from primary aluminium production, and further activities emitting CO_2 were introduced.

In order to reduce GHG and to comply with the goals for 2030, the EU has a comprehensive policy which includes, in addition to the EU ETS, the emissions of households, transport users, agriculture and land use, land use change, and forestry (LULUCF). The latter has been added to reach the EU 2030 targets (Delbeke and Vis 2019, p. 17).

Conclusion

Mexico's commitment to climate change has been influenced by international agreements such as the 1997 Kyoto Protocol and the 2015 Paris Agreement. The former has put emission trading on the international agenda and the latter has placed emission trading as a new instrument for climate change policy around the world.

Additionally, Mexico has undertaken reforms regarding GHG mitigation in line with the 2015 Paris Agreement. The 2015 Energy Transition Law promotes sustainable and efficient use of energy and SENER has recently included the clean energy target of 39.9% by 2033 and 50% by 2050. The 1988 LGEEPA was amended to include the climate change agenda. The economic instruments for mitigating climate change in Mexico are: (i) the carbon tax, (ii) the Climate Change Fund, and (iii) the ETS pilot programme.

The ETS is a new instrument for environmental protection in Mexico. Therefore, it is very important to gain experience in its implementation during the current pilot phase. However, the effective operation of such a regime requires rules for monitoring, reporting, and verification.

As a developing country, the challenge for Mexico is getting technical and economic support from developed nations. In dealing with climate change, Mexico faces a lot of problems because of its lack of infrastructure, capacity building, transfer of technology, and finance, among others. Therefore, international support in these areas is required.

The EU ETS experience will be taken into account in the pilot phase of ETS (2020–2022) according to the LGCC (36 months). It is expected that at the end of the three phases, Mexico will be able to establish an Emissions Trading System, achieving the emissions reductions not only in the energy and industrial sector, but in other sectors as well. After that time, it is expected that this implementation will be mandatory. The ETS in Mexico will play a key role in promoting decarbonization in sectors such as the energy and industry sectors.

The implementation of an Emission Trading System under the cap and trade principle will help Mexico reach and comply with its INDC goals in the future and combat climate change. One of the benefits of emissions trading is that it provides certainty for environmental outcomes and will foster Mexico's sustainable development.

For Mexico, the implementation of an ETS as a cost-effective mitigation measure to achieve the Nationally Determined Contributions of reducing 22% GHG emissions by 2030 (increasing to 36%, if it has international support and financing) and 50% by 2050 is very relevant as a developing country.

The European Union's experience in the Emissions Trading System serves as a good example for implementing the Emissions Trading System in Mexico due to the fact that Mexico has the opportunity to learn and build technology capacities in this matter.

In sum, as a developing country, Mexico needs international support to achieve its targets, including technical, financial, and capacity building. The pilot phase of ETS will help the country to gain experience in this area and collect first experiences with auctioning allowances. Moreover, there are many lessons to be learned from the EU ETS, but the most important thing that Mexico has to take into account is that the GHG emissions reductions, clean energy, climate policy, etc., must be done gradually. The learning-by-doing approach of the EU ETS should be implemented in the Mexican climate change policy.

References

Birnie P, Boyle A, Redgwell C (2009) International law & the environment, 3rd edn. Oxford University Press, p 358

Bodansky D, Brunnée J, Rajamani L (2017) International climate change law. Oxford University Press, p 228

Borghesi S, Montini M (2016) The best (and worst) of the GHG emission trading systems: comparing the EU ETS with its followers. Front Energy Res 3

CISL (2015) 10 years id Carbon Pricing in Europe, Institute for sustainability leadership, Cambridge University Press

Cassese A (2005) International law, 2nd edn. Oxford University Press, p 487

Commission Decision of 27 April 2011 determining transitional Union-wide rules for harmonized free allocation of emission allowances pursuant to Article 10a of Directive 2003/87/EC of the European Parliament and of the Council (notified under document C (2011) 2772) (2011/278/EU). https://eur-lex.europa.eu/legal-content/EN/TXT/PDF/?uri=CELEX:32011D0278&from=EN

Delbeke J, Vis P (2019) Towards a climate-neutral Europe: curbing the trend. Routledge. https://ec.europa.eu/clima/sites/clima/files/toward_climate_neutral_europe_en.pdf

Diario Oficial de la Federación, 01 de octubre de 2019. Acuerdo por el que se establecen las bases preliminares del Programa de Prueba del Sistema de Comercio de Emisiones. Diario Oficial de la Federación, 01.10.2019. https://www.dof.gob.mx/nota_detalle.php?codigo=5573934&fecha=01/10/2019

Dixon M, McCorquodale R, Williams S (2011) Cases and material son international law, 5th edn, p 441

Directive 2003/87/EC of the European Parliament and of the Council of 13 October 2003 establishing a scheme for greenhouse gas emission allowance trading within the Community and amending Council Directive 96/61/EC. https://eur-lex.europa.eu/legalcontent/EN/TXT/PDF/?uri=CELEX:32003L0087&from=EN

Dupuy M, Viñuales J (2018) International environmental law, 2nd edn. Cambridge University Press, p 177

European Union (2020) Emission Trading System, climate policies. https://ec.europa.eu/clima/policies/ets_en

Global Carbon Atlas (2018) http://www.globalcarbonatlas.org/en/CO2-emissions

IETA (2018) Mexico: a market based climate policy case study. https://www.ieta.org/resources/Resources/Case_Studies_Worlds_Carbon_Markets/2018/Mexico-Case-Study-Jan2018.pdf

Intended Nationally Determined Contribution, Mexico, March 30, 2015. https://www4.unfccc.int/sites/submissions/INDC/Published%20Documents/Mexico/1/MEXICO%20INDC%2003.30.2015.pdf

International Carbon Action Partnership ICAP, ETS Mexico (2020a). https://icapcarbonaction.com/en/?option=com_etsmap&task=export&format=pdf&layout=list&systems[]=59

International Carbon Action Partnership ICAP, EU ETS (2020b). https://icapcarbonaction.com/en/?option=com_etsmap&task=export&format=pdf&layout=list&systems%5b%5d=43

Intergovernmental Panel on Climate Change (IPCC) (2012). Glossary of terms. In: Field CB, Barros V, Stocker TF, Qin D, Dokken DJ, Ebi KL, Mastrandrea MD, Mach KJ, Plattner G-K, Allen SK, Tignor M, Midgley PM (eds) Managing the risks of extreme events and disasters to advance climate change adaptation. A Special Report of Working Groups I and II of the Intergovernmental Panel on Climate Change (IPCC). Cambridge University Press, Cambridge, UK, and New York, NY, USA, pp 555–564. https://archive.ipcc.ch/pdf/special-reports/srex/SREX-Annex_Glossary. pdf

Kyoto Protocol to the United Nations Framework Convention on Climate Change (1997). https:// unfccc.int/resource/docs/convkp/kpeng.pdf

Ley General de Cambio Climático, June 6, 2012. http://www.diputados.gob.mx/LeyesBiblio/pdf/ LGCC_130718.pdf

Ley de Transición Energética, December 24, 2015. http://www.diputados.gob.mx/LeyesBiblio/pdf/ LTE.pdf

Malcolm MS (2014) International environmental law, 7th edn. UK, Cambridge University Press, p 617

Paris Agreement (2015) To the United Nations framework convention on Climate Change. https:// unfccc.int/files/essential_background/convention/application/pdf/english_paris_agreement.pdf

PWC (2017) The long view. How will the global economic order change by 2050? February 2017. https://www.pwc.com/gx/en/world-2050/assets/pwc-the-world-in-2050-full-rep ort-feb-2017.pdf

Reglamento de la Ley General de Cambio Climático en materia del Registro Nacional de Emisiones, RENE, 28 de octubre de 2014. http://www.diputados.gob.mx/LeyesBiblio/regley/Reg_LGCC_M RNE_281014.pdf

Reglamento de la Ley de Transición Energética, 04/05/2017. http://www.dof.gob.mx/nota_detalle. php?codigo=5481526&fecha=04%2F05%2F2017

Rio Declaration on Environment and Development to the United Nations framework convention on Environment (1992). https://www.un.org/spanish/esa/sustdev/documents/declaracionrio.htm

Sands P et al (2018) Principles of international environmental law, 4th edn. Cambridge, p 321

SEMARNAT (2019a) AVISO sobre el tope de emisiones del programa del Sistema de Comercio de Emisiones, 27 de noviembre de 2019. https://www.gob.mx/cms/uploads/attachment/file/513 702/Aviso_Tope.pdf

SEMARNAT (2019b) AVISO sobre la asignación sectorial del programa de prueba del Sistema de Comercio de Emisiones, 27 de noviembre de 2019. https://www.gob.mx/cms/uploads/attach ment/file/513701/Aviso_Asignacion_Sectorial.pdf

SENER, Acuerdo por el que la Secretaria de Energia aprueba y publicala actualizacion de la Estrategia de Transición para promover el Uso de Tecnologías y Combustibles más Limpios, en términos de la Ley de Transición Energética, 07/02/2020 http://www.dof.gob.mx/nota_deta lle.php?codigo=5585823&fecha=07/02/2020

Tan Z (2014) Air pollution and greenhouse gases, from basic concepts to engineering applications for air emission control. Springer

Titenberg TH (2020) Emissions trading: principles and practice. Washington, DC, Resources for the Future 2006. UNEP DTU CDM/JI Pipeline/CDM projects by host region. https://www.cdm pipeline.org/cdm-projects-region.htm#top

United Nations Framework Convention on Climate Change (1992). https://unfccc.int/files/essent ial_background/background_publications_htmlpdf/application/pdf/convsp.pdf

United Nations (2020) Treaty collection. https://treaties.un.org/Pages/ViewDetails.aspx?src=TRE ATY&mtdsg_no=XXVII-7-d&chapter=27&clang=_en

World Bank (2020) Carbon pricing dashboard. https://carbonpricingdashboard.worldbank.org/ map_data

Chapter 6
Particularities of the Legal Framework for the Mexican Emissions Trading System

Rosalía Ibarra Sarlat

Abstract This paper examines the legal bases for the mandatory regulation of the emissions trading system in Mexico. They are derived from the main international instruments on climate change: the United Nations Framework Convention on Climate Change (UNFCCC) and its ambitious objective, the quantifiable commitment of the Kyoto Protocol, and its tie to economic instruments. The Paris Agreement, the Nationally Determined Contributions (NDCs) and the market mechanisms regulated in Article 6, the implementation of which is essential to achieve the Agreement's objectives are also part of this broad system. Legally, the international foundations of the emissions trading system are reflected at the national level. For these, the constitutional and legal bases underpin the current regulation of the mandatory market instrument. It aims to effectively reduce, in terms of costs, the greenhouse gas emissions from the most polluting economic activities, without replacing direct control measures. The core aspects of this system are highlighted from a national regulatory analysis, with special emphasis on the importance of a limited cap and its future reduction, as well as the legal nature of allowances that are allocated by the public administration to the regulated industries' facilities.

Keywords Emissions trading · Economic instruments · Emission allowances · Carbon markets · Regulations · Mexico

Introduction

The economic theory considers climate change as a negative externality derived from the burning of fossil fuels for industrial and energy production. To correct this great market failure and to internalize the externalities in the design of environmental public policies, the environmental economy plays a fundamental role as it consists of applying the economic principles to the study of natural resource management.

R. Ibarra Sarlat (✉)
Institute of Legal Research (IIJ) of National Autonomous University of Mexico (UNAM), CDMX, Circuito Maestro Mario de la Cueva s/n. Ciudad Universitaria, c.p. 04510, México City, Mexico
e-mail: ribarra@unam.mx

© The Author(s) 2022
S. Lucatello (ed.), *Towards an Emissions Trading System in Mexico: Rationale, Design and Connections With the Global Climate Agenda*, Springer Climate,
https://doi.org/10.1007/978-3-030-82759-5_6

This is accomplished by integrating the environmental variable into conventional economic analysis so that the structure of environmental decisions is based on cost-effectiveness and cost–benefit considerations. In this regard, to mitigate greenhouse gas emissions, global and national interests have been strengthened by the adoption of economic instruments, especially carbon markets as complementary measures to command and control regulations to ensure global benefits at the lowest possible cost. The Mexican government is no exception. To achieve its emissions reduction goals, it has opted for the mandatory implementation of the emissions trading system in a cap and trade scheme, an instrument to promote energy efficiency, the introduction of new low-carbon technologies, investment in technological innovation, the modification of consumption patterns, etc., all aimed at decarbonizing our economy. The implementation of this instrument has legal backing at the international and national levels, which has been progressive in promoting the use of these systems, focused on improving their regulatory framework and broadening their scope in both jurisdictions.

Due to methodological issues and legal hierarchy, this paper starts from the study of the international instruments on which the regulatory regime on climate change is based, with special emphasis on certain particularities, such as in the case of the ambitious objective of the United Nations Framework Convention on Climate Change (UNFCCC); the quantifiable commitment of the Kyoto Protocol and its link to economic instruments; the Paris Agreement, the Nationally Determined Contributions (NDCs) and the market mechanisms regulated in Article 6 whose scope of application is still under discussion in international forums.

After analyzing the international foundations of the emissions trading system, its application at the national level is studied, which is reflected in constitutional bases and in the regulations pertaining to the economic instruments of the General Law of Ecological Balance and Environmental Protection (Ley General de Equilibrio Ecológico y Protección al Ambiental or LGEEPA) and the General Law on Climate Change (Ley General de Cambio Climático or LGCC), which support the current regulation of the emissions trading system in Mexico.

Finally, the analysis of the national regulations highlights the core aspects of the system, with special emphasis on the cap, as well as the free allocation of allowances and the polluter pays principle, to conclude with the importance of defining and identifying the legal nature of the allowances allocated by the public administration to the facilities of the regulated sectors.

International Legal Basis

The international legal regime for climate change is made up of three binding instruments to which Mexico is a State Party:

– The United Nations Framework Convention on Climate Change (UNFCCC) adopted in New York on 9 May 1992. In effect since 21 March 1994.

- The Kyoto Protocol (Protocol to the UNFCCC) adopted in Kyoto, Japan, on 11 December 1997. In effect since 16 February 2005.
- The Paris Agreement adopted in Paris, France, on 12 December 2015. In effect on 4 November 2016.

The Objective of the UNFCCC

The main objective of the Kyoto Protocol and the Paris Agreement, as the Convention's legal instruments for its effective implementation, is stipulated in Article 2 of the Convention, which states that:

> The ultimate objective of this Convention and any related legal instruments that the Conference of the Parties may adopt is to achieve, in accordance with the relevant provisions of the Convention, stabilization of greenhouse gas concentrations in the atmosphere at a level that would prevent dangerous anthropogenic interference with the climate system. Such a level should be achieved within a time frame sufficient to allow ecosystems to adapt naturally to climate change, to ensure that food production is not threatened and to enable economic development to proceed in a sustainable manner.

The achievement of the UNFCCC's objective is long term and requires the participation and collaboration of the various sectors involved in order to modify behavior patterns in strategic areas, such as energy and industry. The interests are many and diverse, the difficulty of which lies in making them compatible.

A key point is the sustainable development principle, adopted by the UNFCCC in accordance with its Article 3, from which it follows that efforts to prevent and combat the effects of climate change must not run counter to the economic development of countries, especially the most vulnerable ones, as long as it is not detrimental to the environment. This way, Parties must promote and opt for a sustainable economic development model so that the protection of the climate system is for the benefit of present and future generations.[1]

For sustainable development to be a viable proposal, economic, social, and political changes are needed at the international and national levels in which strategies respond to the interests of the majority and where a new order is sought, especially an economic one that promotes the rational and balanced democratic management of natural resources that guarantee peace, stability, and prosperity in the world community.

Under this scheme, sustainable development must be related to the simultaneous and ongoing evolution of the economic, social, environmental, technological, and political sectors at the international and national levels (World Resources Institute 1992). Its implementation implies a development that can be maintained in the long term from environmental and economic points of view through the use of less polluting technologies and with lower or no greenhouse gas (GHG) emissions.

---

[1] The Convention refers to the right to development, recognized as a human right in the Declaration on the Right to Development approved by Resolution 41/128 of the UN General Assembly on 4 December 1986.

Emissions trading through effective implementation is, precisely, an instrument that can contribute to the achievement of this objective.

The Quantified Commitment of the Kyoto Protocol and Its Link to Economic Instruments

Stabilizing greenhouse gas (GHG) concentrations in the atmosphere at a level that would prevent dangerous anthropogenic interference with the climate system involves the adoption of quantified commitments to reduce total emissions by at least 5% relative to 1990 levels for the 2008–2012 period. This is a core obligation of the Kyoto Protocol exclusively for developed countries and those with economies in transition (Article 3). For compliance and in order to promote sustainable development, each of the obligated Parties must implement and/or develop measures and policies for the progressive reduction or phasing out of market deficiencies, fiscal incentives, tax and duty exemptions, and subsidies contrary to the objective of the Convention in all GHG-emitting sectors, and apply market-based instruments (Article 2).

Agreeing on the commitments assumed in the Kyoto Protocol was possible thanks to the inclusion of three economic mechanisms, known as *flexible mechanisms* aimed at making it easier for the obligated Parties to meet the quantified emission reduction commitment at a low cost. These were proposed in previous negotiations by the United States Government:

(1)   The Joint Implementation (JI), regulated in Article 6, allows governments and companies from developed countries grouped in Annex I of the UNFCCC (Annex B of the Kyoto Protocol) to invest in other countries of the same group in the realization of projects for emission reduction or carbon sequestration by promoting sinks, which is compensated by Emissions Reduction Units (ERUs).

(2)   The Clean Development Mechanism (CDM), regulated in Article 12, is an instrument that promotes and regulates public or private investments by countries included in Annex I of the UNFCCC (Annex B of the Kyoto Protocol) in emission reduction or carbon sequestration projects carried out in a non-Annex I country (developing countries) to help it achieve sustainable development and contribute to the ultimate objective of the Convention, which is rewarded with Certified Emissions Reductions (CERs).

(3)   The Emissions Trading (ET), regulated in Article 17, is a system that operates in a market based on an emissions limit and its trade (cap and trade system). The certainty of the results is based on the establishment of a total quota of allocated allowances, which represent the total limit of authorized emissions. Under this regime, UNFCCC countries party to Annex I (Annex B of the Kyoto Protocol) or those legal entities that they have authorized[2] may trade through purchase

---

[2] At the time, the Protocol only perceived emissions trading between States, however, it was later admitted that States could associate their companies, without replacing State responsibility.

and sale depending on whether they have emission surplus or deficit. The different types of accounting units are recognized by the Protocol (Assigned Amount Units-AAU, or those generated by projects, such as CERs and ERUs). The incorporation of this system, for which there are precedents and experiences in several countries, has led to major changes in the economic structure. The Mexican government has based the establishment of its emissions trading system on these lessons learned.

The JI and CDM are project-based compensation mechanisms that enable carbon credits to be obtained with *ex-post* verification of the emission reductions achieved. Additionally, they seek to promote technological and financial transfers through international investment, whereas emissions trading is a reduction mechanism with pre-set emission limits for a period of time determined by the competent authority so that the reduction depends on the size of the cap, distributed in emission quotas, thus building on the *ex ante* allocation of emission allowances that can be traded on the market (Ibarra 2019).

The CDM was an innovative tool that involved developed and developing countries in financing climate action. The CDM is an existing mechanism, in which Mexico has had extensive participation. Currently, in the Latin American region, Mexico ranks second in the number of registered projects with 192, especially in the areas of methane recovery, waste management and recovery, renewable energy, energy efficiency, industrial processes, and agriculture (UNFCCC 2020).

In general terms, the flexible mechanisms have the peculiarity of integrating the environmental variable into economic systems by introducing market rules in favor of the environment, facilitating the fulfillment of commitments at a low economic cost, as an incentive, with the advantage of being voluntary and introducing the private sector without displacing State responsibility. They are therefore linked to national policies and measures of supplementary nature—an intended feature of the Protocol (Articles 6(d) and 17 of the Protocol). This means that under no circumstance should these mechanisms supplement national measures to comply with the obligations assumed and are, therefore, not a means of compliance in *stricto* sensu, but rather an aid so that States comply (Saura 2003; Ibarra 2012).

The Kyoto Protocol helped enable developed and developing country governments to adopt climate policies and laws, and propelled some industries and companies to take into account the environmental variable, specifically climate change, in decision-making about their investments. However, this legal instrument was ineffective in terms of global mitigation (Salinas 2017) because, despite the fulfillment of quantified commitments (European Environment Agency 2017), the steady global increase in the volume of GHG emissions continued, aggravating the climate problem and failing to meet the UNFCCC's target, largely due to the Protocol's original structural deficiencies.[3] This led to the adoption of a new framework of obligations for States: the Paris Agreement.

The Paris Agreement NDCs

The Paris Agreement, in addition to achieving the Convention's objective, aims to strengthen the international response to sustainable development with low GHG emissions with the goal of maintaining the average global temperature increase well below 2 °C with respect to pre-industrial levels, limiting the increase to 1.5 °C (Article 2). To this end, the Agreement changes the strategy at the mitigation level, adopting "voluntary" commitments from developed and developing countries, replacing the top-down scheme with a bottom-up scheme. The commitments are specified in the instrument of Nationally Determined Contribution (NDC) so that emission reduction is not limited to some countries, but to all, implying broad coverage in which each State establishes its own contribution to mitigation and determines the means to achieve it.

In this regard, the Mexican government assumed two types of mitigation commitments in its NDC: unconditional (with its own resources) and conditional (with additional resources and technology transfer through international cooperation). In terms of the former, it committed to reducing 25% of its GHGs and short life climate pollutants (SCCP) by 2030, which implies: (a) A 22% baseline reduction in direct emissions of carbon dioxide, methane, nitrous oxide, and fluorocarbon gases; and (b) A 51% reduction in black carbon particle emissions. This commitment incorporates a peak in net emissions for 2026; thereafter net emissions should start to decline.

Concerning conditional commitments, the objective expressed above extends to a 40% reduction in GHGs and HCVs by 2030, including: (a) GHG reductions of up to 36%; and (b) Reductions in black carbon emissions to 70%. The commitment is conditional on a global agreement, including, for example, provisions for an international carbon price, tariff adjustments for carbon content, financial cooperation, and technology transfer, among others.[4]

---

[3] Under the regulatory scheme of the Kyoto Protocol, the turning point was precisely the non-imposition of quantifiable emissions reduction commitment on the States considered as developing countries (not included in Annex I of the UNFCCC), countries that, as principally China and India are currently among the main emitters. However, this does not mean that developed countries are now the ones who emit the least, since the United States, the 28 countries of the European Union (mainly Germany), Russia, Japan and Canada are among the 10 territories with the most CO_2 emissions including, of course, the territories of the emerging economies of China and India.

[4] Prior to the 21st meeting of the Conference of the Parties to the UNFCCC (COP21), at COPs 19 and 20 in accordance with its first decision (1/CP.19 and 1/CP.20), all Parties were invited to communicate to the UNFCCC Secretariat their Intended Nationally Determined Contribution (INDC), with the aim of being brought for negotiation at COP21. With the implementation of the Paris Agreement, the approved INDCs were considered official and thus acquired the name Nationally Determined Contribution (NDC). Mexico submitted its NDCs on 27 March 2015, becoming the first developing country to do so.

Article 6 of the Paris Agreement

Achieving the objective of the Paris Agreement depends on the effective implementation of the mitigation commitments of the NDCs and their progressive ambition. Although the Agreement does not stipulate an obligation to coercively fulfill such commitments, Parties shall pursue domestic mitigation measures, with the aim of achieving the objectives of such contributions, in accordance with Article 4(2). In this regard, the measures adopted should aim at the decarbonization of the economies, which will lead to radical changes in production sectors, the promotion of energy efficiency, the introduction of new low-carbon technologies, investment in technological innovation, changing consumption patterns, etc. This will require the promotion and/or imposition of diverse measures in national climate policies, where the use of economic instruments has increased due to its flexible and profitable framework. Specifically, emissions trading has proliferated geographically. Its introduction and development was the legacy of the Kyoto Protocol, which some scholars consider to be its great merit (Baron and Philibert 2007; Sanz 2009; Philibert 2005).

Article 6 of the Paris Agreement does not specifically mention the flexible mechanisms of the Protocol, however, it alludes to them implicitly as we will see next. Furthermore, it is assumed that they remain applicable as they do not conflict with the content of the Agreement, in accordance with Article 30 of the 1969 Vienna Convention on the Law of Treaties (Salinas 2017).

The execution of Article 6 is considered fundamental to achieve the Agreement's objectives in compliance with the NDC's mitigation targets and increased ambition. However, to date, the State Parties have not been able to agree on its rules of application, due to the fact that its content has become controversial as it requires common guidelines for the use of cooperative approaches on a voluntary basis, among which are market mechanisms. The above is drawn from the core points of Article 6, paragraphs 2 and 4, to which we will refer without being exhaustive but with the objective of pointing out elements that the Mexican government should address in its participation in the carbon markets and offsets.

Article 6(2) of the Paris Agreement

Article 6(2) states that:

> Parties shall, where engaging on a voluntary basis in cooperative approaches that involve the use of internationally transferred mitigation outcomes towards nationally determined contributions, promote sustainable development and ensure environmental integrity and transparency, including in governance, and shall apply robust accounting to ensure, inter alia, the avoidance of double counting, consistent with guidance adopted by the Conference of the Parties serving as the meeting of the Parties to this Agreement.

This section states that to facilitate compliance with the commitments assumed in the NDCs, units known as Internationally Transferred Mitigation Outcomes

(ITMOs) may be transferred between State Parties through mechanisms such as carbon markets. It is also established that projects and programs that participate voluntarily in cooperative approaches to generate ITMOs, in addition to being profitable, must promote sustainable development and guarantee environmental integrity, as well as transparency.

The principle of environmental integrity implies recognizing the repercussions of human activity on ecological systems, respecting the limits of the regenerative capacity of ecosystems, and avoiding irreversible damage to populations and plant and animal species (actions that lead to irreplaceable losses, such as the extinction of endangered species). At the same time, it implies protecting areas internationally recognized for their ecological, cultural, or historical value (International Institute for Sustainable Development 1994). For environmental integrity to exist, the environmental variable must first be incorporated in decision-making, which is undoubtedly a state responsibility, whose level of incorporation will depend on the degree of maturity of the political and economic systems. That is, in particular, the strength of the planning, management, and resource allocation structure. The strengthening of the rule of law in environmental matters is required; a key element for the protection, conservation, and restoration of environmental integrity, without which environmental governance and compliance with national and international rights and obligations could become arbitrary, subjective, and unpredictable (UICN 2016).

In view of the multiple legal, economic, political, social, environmental, and institutional differences—which are reflected in the diversity of the NDCs presented by the State Parties to the Paris Agreement—the demand for effective enforcement of environmental integrity is of great relevance when approving projects and programs that ensure real and measurable GHG emission reductions traded in international markets.

While the Paris Agreement does not define environmental integrity, Schneider and La Hoz Theuer (2019) identify four main factors to ensure that global GHG emissions do not increase as a result of transfers and that environmental integrity is not undermined:

1. Accounting for international transfers.
2. Quality of units.
3. Ambition and scope of the mitigation target of the transferring country.
4. Incentives or disincentives for future mitigation action.

On the other hand, Article 6(2) urges that two countries should not claim the same emission reductions for the fulfillment of their respective mitigation targets, which would imply that the reduction of one tonne of emissions would be counted in favor of two countries and not just one. Preventing double counting or double claiming requires strong and robust monitoring, reporting, and verification (MRV) systems to demonstrate net emission reductions, which is complemented by the Transparency Framework under Article 13 of the Paris Agreement.

Schneider and La Hoz Theuer (2019) point out that sound accounting is a prerequisite for both ensuring environmental integrity and making accounting for international transfers representative of Parties' mitigation efforts over time. They also establish that in order to prevent cumulative global emissions from increasing, it is essential that there be adequate accounting of the reductions or removals of emissions from activities with temporary results, such as in the land-use, land-use change, and forestry sectors or in the case of geological storage of CO_2. Regarding the other elements, the authors note that the quality of the transfer units will depend on criteria similar to those of the accreditation systems used in the flexible mechanisms of the Kyoto Protocol. In turn, they point out that the quality of the units is linked to the mitigation objectives, the more ambitious the objectives, the higher the quality.

Finally, they highlight the international carbon markets as instruments that can provide both incentives and disincentives for future mitigation actions, since they consider that for the countries acquiring carbon units the market is an incentive that allows for reducing costs in order to achieve more ambitious objectives, but the system could discourage selling countries from setting more ambitious future mitigation targets, in order to sell carbon units at a better price in order to obtain more benefits from the transfer units internationally. This is because the system responds to supply and demand (the lower the volume of carbon units, the higher the price).

Although the mitigation objectives of the NDCs are established outside the market and cannot be modified once they are presented under the framework of the Paris Agreement, in the updates of the future NDCs of the country that sells carbon units could consider the prices of carbon units in the market and if they are high, it is very likely that they do not want to saturate the said market with more units to preserve their price, so their level of ambition in mitigation actions could be lower.

Hence, it is appropriate to link emissions trading systems internationally between countries with similar levels of ambition. Also, these authors suggest that the Parties could decide to establish requirements for participation in international transfers under Article 6 that provide incentives for countries to expand the scope of their NDCs, for example, by limiting these transfers to emission reductions generated by sectors or gases covered by the selling country's NDCs or, alternatively, requiring countries to commit in the future to expand the scope of their NDCs to targets in all economic sectors in order to be able to participate in Article 6.

Regarding the international carbon markets, Zapf et al. (2019) consider that a global market is an adequate instrument for achieving the common commitment of the Paris Agreement. To this end, they point out that, in addition to adopting the cap and trade scheme, the introduction of this system must be uniform and cross-sectoral with a specific politically agreed design. When this is an instrument subject to speculation, the best that can be achieved is that the overall costs of reducing emissions are minimized as much as possible, while ensuring that global CO_2 emissions remain within a certain carbon budget.

Article 6(4) of the Paris Agreement

In connection with the previous section, paragraph 4 states the following:

A mechanism to contribute to the mitigation of greenhouse gas emissions and support sustainable development is hereby established under the authority and guidance of the Conference of the Parties serving as the meeting of the Parties to the Paris Agreement for use by Parties on a voluntary basis. It shall be supervised by a body designated by the Conference of the Parties serving as the meeting of the Parties to the Paris Agreement, and shall aim:

(a)   To promote the mitigation of greenhouse gas emissions while fostering sustainable development;

(b)   To incentivize and facilitate participation in the mitigation of greenhouse gas emissions by public and private entities authorized by a Party;

(c)   To contribute to the reduction of emission levels in the host Party, which will benefit from mitigation activities resulting in emission reductions that can also be used by another Party to fulfill its nationally determined contribution; and

(d)   To deliver an overall mitigation in global emissions.

This section anticipates the development and emergence of a new system based on and as a successor to the CDM, known as the Sustainable Development Mechanism (SDM). It is expected to adopt all the positive experiences of the CDM, but also to address its shortcomings, such as the inadequate use of additionality and the lack of project sustainability (their inequitable geographical distribution and compensation[5] as opposed to a reduction scheme) (Ibarra 2012).

In the implementation of projects under the CDM scheme, additionality in relation to the baseline scenario, also called business-as-usual (BAU), linked to the need to ensure that real, measurable, and certifiable GHG reductions or removals are generated, has been one of the most discussed issues (Greiner and Michaelowa 2003; Schneider 2009; Cames et al. 2016; Fischer 2005) as it is fundamental to ensure environmental integrity and thus the effectiveness of global climate action since additionality requires that any mitigation or removal activity considered for a market-based mechanism must demonstrate that the reduction of GHG emissions or the increase in GHG removals are greater than what would occur if the project activity were not carried out.

Additionality applied to the new Paris Agreement's liability scheme implies ensuring that activities based on market mechanisms are truly additional to host countries' CDMs. It also implies that those activities should generate real GHG reductions. Because countries and other participating entities use dummy units to meet their mitigation obligations, through offsets, an overall increase in emissions will be generated instead of a reduction, altering the environmental integrity of the Agreement, hence the importance of ensuring additionality. However, its verification is one of the most difficult methodological problems, but it is essential to address it within the institutional and legal framework of the verification system (Michaelowa et al. 2019).

Following this order of ideas, additionality is an integral concept to ensure sustainable development, since it cannot be determined by the economic factor alone, that is

---

[5] The CDM is not considered to be an instrument of global emission reduction, but rather an instrument of compensation, since the emissions made in excess in the territory of developed countries are compensated with those avoided in the projects implemented in developing countries.

to say, by cost–benefit criteria of monetary value, since the implementation of activity for a market mechanism should also promote environmental and social benefits in addition to being an important means of technology transfer.

In this regard, the experience developed in CDM projects is relevant, since many projects lacked sustainability (Schneider 2007) because they generated negative impacts, either social (conflicts over land tenure, restrictions on access to goods and services for local communities, population displacement, evictions, and expropriations, among others, with serious human rights implications), environmental (loss of biodiversity, habitat destruction, alteration of ecosystems, soil erosion, increased risk of fire, etc.), economic (loss of long-term benefits, increased illegal economic activities), cultural, and/or religious (Ibarra 2012). The concept of sustainable development in practical terms is highly debatable in the CDMs, especially in determining the extent to which it has been achieved. This led to the acceptance of projects with negative effects that were especially economically beneficial to developed countries by generating CERs (Muller 2007).

Currently, formal modalities and procedures for the new mechanism proposed in Article 6(4) are being negotiated at the Conference of the Parties to the Country Agreement, for which the experience gained from the flexible mechanisms of the Kyoto Protocol should undoubtedly be valued and go even further in obtaining results. Michaelowa et al. (2019) point out that in the face of the new paradigm changes, the activities proposed under the scheme of the new mechanism, could not only be evaluated in terms of their sustainability, but rather in terms of their transformative potential. This is to say, toward a transformational development which could complement or even replace the proof of additionality. The criterion of transformational change is a great challenge (Michaelowa et al. 2019), however, it is necessary to confront it given the climate emergency we are facing.

In general terms, Article 6 requires a great deal of negotiation between the Parties to establish international rules for the implementation of the new mechanisms. At the same time, it determines their transition with the existing mechanisms under the Kyoto Protocol, such as with the regional and national carbon markets that are increasing in number and in operations, which can undoubtedly contribute to implementing the new mechanisms of international cooperation.

National Legal Basis

Regarding climate change, over the years the Mexican government has adopted a legal and institutional framework to comply with its international commitments. In this process of gradual and progressive advancement, the emissions trading system is currently regulated as a mandatory instrument, which is expected to contribute to achieving the mitigation objectives of the CDMs presented in the provisions of the Paris Agreement.

However, the application of the cap and trade scheme is not a novelty in the Mexican system. From 2001 to 2005, Petroleos Mexicanos (PEMEX) voluntarily implemented an internal market for carbon emission permits with the technical support of Environmental Defense, a non-governmental organization from the United States.

PEMEX began to quantify its emissions in 1997 and established goals for reducing CO_2 emissions by 1% with respect to 1999 levels for the 2001–2003 period. To achieve this, it operated the internal commercialization of emission permits in which 25 Business Units (PEMEX industries with the highest carbon emissions) participated and to which permits equivalent to one ton of CO_2 were granted. For the coordination and development of the market, the Corporate Environmental Protection Audit was created; online market operations were carried out in the Transaction Registration System developed by PEMEX. The mechanism stimulated competition among the Business Units within the market to reduce their emissions and reach the goal imposed by the same company, in which each Unit developed better operational practices and executed projects that were cost-effective and cost-beneficial in the reduction of emissions to obtain a surplus in the permits granted, sell them on the market and make a profit, as well as obtaining business experience for networking purposes with future global markets (Gómez 2004). The results were favorable since, in 2002, an accumulated emissions reduction of 11% was achieved with respect to the base year (1999). This is equivalent to almost 5 million tons of CO_2 (Fernández and Martínez 2003). However, with the Kyoto Protocol entering into force in 2005, the market ceased to function and was replaced by CMD projects; the company only registered one project in 2012, which to date has not entered the stage of soliciting the corresponding certificates[6] (UNFCCC 2020).

Economic Instruments in the LGEEPA

Legally, the economic instruments were introduced in the General Law of Ecological Balance and Environmental Protection (Ley General de Equilibrio Ecológico y Protección al Ambiental or LGEEPA)[7] by means of reforms published in the Official Gazette of the Federation (DOF) on 13 December 1996; until then, environmental enforcement had been based on a system of direct regulation (command and control): permits, inspections, and sanctions. However, it was considered that it should be complemented by other types of instruments to achieve the environmental policy objectives.

According to the explanatory memorandum of the 1996 reform initiative, the indirect regulation based on economic instruments was considered ideal for giving effect to two fundamental principles of environmental policy:

(1) *Whoever pollutes, makes excessive use of natural resources, or alters the ecosystems must assume the costs inherent to their conduct.*

---

[6] Project 4966: Waste Energy Recovery Project at PEMEX TMDB.
[7] Published in the Federal Official Gazette (DOF) on 28 January 1988.

The Polluter Pays Principle was introduced in 1972 by the members of the Organization for Economic Cooperation and Development (OECD), which postulates that the polluter should bear the full costs of any environmental damage caused by the production of goods and services, which implies the internalization of environmental degradation costs as a negative externality (OECD 1975).[8] By means of this principle, it is intended to compel the producer of goods and services to internalize the externality, absorbing the costs of production, as well as the environmental ones (Butze 2006).

(2)   *Whoever conserves resources and invests in ecological conservation, reconstructing the environmental capital of the nation, should receive a stimulus or compensation.*

In *contrario* sensu, whoever protects and conserves obtains an economic benefit when referring to who should be stimulated or rewarded for protecting the environment.

In consideration of the above, the LGEEPA regulates in Section III the economic instruments and defines them in its Article 22, first paragraph, as those of:

Regulatory and administrative mechanisms of a fiscal, financial, or market nature, through which people assume the environmental benefits and costs generated by their economic activities, encouraging them to carry out actions that favor the environment.

The Law distinguishes between fiscal, financial, and market instruments, with respect to the latter, the fourth paragraph establishes that they are:

The concessions, authorizations, licenses, and permits that correspond to pre-established volumes of emissions of pollutants into the air, water, or soil or that establish limits to the exploitation of natural resources [...].

The prerogatives derived from economic market instruments shall be transferable, non-taxable, and remain subject to the public interest and the sustainable use of natural resources.

The emissions trading system has its legal basis in this provision, which allows for the establishment of permits or transferable rights in the market through the prior establishment of an air pollutant emissions limit by the State. The provision is also extended to the compensation system. However, the Law only recognizes this instrument legally but without giving it a mandatory nature. Additionally, the Law, although it provides for the prevention and control of air pollution (Articles 110 to 116), does not focus on climate change and its complex problems, hence the need for specific legislation.

---

[8] The controversy of this principle has been around its adequate application and its link to the valuation of the real costs of environmental degradation; distorting on multiple occasions its pragmatic background by interpreting it as "a right to pollute" through payment. On the other hand, acceptance and application of the principle in international relations still generates discord among States.

The Economic Instruments in the LGCC

Mexico was the first developing country to adopt a law on climate change, published in the DOF on 6 June 2012, and in force as of 10 October of the same year in accordance with its First Transitory Article.

The General Law on Climate Change (Ley General de Cambio Climático or LGCC), in accordance with its Article 1, regulates the provisions of the Political Constitution of the United Mexican States (Constitución Política de los Estados Unidos Mexicanos or CPEUM) on environmental protection, sustainable development, and the preservation and restoration of ecological balance.

The LGCC has among its objectives, in accordance with its stipulations in Article 2:

- Guarantee the right to a healthy environment.
- Regulate the emissions of greenhouse gases and compounds.
- Regulate actions for the mitigation of climate change.
- Promote education, research, development, and transfer of technology and innovation, and dissemination in the field of climate change mitigation.
- Promote the transition to a competitive, sustainable, low-carbon economy that is resilient to the extreme hydrometeorological events associated with climate change.
- Establish the bases so that Mexico contributes to the fulfillment of the Paris Agreement.

Likewise, the Law determines an indicative target or aspirational goal for the reduction of emissions in the Second Transitory Article, consisting of a 30% reduction of GHG emissions by 2020 with respect to the baseline, and conditioned to economic and technological support, a reduction of 50% by 2050 based on what was emitted in 2000. At the same time, the Third Transitory Article establishes that the actions taken in the matter of mitigation should reach the following aspirational goals and indicative deadlines: Promotion of electricity generated by means of clean energy sources by at least 35% by the year 2024; which constitutes an essential component to reach the emissions reduction goals committed to in the NDC of Mexico, which are in line with the conditional goal set forth by this law.

Regarding the NDCs, it should be noted that those presented by the Mexican government in 2015 were based on the following instruments:

- General Law on Climate Change (LGCC, adopted in 2012)
- National Climate Change Strategy, 10–20-40 Vision (Estrategia Nacional de Cambio Climático, Visión a 10–20-40, adopted in 2013)
- Carbon Tax (Impuesto al Carbono, implemented in 2014)
- National Emissions Register (Registro Nacional de Emisiones or RENE, created and launched in 2014)
- Energy Reform (Reforma Energética, laws and regulations adopted in 2013)
- National Inventory of Greenhouse Gas and Compound Emissions (Inventario Nacional de Emisiones de Gases y Compuestos de Efecto Invernadero, update

presented in 2015 by the National Institute of Ecology and Climate Change (INECC) in compliance with Article 74 of the LGCC).

In view of the objectives, the LGCC determines the scope and content of national climate change policy, for whose orientation and implementation, in accordance with Article 58, the Special Climate Change Program, as well as the Nationally Determined Contributions (NDC), are among the planning instruments for the National Climate Change Strategy.

The LGCC establishes two guiding principles for action: adaptation and mitigation, both with medium- and long-term focuses. In terms of mitigation, Article 31 states that through planning and economic instruments, the policy should include a diagnosis, planning, measurement, monitoring, reporting, verification, and evaluation of national emissions; in response to which the National Inventory of Emissions of Gases and Greenhouse Compounds and the National Emissions Registry are instituted, the latter regulated by the LGCC's Regulations on the matter of the National Emissions Registry, published in the DOF on 28 August 2014.

The national policy on climate change determined, in accordance with Article 26, section IX, that the use of economic instruments in mitigation encourages the protection, preservation, and restoration of the environment, and the sustainable use of natural resources, in addition to generating economic benefits for those who implement them.

The economic instruments are regulated in Chapter IX in Articles 91 to 95. They are defined in Article 92 and based on the concept of the LGEEPA, the variant is due to the objective of the LGCC, with respect to which it is indicated that through these instruments of fiscal, financial, or market character people assume the benefits and costs of the mitigation and adaptation to climate change, with the objective of encouraging them to carry out actions that favor achieving the objectives of the national policy on climate change.

The fourth and fifth paragraphs of the same Article, in accordance with the LGEEPA, state that market instruments are:

The concessions, authorizations, licenses, and permits that correspond to pre-established volumes of emissions, or that provide incentives to carry out actions to reduce emissions by providing alternatives that improve their cost-efficiency ratio.

The prerogatives derived from market economic instruments shall be transferable, non-taxable, and will be subject to the public interest.

In this sense, the Law shows the suitability of privileging the actions with the greatest potential for mitigation in a cost-efficient manner and that at the same time generate a collective benefit. This scheme contemplates the emissions trading system, the use of which was always considered in Article 94, but on a voluntary basis. It was not until the 2018 reforms to the LGCC that its implementation was stipulated as mandatory at the federal level. It should be noted that the Law gives the federation the power to create, authorize, and regulate emissions trading (Article 7, section IX), while the National Institute of Ecology and Climate Change is empowered to participate in its design (Article 22, section III).

Economic instruments in general, and specifically market instruments, regulated by the provisions of the LGEEPA and the LGCC have their basis in Articles 4, 25, and 27 of the CPEUM, to which we refer below.

Article 4, in its fifth paragraph, stipulates the right to a healthy environment, which is the guiding principle of Mexican environmental policy, and therefore, of any instrument or measure adopted to achieve its objectives. Such is the case of market instruments. It should also be noted that maintaining the atmosphere in optimal conditions and free of excess pollutants, will guarantee the enjoyment of this right.

The first paragraph of Article 25 states that the Mexican government should guarantee comprehensive and sustainable economic development, in response to which the seventh paragraph establishes that, under the criteria of social equity, productivity, and sustainability, it will support and encourage businesses in the social and private sectors of the economy. But they will be subject to the modalities dictated by the public interest and the use of productive resources will be for the general benefit, seeking to preserve them and the environment.

Preservation of the chemical composition of the atmosphere through stabilization of the concentrations of GHG emissions to a level that avoids dangerous anthropogenic interference in the climate system is of public interest. Hence, the establishment of emission limits by the State for the productive sectors, through market instruments, is in the interest of the community and linked to the postulate of guaranteeing sustainable development, whose achievement requires in-depth structural changes where the general interest prevails over the individual.

As for Article 27, in its third paragraph, the principle of the social function of private property is consecrated, which is subject to the limitations dictated by the State in benefit of environmental protection. Market instruments restrict (does not imply deprivation) rights under the protection of the general interest (Sanz 2007). Rosembuj (2005) points out that, in the establishment of social order, there exist two opposing principles: the principle of individual freedom, supported by the right to property; and the second, the principle of equality of all humans, where the beholder of social wellbeing is the State. It is precisely in the adequate concordance of both principles where the ideal environmental protection lies.

The Emissions Trading System in the LGCC

On 13 July 2018, the Decree in which various provisions of the LGCC are amended and added was published in the DOF, including the reforms to Articles 94 and 95, through which the creation, authorization, and regulation of the emissions trading system were consolidated.

In response to these modifications, Article 94 mandates the establishment of a progressive and gradual emissions trading system to promote the reduction of emissions at the lowest possible cost, in a measurable, reportable, and verifiable manner and without compromising the competitiveness of the participating sectors facing international markets. In turn, Article 95 stipulates the possibility to network with carbon markets from other countries in order to carry out operations and transactions.

Likewise, it was established in the Second Transitory Article that the system in its operational phase would remain subject to the adoption of the preliminary bases for a test program with a validity of thirty-six months, without economic effects for the participating sectors and without harming their business competitiveness. This provision was the guideline for formalizing the system's pilot phase.

The preliminary bases for the Test Program of the Emissions Trading System were established by means of an Agreement published in the DOF on 1 October 2019; subsequently, in compliance with the provisions of the Agreement, the Secretariat of the Environment and Natural Resources (SEMARNAT) distributed notices on 27 November 2019 in regards to the emissions cap and the sectoral allocation of allowances. In accordance with these instruments, the core elements of the system are stipulated in its preparatory phase (Table 6.1).

Table 6.1 Mexican emissions trading system. preparatory phase

Market type	Cap and trade
Object	Emission allowances and offsets
Allocator	The State through SEMARNAT
Allocation	100% free allocation in test phase (grandfathering criterion)
Actors	Transactions between participating sectors
Sectors to support emissions with quotas	Energy and Industry
Participation threshold	Applicable to facilities whose annual emissions are equal to or greater than 100,000 tons of direct carbon dioxide emissions from fixed sources, as reported in the National Emissions Registry (RENE)
Phases	Preparatory or pilot phases (test program) Operational (after the test program)
Territorial Scope	Federal
Time frame	Thirty-six-month pilot phase, which will start on 1 January 2020 and end on 31 December of 2022, divided into two periods (1 January 2020 to 31 December 2021 will correspond to the pilot phase; between 1 January 2022 to 31 December 2021 will be the transition phase from the Test Program to the Operational Phase of the Emissions Trading System)
Gases	In the pilot phase only CO_2, direct emissions from fuel consumption and industrial processes
Limit of use of offsets	Up to 10%
Coercivity	Non-economic in pilot program
Annual cap	271.3 million allowances for 2020 273.1 million allowances for 2021

In order to comply with the goals indicated in its NDC, the Mexican government opts for the obligatory execution of the emissions trading system with the goal of promoting the reduction of emissions at the lowest possible cost. This emissions trading system, like other economic instruments, does not substitute the direct control measures related to the participating facilities, rather, it is complementary.

It is important to note that the emissions market does not privatize the collective ownership of the atmosphere (a natural resource of common use), but rather, legally creates, with respect to this asset, usage rights (emission allowances) that are marketable under the control of public administration in order to safeguard the function and quality of the protected good. The system self-imposes a legal obligation on the participating sectors to support their economic activity with these allowances during a determined compliance period. The emissions trading legalized in our system, therefore, allows limits to be placed on the freedom to emit CO_2 into the atmosphere at the facilities of the sectors indicated by the legal standard (energy and industry, according to the classification provided for in the RENE Regulation), who may continue exercising their polluting activity, but only in the amount established by the number of allowances allocated or obtained in the market. The system does nothing more than transforming the freedom of use of the atmosphere into a right to use it with its respective restrictions and obligations for its holder; with which the quantitative problem of emissions is addressed to qualitatively protect the natural resource.

The Importance of a Reduced Cap

Emissions trading seeks to flexibly incentivize the participation of the sectors involved in order to achieve climate objectives. Its establishment constitutes a direct regulation, while eliminating rigidities. Hence, the system as an emissions control technique forms part of the administrative management sphere, where the market does not take precedence over regulation, but rather is born thanks to it (Sanz 2014). The system is in itself an economic instrument wrapped in a legal system since it not only complies with market rules, but also involves legal regulation, whose ultimate aim is the protection of atmospheric quality on the basis of sustainable development. Public intervention in the system's design is necessary to achieve the climate objectives. Reaching the quantified result depends on the authorized emissions cap and the allowances allocated by the State. In this sense, it is indispensable in the legal system to ensure that it is possible to trade below this cap but not exceed the maximum limit of permitted emissions, with which it will be possible to have certainty over the proposed results.

In this regard, the cap established by SEMARNAT is not in line with the structural objective of the system, as can be seen (Table 6.1) from 2020 to 2021, as it increases instead of decreasing. The cap as a core element must ambitiously decrease in order for it to be effective and have environmental integrity while the transition toward low-carbon growth is encouraged. Otherwise, the system will only have economic effects.

These types of markets operate with absolute objectives by establishing a perfectly fixed and assured maximum limit on emissions that prevent them from increasing (Lefevere 2005). A cap with an increasingly reduced limit is the means by which the first condition for a market is generated: the scarcity of emission allowances, which is created by the public administration at the moment of setting the maximum offer by establishing an emissions limit (Sanz 2014) and with which the allowances acquire a major economic value that internalizes the environmental cost.

The system designed in Mexico with a cap that increases instead of decreasing is due to the fact that the climate mitigation targets are relative to a BAU scenario in which emissions increase, that is, they are not established in absolute terms. On the other hand, the increased cap of the Mexican system is linked to the expectations of economic growth in which the expansion of the production of some participants is expected, as well as the incorporation of new participants to the market, which is related to the peak of net emissions for the year 2026 indicated in the unconditional mitigation commitments provided in the Mexican government's NDC.

The above is positioned within the scope of the discussion of the right to development. However, it should be noted that the current world economic development is not sustainable and cannot be a model to follow, countries like Mexico cannot repeat the same mistake, they must develop but under the scheme of sustainability and the guarantee of environmental integrity that entails changing the current patterns of the prevailing economic development model. Hence, it is essential to look for development alternatives that generate radical changes linked to the reduction of CO_2 emissions and therefore both to the progressive reduction of the cap in an emissions trading system.

Free Allocation and the Polluter Pays Principle

The allocation of allowances to regulated facilities is a transcendent issue since each allowance has an economic value that forms part of the assets of the operation/facility, hence a smaller or larger allocation of allowances has major economic implications, where the allocation method generates different impacts on the facilities, sectors, and national economy (Del Río González 2004).

During the trial period in the Mexican system, the allocation of allowances was stipulated free of charge through the grandfathering method, that is, based on historical information reported to RENE. In this regard, we can refer to the fact that acquired allowances to use the atmosphere were recognized.

The free allocation of allowances granted by the State, in a preparatory phase, is more widely accepted among facilities subject to the obligation to support their emissions with quotas. However, Antunes (2006) considers that an emissions market based on this type of allocation is a real violation of the Polluter Pays Principle since it gives polluters an economic advantage by assigning them a valuable title free of charge, which will have a certain price on the market and with which several polluters would consequently be able to profit from obtaining certain economic gains.

In this regard, Del Río González (2004) points out that regardless of what the initial allocation of allowances was, what matters is the optimal final result of the purchase and sale of allowances among the participants, that is, in a comprehensive view. In this regard, Sanz (2007), with whom we agree, considers that emissions trading in itself is a system that if the Polluter Pays Principle is applied, since any polluter who needs more allowances, apart from those allocated, will have to resort to the market to obtain, by means of payment, additional allowances that will enable them to use the atmosphere to dump their emissions while still respecting the established limit. Or, in the event that the purchase of allowances is more onerous, or if they are in short supply on the market, they must make the necessary adjustments so that their emissions match the allowances received, which implies a cost assumed by the facility concerned. In this case, the polluter bears the costs of carrying out the measures imposed by the public authority to meet the emission reduction targets.

Legal Nature of Emission Allowances

The third Article, section IV, of the preliminary bases of the Emissions Trading System Pilot Program defines the emission allowance as:

> The administrative instrument issued by the Secretariat that grants the right to emit one ton of carbon dioxide during a given compliance period.

Legally, it is an administrative act that grants a subjective right. A holder has the power to use one ton of CO_2, as opposed to the obligation to reduce their emissions; whose extraordinary reduction allows them to not use that allowance, to which they could obtain an economic benefit at the time of selling it to those who require extra allowances in order to comply with their obligations. In this way, the right holder obtains two options of use, depending on whether an action or an omission is carried out. In any case, a benefit is obtained, since if they emit the permitted amount, they are simply meeting their acquired allowance. If they choose not to emit and exercise their right to transfer the unused designated quota, the benefit is greater, since it allows for the possibility of obtaining an economic asset with respect to the good for sale on which the negotiated allowances fall (the emission). The price for which it is sold should be high given its scarcity when carrying out production activities due to the emissions limit established for the compliance period. This is in benefit of the ultimate goal, which is the protection of atmospheric quality (Ibarra 2012). In this context, allowances can be classified as subjective "administrative" rights, because they are rights created by the legal norm and assigned through an administrative act, with respect to which the "authentic interpretation" that establishes the preliminary bases themselves is considered to be unsuitable, and will, therefore, be subject to the interpretation established in practice.

On the other hand, it is important to emphasize that this right recognized in the legal system does not imply, nor can it imply, a privatization or appropriation of the air (Pâques and Charneux 2004). In fact, at no time do the regulatory norms recognize the appropriation of the atmosphere. There is only one property right over the allocated emissions quota. However, it is a ownership regulated and limited by the administration (Sanz 2007) in terms of the following elements:

- Expiration: participants who properly comply in a timely matter with the obligation provided for in the previous paragraph, may use the surplus emission allowances they have in their account to carry out transactions or comply with their obligations in subsequent compliance periods during the Pilot Program (Article 22).
- Timing: allowances issued will only be valid for the Pilot Program (Article 23).
- Conditions for their management and monitoring in the market (Articles 25 to 32)

Conclusion

The implementation of emissions trading as a complementary tool to achieve GHG emissions reduction targets has become more popular due to its flexible and cost-effective scheme. At the international level with the new Paris Agreement strategy in the plans of the mitigation commitments, replacing the top-down scheme with the bottom-up scheme, it is expected to lead to a better and more comprehensive result in the implementation of carbon markets. However, in the framework of Article 6 of the Agreement, operational rules with common guidelines for the use of cooperative approaches on a voluntary basis are still being discussed among the States Parties. Provisions that the Mexican government should take into account in its participation in the carbon markets and offsets. In this regard, the demand for the effective application of environmental integrity will result in great relevance when approving projects and programs that ensure real and measurable GHG emission reductions, that can be traded in international markets. For this, it is essential to take into account the experience gained in the flexible mechanisms under the Kyoto Protocol.

At the national level, the mandatory nature of the emissions trading system is relevant, since it is aligned with international provisions and requirements for a common front against the continuous and constant rise of GHG emissions. The preliminary bases of the system in its pilot phase determine its core elements, which should be stricter in order to achieve international and national commitments to real emission reductions. In this regard, the cap is particularly important as a fundamental element, which should establish decreasing limits so that it has effectiveness, environmental integrity, and to encourage the transition to low-carbon growth, otherwise, the system will only have economic effects. This is important since the quality of the allowances is linked to the mitigation objectives, the more ambitious the objective, the greater the quality. This will have international repercussions since it will be more convenient to relate to emissions trading systems with similar levels of ambition.

References

Antunes T (2006) O Comércio de Emissões Poluentes à luz da Constituição da República Portuguesa. AAFDL, Lisboa

Baron R, Philibert C (2007) Climat, énergie et marchés de quotas échangeables. Revue De L'énergie 575:17–25

Butze W (2006) Permisos de contaminación negociables: Un instrumento de mercado para la regulación ambienta. Análisis Económico, XX I(48):257–288

Cames M et al (2016) How additional is the clean development mechanism? Analysis of the application of current tools and proposed alternatives. Öko-Institut e.V, Berlin

Del Río González P (2004) Ventajas e inconvenientes de los métodos de asignación de derechos de emisión de CO_2 en el contexto de la Directiva Europea de Comercio de Emisiones. Revista Interdisciplinar De Gestión Ambiental 6(71):16–30

European Environment Agency (2017) Annual European union greenhouse gas inventory 1990–2015 and inventory report 2017. EEA, Luxemburgo

Fernández A, Martínez J (Coords.) (2003) Avances de México en materia de cambio climático 2001–2002. SEMARNAT, INE, México

Fischer C (2005) Project-based mechanisms for emissions reductions: balancing trade-offs with baselines. Energy Policy 33(14):1807–1823. https://doi.org/10.1016/j.enpol.2004.02.016

Gómez S (2004) Mercado interno de permisos de emisiones de carbono. Estudio de Caso, Pemex. In: Martínez J, Fernández A (Comps.) Cambio climático: una visión desde México. México, Instituto Nacional de Ecología, Secretaría del Medio Ambiente, pp 447–453

Greiner S, Michaelowa A (2003) Defining investment additionality for CDM projects—practical approaches. Energy Policy 31(10):1007–1015. https://doi.org/10.1016/S0301-4215(02)00142-8

Ibarra R (2012) El Mecanismo de Desarrollo Limpio. Estudio crítico de su régimen jurídico a la luz del imperativo de sostenibilidad. Aranzadi, Pamplona

Ibarra R (2019) De la Convención Marco de las Naciones Unidas al Acuerdo de París: Una larga trayectoria científica, política y económica. In: Ibarra R (Coord.) Cambio Climático y Gobernanza: Una visión transdisciplinaria. Instituto de Investigaciones Jurídicas, UNAM, México, pp 3–36

International Institute for Sustainable Development (1994) Trade and sustainable development. Principles. IISD, Canada

Lefevere J (2005) Greenhouse gas emissions trading: a background. In: Bothe M, Rehbinder E (eds) Climate change policy. Eleven International Publishing, Utrecht, pp 103–129

Michaelowa A et al (2019) Additionality revisited: guarding the integrity of market mechanisms under the Paris Agreement. Climate Policy 19(10):1211–1224. https://doi.org/10.1080/14693062.2019.1628695

Muller A (2007) How to make clean development mechanism sustainable-the potential of rent extraction. Energy Policy 35(6):3203–3212. https://doi.org/10.1016/j.enpol.2006.11.016

OECD (1975) Le principe polleur – payeur. Definition, analyse, mise en oeuvre. OCDE, París

Pâques M, Charneux S (2004) Du quota d'émission de gaz à effect de serre. Revue Européenne De Droit De L'environmmement 3:266–278

Philibert C (2005) Transformer Kyoto. Esprit 318(10):28–42

Rosembuj F (2005) El precio del aire; Aspectos jurídicos del mercado de derechos de emisión. El Fisco, Barcelona

Salinas S (2017) El esfuerzo de mitigación de emisiones en el marco del régimen internacional contra el cambio climático. Estado de la cuestión tras el Acuerdo de París. In: García M, Amaya O (eds) Retos y compromisos jurídicos de Colombia frente al cambio climático. Universidad Externado de Colombia, Colombia, pp 23–53

Sanz I (2014) ¿Mercados para la protección del medio ambiente? Veredas Do Direito 11(21):11–30

Sanz I (Dir.) (2007) El mercado de derechos a contaminar. Régimen jurídico-público del mercado comunitario de derechos de emisión en España. Lex Nova, Valladolid

Sanz I (2009) Los derechos de emisión y su aplicación en España. In: Martín JJ (Dir.) Hacia una política comunitaria europea en cambio climático y sus consecuencias para España. Universidad de Burgos, España, pp 197–218

Saura J (2003) El cumplimiento del protocolo de Kioto sobre cambio climático. Publicacions de la Universitat de Barcelona, Barcelona

Schneider L (2009) Assessing the additionality of CDM projects: practical experiences and lessons learned. Climate Policy 9(3):242–254. https://doi.org/10.3763/cpol.2008.0533

Schneider L, La Hoz Theuer S (2019) Environmental integrity of international carbon market mechanisms under the Paris Agreement. Climate Policy 19(3):386–400. https://doi.org/10.1080/14693062.2018.1521332

Schneider L (2007) Is the CDM fulfilling its environmental and sustainable development objectives? An evaluation of the CDM and options for improvement. Öko-Institut e.V, Berlin

UICN (2016) Declaración Mundial de la Unión Internacional para la Conservación de la Naturaleza (UICN) acerca del Estado de Derecho en materia ambiental. Word Commission on Environmetal Law, Rio de Janeiro

UNFCCC (2020) CDM project search. https://cdm.unfccc.int/Projects/projsearch.html. Accessed 26 April 2020

World Resources Institute (1992) World resources 1992–93. Oxford University Press, New York

Zapf M, Pengg H, Weindl C (2019) How to comply with the Paris agreement temperature goal: global carbon pricing according to carbon budgets. Energies 12(15). https://doi.org/10.3390/en12152983

Chapter 7
The Political Economy of Carbon Pricing: Lessons from the Mexican Carbon Tax Experience for the Mexican Cap-and-Trade System

Juan Carlos Belausteguigoitia, Vidal Romero, and Alberto Simpser

Abstract Price-based climate change policy instruments, such as carbon taxes or cap-and-trade systems, are known for their potential to generate desirable results such as reducing the cost of meeting environmental targets. Nonetheless, carbon pricing policies face important economic and political hurdles. Powerful stakeholders tend to obstruct such policies or dilute their impacts. Additionally, costs are borne by those who implement the policies or comply with them, while benefits accrue to all, creating incentives to free ride. Finally, costs must be paid in the present, while benefits only materialize over time. This chapter analyses the political economy of the introduction of a carbon tax in Mexico in 2013 with the objective of learning from that process in order to facilitate the eventual implementation of an effective cap-and-trade system in Mexico. Many of the lessons in Mexico are likely to be applicable elsewhere. As countries struggle to meet the goals of international environmental agreements, it is of utmost importance that we understand the conditions under which it is feasible to implement policies that reduce carbon emissions.

Keywords Political economy · Carbon pricing · Mexico · Fiscal reform · Climate policy

J. C. Belausteguigoitia (✉)
Centro ITAM de Energía y Recursos Naturales, Avenida Camino a Santa Teresa No. 930, Colonia Héroes de Padierna, 10700 Ciudad de México, Mexico
e-mail: juan.belausteguigoitia@itam.mx

V. Romero · A. Simpser
Department of Political Science, ITAM, Rio Hondo 1, 01080 Mexico City CDMX, Mexico
e-mail: vromero@itam.mx

A. Simpser
e-mail: alberto.simpser@itam.mx

© The Author(s) 2022
S. Lucatello (ed.), *Towards an Emissions Trading System in Mexico: Rationale, Design and Connections With the Global Climate Agenda*, Springer Climate,
https://doi.org/10.1007/978-3-030-82759-5_7

133

Why Is It so Difficult to Implement Effective Climate Change Policies Based on Carbon Pricing?

Carbon pricing is a policy concept used at the national and subnational levels across the world to incentivize carbon dioxide emission reductions. Carbon pricing entails charging emitters for the carbon dioxide emissions for which they are responsible. Such emissions result from burning fossil fuels for essential activities such as electricity generation, industrial production, transportation, and heating and air-conditioning in residential and commercial buildings.

Since Pigou (1932) formalized the idea that with the right fiscal intervention, prices could lead to socially efficient (desirable) outcomes, correcting market failures caused by externalities, economists have advocated pricing signals as the main driver of environmental policies. Climate change is no exception. Economic instruments in the form of pricing signals, such as taxes and cap-and-trade systems, have an outstanding theoretical performance with regard to important evaluation criteria such as efficiency, effectiveness and cost-effectiveness.[1] Flexible responses by economic agents are essential to explaining efficiency and cost-effectiveness, as carbon pricing allows firms to decide on the most efficient ways to mitigate emissions in response to carbon prices. Similarly, equalizing marginal cost of mitigation across firms and sectors is a necessary condition for cost-effectiveness, i.e. minimizing overall costs of mitigation. This is accomplished if a uniform carbon price applies across all firms and sectors, regardless of the emissions source.

If economists had their way, price signals would constitute the pillar of environmental policy. With regard to climate change, there may be disagreements among economists as to whether a carbon tax outperforms a cap-and-trade system, or vice versa, but most economists agree that pricing carbon emissions should be an essential component of emissions-mitigation policy. Weitzman (1974) was the first to explain the basic difference between taxes and cap-and-trade systems. Under perfect certainty, in principle, both lead to the same socially efficient outcome. However, when the marginal benefits and marginal costs functions of mitigation are not known, taxes are preferred if the main policy objective is to provide price certainty, as regulated agents know the price per ton emitted. Notice that the tax does not guarantee that the desired emission reduction will take place, as the exact reaction of the regulated entities is uncertain. On the other hand, cap-and-trade programmes provide emission reductions certainty but uncertainty about the price per ton of emissions. Taxes and cap-and-trade methods also differ with regard to the political dynamics to which they give rise. Stavins (2012) argues that "the key difference is that political pressures on a carbon tax system will most likely lead to exemptions of sectors and firms, which reduces environmental effectiveness and drives up costs, as some low-cost emission reduction opportunities are left off the table;" whereas "…[the] political pressures

---

[1] For an excellent discussion of both theoretical and practical aspects of a cap and trade system see Tietenberg (2006); for a detailed analysis of cap and trade systems from both the legal and economic perspectives and a comparison of their characteristics see Borghesi et al. (2016).

on a cap-and-trade system lead to different allocations of the free allowances, which affect distribution, but not environmental effectiveness, and not cost-effectiveness."

Whether in the form of a tax or a cap-and-trade system, a carbon price would provide consumers and producers incentives to make changes, sometimes marginal but sometimes significant, to their behavioural patterns leading to less carbon dioxide emissions. This would, in turn, lead to a more efficient, or at least a cost-effective, outcome. However, despite the growing importance that carbon pricing has had in the last decade—almost 23% of global emissions today are affected by a carbon pricing scheme[2]—prices still do not play the role they will need to play if they are to limit emissions in a manner consistent with climate change goals.

Why is it so difficult for governments to implement the right policies? The nature of climate change as a global-public-bad (i.e. the non-rivalry and non-excludability of its impacts) and the temporal asymmetry of the relevant costs and benefits represent major obstacles to the design and implementation of effective public policies. On the one hand, the benefits of emissions-mitigation actions are distributed globally, while mitigation costs are borne entirely by those who take such actions. This is an instance of the famous free-rider problem.[3] Importantly, the free-rider problem arises both between countries as well as within countries, across sectors of economic activity, and even across firms.

At the same time, the temporal dimension of costs and benefits further complicates the establishment of emissions-mitigation policies, since the costs are borne in the present, while most of the benefits are obtained in the future. The existence of high initial costs and deferred benefits creates a political challenge. Politically, it is much easier to offer benefits in the present and pass the costs on to future generations. Free riding and the intertemporal asymmetry of the costs and benefits of emissions-mitigation actions hinder the implementation of mitigation policies and highlight the importance of considering both political and economic incentives in policy design and execution.

The next section provides a brief history of climate change policy in Mexico as background to understand the difficulties faced by efforts to introduce price-based mechanisms. The section after that analyses the conditions that provided a favourable scenario for the Mexican Treasury (*Secretaría de Hacienda y Crédito Público* or SHCP) to consider introducing a carbon tax. Next, we describe the tax design process as well as the negotiations and lobbying that transformed the proposed tax into the tax that was eventually approved by the Mexican Congress; next, we analyse political opposition to the tax; and finally, we draw some lessons and offer conclusions.

---

[2] World Bank (2020).

[3] The free-rider problem refers to the incentives that self-interested individuals have to enjoy the benefits of a non-rival good without contributing, given that exclusion is unfeasible.

A Brief History of Climate Change Policy in Mexico Before the Carbon Tax Was Passed

Despite a very active and positive international agenda and a determined domestic institutional strengthening programme, prior to the 2013 carbon tax, Mexico had not taken any policy measures that would impose costs on important economic sectors for emitting carbon dioxide.

The Mexican government began analysing climate change in the late 1980s, focusing on its potential impacts on the country. The ministry charged with this task was the Ecology under-Secretariat of the Department of Ecology and Urban Development (*Secretaría de Ecología y Desarrollo Urbano y Ecología* or SEDUE). This made it possible for Mexico to participate in the meetings that led to the adoption of the United Nations Framework Convention on Climate Change (UNFCCC) in 1992. That same year, SEDUE was transformed into the Ministry of Social Development (*Secretaría de Desarrollo Social* or SEDESOL). In addition to environmental and urban development policies, the new ministry was responsible for poverty alleviation. Meanwhile, two new environmental agencies were created: The National Institute of Ecology (*Instituto Nacional de Ecología* or INE),[4] responsible for policy formulation and implementation, and an environmental enforcement agency (*Procuraduría Federal de Protección al Ambiente* or PROFEPA). Despite this surge in institutional capacity focusing on the environment, climate change itself continued to be analysed—like at SEDUE and SEDESOL—mainly by a small staff of advisors to the president of the INE.

In December 1994, responsibility for environmental and natural resource policy moved from SEDESOL to a new specialized ministry, the *Secretaría del Medio Ambiente, Recursos Naturales y Pesca* (SEMERNAP) at the beginning of President Ernesto Zedillo's administration. INE and PROFEPA were moved from SEDESOL to SEMARNAP and a special climate change office was created at INE. Both in Mexico and globally, 1997 was a special year for climate change-related institutions. That year, Mexico launched the Climate Change Inter-Departmental Committee, led by SEMARNAP, in order to improve policy coordination within the federal government and to facilitate meeting its commitments under the UNFCCC. The Kyoto Protocol was also adopted in December 1997. During the process, Mexico was pressed to become an Annex B country due to its recently acquired status as an OECD member. In 1999, the federal government issued the Climate Action Programme (*Programa de Acción Climática*). The programme helped Mexico resist the pressure to become an Annex 1 country or to commit to more ambitious goals than the ones established under the UNFCCC. Before 2000 (the last year of President Zedillo's administration), when Mexico ratified the Kyoto Protocol, the government focused its climate change-related actions on scientific research and important support studies such as the greenhouse gas inventory.

---

[4] Not to be confused with the current *Instituto Nacional Electoral* or INE, charged with electoral matters.

The first mechanism based on economic instruments to mitigate emissions in Mexico was the Clean Development Mechanism or CDM (*Mecanismo de Desarrollo Limpio*). The Clean Development Mechanism was a Kyoto Protocol instrument that allowed Annex 1 countries—those with binding mitigation commitments under the protocol—to invest in mitigation projects in foreign developing countries as an alternative to investing in more-expensive mitigation projects in their own countries. The CDM raised expectations among environmentalists and industry in Mexico because it was considered to be an excellent vehicle to finance emissions reductions in various industrial sectors. However, those expectations never came to fruition, partly because of the large transaction costs of the mechanism, largely stemming from the inherent practical difficulty of verifying emission baselines and reductions.

Under President Felipe Calderón's administration (2006–2012), Mexico enhanced its leadership in climate policy. On the domestic front, in 2007 the government passed the Climate Change National Strategy (*Estrategia Nacional de Cambio Climático* or ENCC) and, in 2008, the Climate Change Special Programme (*Programa Especial de Cambio Climático* or PECC). Additionally, the Mexican congress enacted the General Law on Climate Change (*Ley General de Cambio Climático*) in 2012. This law established, conditional on financial and technological support from the international community, mitigation targets for greenhouse gas (GHG) emission reductions: a 30% reduction in emissions by 2020 with respect to a business-as-usual scenario and a 50% reduction by 2050 with respect to the year 2000 emissions.

In the international arena, Mexico was one of the countries that proposed the creation of the Green Climate Fund at the G8 meeting in Rome in 2008. However, Mexico is best remembered for its role in re-establishing trust in the multilateral process, when it hosted the 16th Conference of the Parties (COP16) in Cancún. Despite all its domestic and international accomplishments, President Calderón did not use carbon pricing to promote emissions-mitigation actions during his administration.

President Peña Nieto's administration began in December 2012, and climate change was not one of its priorities. He proposed a very ambitious set of reforms including, among others, major changes in the labour, antitrust, telecom, energy, and fiscal sectors. Even though President Peña Nieto did not have the same personal passion for climate change as his predecessor, it can be argued that his reform of the electricity sector (through the promotion of renewable energy) and fiscal reform (through the introduction of the carbon tax) had the potential to have a very significant impact on GHG emissions. Why was there a fiscal reform? And why and how was a carbon tax part of it? We now turn to these questions.

Lack of Resources and Oil Dependency in Mexican Public Finances

The fiscal reform of 2013 had two important drivers. The first was to increase government revenue. Due to lack of resources, the government of Mexico was not able to meet the most basic needs of its population, including health, education, infrastructure and security. Excluding oil revenues, tax collection represented only 10% of GDP in Mexico in 2012, compared to 19 and 25% for Latin American and OECD countries, respectively. The second driver of the fiscal reform was the imperative of reducing the degree to which government revenue depended on oil. Due to the inherent volatility of the price of crude oil and the production platform, government dependence on oil production was a major source of uncertainty, variability and risk for Mexican public finances.

After the discovery of Cantarell, a giant oil-field, at the beginning of the 1980s, the Mexican economy became heavily oil-dependent. In 1982, oil accounted for around two thirds of exports and one-third of federal government revenue, and taxes paid by PEMEX—the national state-owned oil monopoly—amounted to a full 10% of GDP. The 1994 North American Free Trade Agreement diversified and increased non-oil-related exports, but government revenues continued to be oil-dependent. Oil-related revenue reached 44% of total government revenue in 2008. However, oil prices collapsed in 2009 (from US\$92 to US\$54 per barrel), clearly signalling to the federal government the risks of its oil dependency.

Seeking to reduce government dependence on oil revenue, in December 2012, the Peña Nieto administration began promoting a series of structural reforms. Energy reform was one of the most conspicuous because it went against a long history of nationalistic policies in the sector. The reform would require PEMEX to compete with private companies, but it would not be able to do so successfully unless its tax scheme changed.

The Mexican treasury (SHCP) issued a fiscal reform project in 2013 aimed at increasing government revenue. Environmental taxes were among the new proposed sources of government revenue. After discussing the pros and cons of the proposed list of environmental taxes with the Ministry of Environment and Natural Resources (SEMARNAT), SHCP decided to include only two environmental taxes in the fiscal reform plan (alongside many other taxes unrelated to the environment): a carbon tax and a tax on pesticides.

Overall prospects for introducing carbon taxes in Mexico in the early 2010s appeared bleak. For one thing, in 2013, as a non-Annex-1 country, Mexico did not have binding international obligations to mitigate GHG emissions. Additionally, given that the energy matrix in Mexico is highly dependent on fossil fuels and that the country has a large history of subsidizing fuel on distributional and competitiveness grounds, a carbon tax appeared to go against traditional economic interests. Indeed, in 2012, oil was used to meet 53% of the country's total energy demand, natural gas represented 36%, coal 5%, hydro-electricity 4%, nuclear 1%, and non-hydro renewables 1%. One might have been surprised, therefore, when the Mexican congress passed a tax on GHG emissions at the points of importation and sale of the fuels on 30 October 2013.

Design of the Proposed Carbon Tax

We now describe the process of designing the draft carbon tax before it became law. As mentioned previously, the Mexican carbon tax had both environmental and fiscal objectives. On the one hand, it aimed to induce fuel efficiency and emissions reductions and, inasmuch as it was a unilateral policy measure, at gaining international legitimacy and placing Mexico in a favourable position for climate change and related international environmental negotiations. At the same time, from the point of view of the treasury, the carbon tax had a clear revenue-generating objective.

The team that designed the carbon tax included experts from SEMARNAT and the Mario Molina Center—a not-for-profit think-tank based in Mexico City. The design team decided that, given the technical difficulties and administrative costs, the tax ought to be an indirect one. Specifically, the tax should be levied on fossil fuels and should be proportional to each fuel's potential to generate carbon dioxide emissions. In order to determine the tax rate, the team considered social-cost-of-carbon estimates and analysed carbon prices in jurisdictions outside of Mexico. At the time, the best-known social-cost-of-carbon estimates ranged from US$20 to US$311 (depending on climate change impact scenarios and discount rates) and 30 countries as well as 18 subnational jurisdictions had either a tax or a cap-and-trade system in place.

The tax design team considered that the social-cost-of-carbon estimates should account for medium- or long-run targets, as even the lowest value in the range was well beyond what was considered politically feasible. The team thus looked at carbon prices in American regional markets, mainly California and the Regional Greenhouse Gas Initiative. The weighted price per ton of carbon dioxide (using market volume as weight) was US$5.70, equivalent at the time to $70.68 Mexican pesos. Intergovernmental Panel on Climate Change default values were used to determine the carbon content of fuels. Table 7.1 shows the proposed tax per volume for different fuels as well as the percentage change in prices.

Figures 7.1 and 7.2 display the expected emission reductions (for the year after the tax was introduced) and the expected tax revenue for 2014, respectively. These figures display estimates for both the proposed tax and the tax that ultimately was implemented.

The design team also analysed specific political economy aspects of the tax. In particular, the team considered impacts on income distribution, specifically the impact on low income groups, as well as on firms' costs, and evaluated the support of the general population and environmental NGOs for the tax. With regard to the distributional impact, the team analysed gasoline consumption by income decile and expenditure increases on liquid petroleum gas (LPG) by decile. Figure 7.3 shows percentage consumption of gasoline by decile. The two highest income deciles consume 52% of gasoline, whereas the four lowest income deciles consume 11%.

Table 7.1 Mexican executive initiative

Initiative of the Executive	*Proposed* tax of MX$ 70.68 pesos per ton of CO_2 (US$ 5.7 dollars)		
	Executive Proposal		
Fuel	Tax (MXN)	Implicit price per ton of CO2 (MXN)	% change in price
Natural Gas	11.94 cents per m3	70.68	4.1%
Propane	10.50 cents per liter	70.68	
Butane	12.86 cents per liter	70.68	
Gasoline	16.21 cents per liter	70.68	1.4%
Jet fuel and other kerosene	18.71 cents per liter	70.68	1.6%
LP Gas	11.42 cents per liter	70.68	1.7%
Diesel	19.17 cents per liter	70.68	1.6%
Fuel Oil	20.74 cents per liter	70.68	2.6%
Petroleum Coke	18.99 cents per kg	70.68	16.1%
Coal Coke	19.30 cents per kg	70.68	17.0%
Mineral Coal	17.83 cents per kg	70.68	
Others		70.68	

- Estimated reduction: 5.8 million tons of CO_2 in 2014.
- Income of $26.7 billion pesos (2 billion US dollars) (1.8% of tax revenues of Federal Government in 2012)

Source Belausteguigoitia (2014)

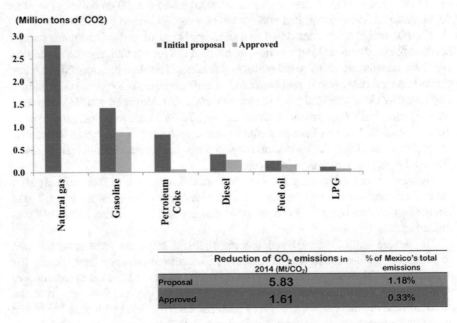

Fig. 7.1 Expected reduction of emissions for 2014. *Source* Belausteguigoitia (2014)

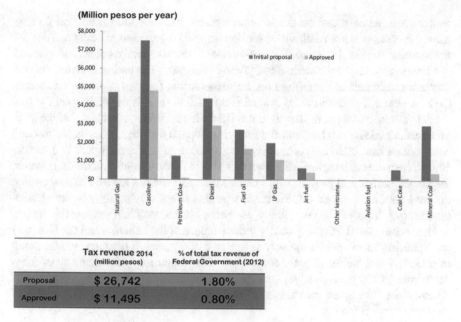

Fig. 7.2 Expected tax revenue for 2014. *Source* Belausteguigoitia (2014)

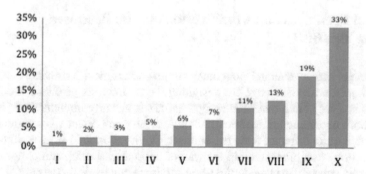

Fig. 7.3 Gasoline consumption by decile (percentage). *Source* CMM with data from *Encuesta Ingreso Gasto de los Hogares* (2018)

With regard to LPG, the increase in expenditure by the two lowest income deciles was $27 Mexican pesos per month per household, by all accounts a small amount. However, according to the IMF, in most countries, the indirect impact of higher general prices represents one-third to one-half of the burden of increased energy prices on households. Because the estimated distributional impact could be significant, the team considered the potential need of compensating the lowest income deciles.

With regard to the tax's potential impact on competitiveness, cost increases were analysed for energy-intensive sectors. Carbon dioxide emitting firms suffer a private welfare loss when a carbon tax is implemented. The intended transition to low-carbon processes is not frictionless; it often imposes significant private cost, despite the increase in social welfare. Energy-intensive sectors, such as steel and cement, can be severely affected under carbon pricing schemes.[5] The team estimated the cost increase as a result of the carbon tax implementation for energy-intensive sectors. Cost increases for cement firms ranged from 0.26 to 0.43%, depending on the fuel used. In the case of steel, the increase was 0.2%. In retrospect, it is clear that the team underestimated the reaction and the lobbying capacity of these sectors. A "second best" option that included some sort of compensation to the most affected sectors could have been contemplated. The fact that Mexico's most important trading partner, the United States, and its most important competitors did not have a national carbon tax (or a national cap-and-trade system) was also effectively used by business associations that opposed the tax. Finally, as for the general public's support for the tax, the team considered that the globally diffuse nature of the benefits, and the difficulty of explaining to the public the way in which carbon pricing operates, would result in relatively low public support for the tax. Given that the great majority of environmental NGOs only consider that carbon pricing is environmentally effective if revenues are earmarked for climate change-related actions, the team did not expect NGOs to support the tax.

Political Pressures and Modifications to the Proposed Carbon Tax Bill

Tax reform is a tricky matter. Under many circumstances, decision-makers may have sufficient technical knowledge for designing the optimal tax given a set of goals. Yet, it is not very often that such an optimal tax is actually implemented (Jenkins 2014). Both economic and political factors can play important roles in rendering the possible quite different from the optimal. On the economic side, various types of transaction costs often stand in the way of optimality. Deficient monitoring and enforcement capacity may render first-best solutions infeasible. In the case of carbon taxes, it is notoriously difficult for governments—and even for international bodies—to properly verify baselines and levels of GHG emission reductions. Uncertainty about the various effects of taxes, the amount of expected revenue, price parameters, or market behaviour also make it possible to err in the design.

When it comes to taxes, though, it is the *political* constraints, pressures and conflicts that loom largest. Economic criteria such as efficiency or optimality—a well as normative criteria such as the "common good"—stand in stark contrast to the interests of specific economic sectors, companies and even individual citizens. From the perspective of business, taxes directly impact profitability and income and they alter the competitive playing field, often benefitting some while harming others.

---

[5] For an excellent discussion of the impact of carbon pricing on competitiveness see Grubb and Neuhoff (2006).

Moreover, taxes are strongly disliked by all—particularly so in low-trust environments where citizens have little faith that tax revenues will be put to good use by their governments.

Dislike of taxes does not, however, automatically translate into effective opposition. The better organized, better connected, and more powerful industries, companies, and persons tend to generate more effective challenges to tax projects that run against their interests. Generally speaking, thus, the feasibility of a tax project hinges in great measure on possibilities for bargaining with, and compensating, powerful actors and veto players. Even when a tax can be forcibly imposed, compliance with it may be affected by the degree to which powerful actors feel adequately treated or compensated.

Yet, not all desirable compensation schemes are feasible. In Mexico, for example, the law forbids earmarking particular tax revenues for specific kinds of expenditures. Moreover, in environments where the legal system is not very effective or corrupt and interpersonal trust is generally low, compensation schemes that promise a future benefit in exchange for a present cost may simply not be credible. In such a context, the feasibility of a tax project may require ex-ante concessions, provisions, and special exceptions to be enshrined in the law itself, often cancelling part of the revenue-generating and environment-protecting potential of the tax project. In other cases, government may simply renege on promises to compensate today's losers tomorrow, when tomorrow comes.

The Mexican carbon tax was indeed importantly shaped by political constraints. To be sure, some such constraints were evident from the start. As mentioned previously, even before the Mexican carbon tax was sent to congress for approval, it already reflected the design team's perceptions of political feasibility of the emissions price per ton on which the tax would be based. Other political pressures made themselves felt as the process of writing the initial tax proposal, and then obtaining the votes in congress, unfolded.

The fiscal reform sent by President Peña Nieto to congress for approval consisted of a tax of $70.68 Mexican pesos per ton of carbon dioxide, for each and every one of the taxed fuels, as shown above in Table 7.1. Table 7.2 describes the tax bill that was ultimately approved. In the approved bill, the implicit price of carbon dioxide emissions is lower across the board. More importantly, that price varies markedly across fuel types. Moreover, both natural gas and jet fuel are, for practical purposes, exempt, as their tax rate was set to zero. In the case of airplane fuel, the reason for the exemption stems from the fact that Mexico is bound by the Chicago Convention, which requires the exemption. The rest of the distortions, however, have their origin in the political process.

By the account of people close to the process of design and approval of the carbon tax in Mexico, there were two major political forces behind the differences between the proposed and the approved bills. The first was a broad-based demand of private sector companies that used natural gas in their processes. A noteworthy role was played by privates who had invested in natural gas-based processes and those industries that are intensive in the use of natural gas, who complained that the government was undermining its own economic growth goals by simultaneously reforming the

Table 7.2 Approved carbon tax's characteristics

Approved version by Congress introduces changes
that will generate inefficiencies in price signals

Approved carbon tax, 2013 Tax Reform

Fuel	Tax (MXN)		Implicit price per ton of CO$_2$ (MXN)	% Change in price
Natural Gas	0.00	cents per m3	0.00	0.0%
Propane	5.91	cents per liter	39.78	
Butane	7.66	cents per liter	42.10	
Gasoline	10.38	cents per liter	45.26	0.9%
Jet fuel and other kerosene	0.00*	cents per liter	46.84	1.1%
LP Gas	6.59	cents per liter	40.68	1.0%
Diesel	12.59	cents per liter	46.42	1.1%
Fuel Oil	13.45	cents per liter	45.84	1.7%
Petroleum Coke	1.56	cents per kg	5.80	1.3%
Coal Coke	3.66	cents per kg	13.4	
Mineral Coal	2.75	cents per kg	10.92	2.6%
Others			39.80	

- CMM emphasized the importance of a uniform price per ton of CO$_2$ for all fuels.
- However, the discussions in Congress led to a different implicit price per fuel.

Impact in the right direction, but inefficiencies due to different implicit prices

The carbon tax does not apply to air fuel because Mexico is a signatory of the Convention on International Civil Aviation (also known as the Chicago Convention) since 1946, which exempts commercial aviation fuel from taxation.

Source Belausteguigoitia (2014)

energy sector, while raising taxes on natural gas. Such investors proposed regarding the level of GHG emissions of natural gas as the floor (to be taxed at a zero rate) on the basis of which the tax rates for other, "dirtier," fuels would be calculated. In the end, the pressure for this change to the tax bill was strong and the change was implemented (see Fig. 7.2).

Private sector players also argued for mechanisms that would make it possible for them to offset some of the tax burden by allowing for partial deductibility of the carbon tax with respect to their overall tax bills, and by making it possible to reduce the carbon tax by offering emission reduction certificates obtained either in Mexico or abroad. The private sector adduced international comparisons to argue for deductibility and for tax offsetting mechanisms. It also emphasized the fact that the United States—by far Mexico's largest trade partner—did not have a general carbon tax and therefore Mexico's tax would reduce the ability of Mexican companies to compete in the international market.[6] Nevertheless, deductibility and the possibility of offsetting emissions by purchasing certificates both ran directly afoul of the government's revenue-raising goals and were severely restricted.

---

[6] This, and the previously mentioned jet fuel treaty exemption, are two examples of the importance of international political factors in domestic environmental policies.

The second major force was the coal sector. For historical reasons beyond the scope of this chapter, coal mine owners have strong friends across the political spectrum. The coal sector used this power to argue that coal is not always burned (e.g. to generate electricity), but it also has alternative uses with much lower emissions, such as steelmaking. To simplify the administration of the carbon tax, the design team initially wanted to tax coal out of the mine (or at the border) regardless of its future use. Eventually, the bill was modified to accommodate the coal sector's demand, and it became necessary for the tax authority to audit the ultimate use of coal. For the rest of the fuels, however, the tax remained fully indirect. Additionally, and perhaps more remarkably, the implicit price of carbon dioxide emissions in the calculation of the tax on coal was reduced as a result of political pressure, to a level well below the analogous price for other fuels.

Designers of the bill were strongly opposed to having different tax rates for different fuels, viewing these as distortions of a formerly clean, cost-effective and simple tax scheme. From the environmental economics point of view, the differentiated treatment reduces the environmental effectiveness of the tax and undermines its cost-effectiveness. Just as importantly, the uneven treatment of different fuels could potentially create legal problems for the viability of the carbon tax as a whole. The tax had been justified on the basis of the externalities of GHG emissions, and this justification was inconsistent with differential treatment of fuels per volume of emissions (in fact, this legal issue was utilized as an argument to partially push against the coal sector's demands). To date, the Mexican carbon tax has not been challenged in the courts, which points to a bargaining outcome in which the main actors' interests were taken into consideration in the final bill. Nevertheless, changes in markets or technology, or the arrival of new players may change actors' incentives to promote legal actions against the current carbon tax.

In sum, the government's imperative to raise revenue was likely the chief motor behind the creation of the carbon tax. Regardless of the initial motive, the design team incorporated experts in the technical aspects of emissions and in environmental economics, and the proposed bill reflected the revenue-raising imperative, anticipated some political constraints, and provided a schedule of economically and environmentally sound incentives to curb carbon dioxide emissions.

Nevertheless, coal mine owners, industrial associations, and the cement and steel industries organized well and had a very active reaction against the tax. Their lobbying capacity resulted in significant changes to the tax bill. The most important were the following: (i) natural gas was for all practical purposes exempted (formally it got a zero percent tax rate); (ii) the indirect price of carbon was significantly lowered and (iii) different tax rates were levied on different fuels, so that some of the dirtiest got very low rates. As shown in Figs. 7.1 and 7.2, the approved tax bill fell quite short of the revenue-raising and emissions-curbing potential of the original bill. Moreover, the inequities it introduced across fuels distorted fuel-usage incentives and created legal risks for the viability of the tax itself.

Political Economy Lessons from the Mexican Carbon Tax

Perhaps the main general lesson is that the feasibility of emissions-reductions policies cannot be understood without regard to the political context. While every country has its idiosyncrasies, we believe the Mexican experience can speak to many other jurisdictions across the globe. Like in Mexico, it is the case everywhere that emissions-reduction policies are by nature difficult due to the external, diffuse, uncertain, and future nature of environmental benefits and the concentration of costs in a few industries, which tend to be politically powerful.

Other aspects of the Mexican experience are perhaps more specific to middle- and low-income countries. In such contexts, institutional weakness, economic inequalities, wealth concentration and low levels of trust render intertemporal bargains and commitments more problematic than in the wealthy parts of the world. It is certainly true in Mexico that large businessmen have a very strong bargaining position vis a vis the government, in part due to the concentration of wealth and ownership, and in part due to the mobility of capital (Elizondo 1994). Those conditions notably constrained the space of feasible carbon taxes in Mexico in ways that departed significantly from those that would have best curbed carbon dioxide emissions.

In fact, environmental concerns were not a top priority for either the government or the private sector. The carbon tax was part of a large package of tax reforms aimed at increasing government revenue. The private sector's concern mirrored that of the government in seeking to boost its bottom line, in what might be viewed as a zero-sum game. Neither environmental NGOs nor private organizations who stood to benefit from the carbon tax were strong or organized enough to play a major role. From this perspective, the fact that Mexico's carbon tax came into existence, when reducing emissions was not a top priority of any of the major players—granting that some are sincerely interested in environmental issues—might be viewed as a minor miracle, rather than as the expression of a coherent, long-term vision about environmental policy.

Mexico's recent history suggests that tax reforms are driven by economic crises, when the government's need for revenue increases sharply and suddenly. But designing fiscal policy with short-term goals in mind implies that the long-term incentive effects of taxes are no more than a second thought, and those who end up being burdened with taxes are those with the least ability to evade them (Magar et al. 2009; Romero 2015). These issues ought to be important concerns throughout the developing world, even while Mexico seems to stand out among Latin American countries for its low levels of tax collection and a negligible distributional effect of taxes (Mahon 2012; Mahon et al. 2015).

The above considerations underscore the importance of finding compensation schemes that render carbon taxes politically feasible, yet avoid undermining the integrity of the behaviour incentives created by the tax.

Conclusions

In this chapter, we have explained the main constraints that Mexican decision-makers face in implementing an effective policy of carbon emissions reduction based on carbon pricing. It is not a trivial endeavour. Yet, there are specific actions that would increase the likelihood of success of an emissions trading system in Mexico. In the following paragraphs we outline our recommendations for Mexico, based on what we have learned from the previous tax reform. Other countries, especially in the developing world and with economic structures similar to Mexico, should also be able to learn from this experience.

1. Carbon pricing should be an essential component of Mexican mitigation policy. At the same time, Mexico needs fiscal reform and a carbon tax could be a pillar of that reform. According to the International Monetary Fund (IMF), in order for Mexico to have an effective contribution to limiting global warming to 2 °C or less, the carbon tax should follow a path that leads to US$75 per ton of carbon dioxide in 2030. Under that scenario, the prices of coal, natural gas, electricity and gasoline would rise 226, 132, 74 and 18%, respectively. If a cap-and-trade-based policy was chosen, the cap would cause similar price increases.

   Decarbonizing the Mexican economy will not only transform production processes, but it will also significantly alter the lives of most people and communities. Such a significant change cannot be driven by the environmental authority (SEMARNAT) on its own. The entire federal government must be involved and the treasury (SHCP) should be the main institutional driver for changes of this magnitude. Total revenue generated by a $75 tax, or by auctioning tradable emission credits, would be equivalent to close to two percent of GDP. The treasury should be interested in considering a carbon tax as an essential element of a new fiscal reform.

   However, in "normal" times, as we have discussed in this chapter, a cap-and-trade reform seems unlikely. Yet, atypical times are in the horizon. The current economic crisis that has deepened the government's need for revenue along with significant pressure from subnational governments for a significant redrafting of their relationship with the central government opens a window of opportunity for a significant fiscal reform that could consider novel instruments that would reduce transaction costs in the bargaining of environmental policies between the government and businessmen.

   An important aspect to consider is President López Obrador's reluctance to consider fuel price increases to consumers (either because of taxes or price liberalization schemes). Goal 5 of the 2020–2024 Energy Programme, titled: "Ensure universal access to energy, so that it is available to all Mexicans for their development," states that:

   > Access to energy is essential for the social and economic development of people and their communities. However, in Mexico there is inequality in access to energy, which is mainly derived from the geographical location and the economic situation of the

people. For this administration, it is of central interest that all Mexicans have access to energy in its various forms, be it electricity, gasoline, diesel, natural gas, among others, to eliminate restrictions on development.[7]

The struggle between the need for more revenue to stabilize the Government's financial situation and the president's commitment not to raise fuel prices will determine whether carbon taxes or auctions of tradable permits will be introduced during President López Obrador's administration.

2. Apart from the driving force of the treasury, building a broad coalition that supports carbon pricing (in either form) will be needed to overcome political challenges. This was completely overlooked in the 2013 reform. The coalition should include environmental NGOs, green business' interests and regular people interested in climate change.

    Given the Mexican government's relative weakness for imposing taxes on powerful industries, external pressure and support from international organizations and countries may be a good way to circumvent this constraint. While the current context is not the most auspicious, mainly because of the COVID-19-related world economic crisis, a new, more propitious, world context may soon potentially arise with the Joe Biden Administration in the United States, and innovation significantly lowers the costs of switching to cleaner technologies. In the meantime, Mexico can learn from the carbon pricing experiences of some United States states (e.g. California), Canada, Europe and China.

3. It should be considered that a reform based only on the "common good" or analogous arguments is doomed to fail. For an effective carbon tax or an effective cap-and-trade system to be politically viable, policymakers need to consider allocating part of the revenue to help some economic sectors make the transition, to compensate regressive distributional effects, and to support displaced workers and hard-hit regions. Unfortunately, promises today about future compensation are not always credible.

4. The long-term tax trajectory or, equivalently, the long-term cap trajectory should be clearly established by law to provide the needed certainty for consumers and producers to efficiently adjust to the new conditions. This would also provide the right incentives for industries to comply, significantly reducing transaction costs.

5. Finally, the government should conduct multiple consultations with affected stakeholders and launch a communications campaign that explains the rationale of the policy, provides the facts supporting the case for carbon pricing, and addresses possible misconceptions. Publicizing the effort would also help potential beneficiaries of the reform to construct a focal point for organizing in support of the reform.

---

[7] Source: Secretaría de Gobernación (2020). Authors' translation from the Spanish-language original.

References

Belausteguigoitia JC (2014) Economic analyses to support the environmental fiscal reform. Mario Molina Center. https://www.thepmr.org/system/files/documents/Economic% 20Analyses%20to%20Support%20the%20Environmental%20Fiscal%20Reform.pdf. Accessed 3 Sept 2020

Borghesi S, Montini M, Barreca A (2016) The European emission trading system and its followers: comparative analysis and linking perspectives. Springer, New York

Elizondo C (1994) In search of revenue: tax reform in Mexico under the administrations of Echeverria and Salinas. J Lat Am Stud 26:159–190

Grubb M, Neuhoff K (eds) (2006) Emissions trading & competitiveness: allocations, incentives and industrial competitiveness under the EU emissions trading scheme. Earthscan Publications, London

Jenkins JD (2014) Political economy constraints to carbon pricing policies: what are the implications for economic efficiency, environmental efficacy, and climate policy design? Energy Policy 69:467–477

Magar E, Romero V, Timmons J (2009) The political economy of fiscal reforms: Mexico. Background paper IADB

Mahon JH (2012) Tax incidence and tax reforms in Latin America. Woodrow Wilson Center Update on the Americas. Wilson Center

Mahon J, Bergman M, and Arnson CJ (2015) Introduction. In: Mahon J, Bergman M, Arnson CJ (eds) The political economy of progressive tax reforms in Latin America. Woodrow Wilson International Center for Scholars, Washington, pp 1–29

Pigou AC (1932) The economics of welfare. Macmillan, London

Romero V (2015) The political economy of progressive tax reforms in Mexico. In: Mahon J, Bergman M, Arnson CJ (eds) The political economy of progressive tax reforms in Latin America. Woodrow Wilson International Center for Scholars, Washington, pp 161–188

Secretaría de Gobernación (2020) Programa Sectorial de Energía 2020–2024. Diario Oficial de la Federación. https://www.dof.gob.mx/nota_detalle.php?codigo=5596374&fecha=08/07/2020. Accessed 3 Sept 2020.

Stavins RN (2012) Cap-and-trade, carbon taxes, and my neighbor's lovely lawn. Harvard Kennedy School, Belfer Center, Technology and Policy. https://www.belfercenter.org/publication/cap-and-trade-carbon-taxes-and-my-neighbors-lovely-lawn. Accessed 3 Sept 2020

Tietenberg TH (2006) Emissions trading: principles and practice. Resources for the Future, Washington, DC

Weitzman M (1974) Prices vs quantities. Rev Econ Stud 41(4):291–477

World Bank (2020) Carbon pricing dashboard. https://carbonpricingdashboard.worldbank.org/. Accessed 3 Sept 2020

Chapter 8
Carbon Finance and Emission Trading in Mexico: Building Lessons from the CDM Experience and FOMECAR (Mexican Carbon Fund)

Simone Lucatello and José Eduardo Tovar Flores

Abstract A more general lesson from the past decade is that climate policy and carbon initiatives such as ETS and carbon pricing are not static concepts, but are instead constantly evolving and building upon previous experiences. The vision of a single, top-down global trading system has shifted toward the reality of various single and regional trading system programmes. Building a national emission trading system in Mexico will surely pass through processes and experiences that the country has somehow undertaken from the Kyoto Protocol (KP) in 2005, particularly with the Clean Development Mechanism (CDM), the Mexican Carbon Fund (FOMECAR) and their legacy. Additional design elements or provisions must be prepared under the new ETS in Mexico: regulation will possibly include definitions, scope, compliance obligation, legal procedures and other necessary provisions such as the allocation of permits. However, in order to start the process, important questions on financing the initiative and accompanying the development of an ETS will go through a finance support scenario. Thus, who is going to finance the starting process for allocating emissions, financing bonds and other design issues for the implementation of the Mexican ETS? Who will be financing and offering technical cooperation to follow up on eligible projects for the ETS and who will be supporting education and information activities about ETS implementation? Those and other questions will be addressed in this article, in the light of international and regional experiences.

Keywords Carbon finance · International cooperation · Clean development mechanism · FOMECAR

S. Lucatello (✉)
Instituto Mora/CONACYT, Mexico City, Mexico
e-mail: slucatello@institutomora.edu.mx

J. E. T. Flores
Eosol Energy de México, S.A.P.I. de C.V., Mexico City, Mexico
e-mail: eduardo.tovar@eosolenergy.com

© The Author(s) 2022
S. Lucatello (ed.), *Towards an Emissions Trading System in Mexico: Rationale, Design and Connections With the Global Climate Agenda*, Springer Climate,
https://doi.org/10.1007/978-3-030-82759-5_8

Introduction: The Stages of the Carbon Markets. Evolution and Lessons Learned

The implementation of a future Mexican ETS will go through different phases, and it will be working in parallel with market mechanisms rules for climate change mitigation that were adopted during COP-25 in 2019. As carbon market actions were built upon many years of intense negotiations and have evolved overtime, it is crucial to understand the legacy of the market mechanisms and learn from past international experiences to better understand the Mexican ETS outreach.

In a synthetic attempt to understand carbon markets' recent history, we can identify four clear stages and periods of evolution. Building on Michaelowa's time-lapse model, the first period refers to the birth and growth of carbon markets in the mid-90s. The second one, named the "gold rush", stretches from entry into force of the Kyoto Protocol (KP) in 2005 until 2012. Then, a fragmentation period occurs during the Second commitment period of the KP until the Paris Agreement in 2015 (2009–2015). And finally, the ultimate period defined is the *post-Paris* implementation, occurring during the period 2015–2024.

Each of these stages was clearly defined by key market agreements as well as regulatory procedures constantly interacting with international negotiations and the various COPs outcomes. The following section will briefly revise each one of the stages and provide some comments on lessons learned.

Concerning the first stage, from 1997 to 2004, the international community saw the genesis of the carbon markets and their design. Carbon markets represent a technical tool that can help governments to achieve their commitments to reducing GHG emissions and fight climate change impacts. Recently, the UNFCCC and the Kyoto Protocol have served as the international legal references under which the idea of carbon markets was born and developed. The starting point was Article 4.2 of the UN Framework Convention on Climate Change (UNFCCC) with its rule on "Joint Implementation" (JI) for Greenhouse Gas (GHG) mitigation by several countries. Taking up previous experiences in the US and South-East Asia, UNFCCC negotiators and experts from different countries began to come up with recommendations for market mechanisms. Even though differences among the way of dealing with climate change impacts and financing climate actions among developed and developing nations, the first conference of the parties in 1995 decided to start a pilot phase of the "Activities Implemented Jointly" (AIJ) lasting until 2000 without generating credit issuance (Michaelowa et al. 2019). This action paved the way for testing different market design options in many countries around the world and the first developing country to implement the AIJ was in Latin America, more precisely Costa Rica in the issues of wind, hydro and reforestation.

This period also saw the development of an important milestone for the carbon markets, meaning a tool to standardize units of GHG emission reductions that can be traded, sold, retired or transferred (EU-ETS 2015). In other words, offsetting and the issuance of a carbon credit measured in CO_2 tons was put in the table of the negotiations. The use of carbon credits within different schemes, like Certified

Emission Reductions (CERs), could be used as offsets under ETS or domestic carbon pricing (World Bank 2010). The foundations of the carbon markets were laid together with other measures adopted during international negotiations, such as technology transfer, innovation, research and development on low-carbon technologies/measures through increasing domestic carbon prices, among others.

With the adoption of Kyoto Protocol (KP) in 1997, the international community adopted some important procedures for developed and emerging economies like the introduction of emissions reduction targets and mitigation goals through emissions allowances—in amount units (AAUs)—distributed to several countries or using market mechanisms to exchange AAUs through emission reduction projects. The market mechanisms were defined as the Joint Implementation (JI) for Annex B countries and the Clean Development Mechanism (CDM) in non-Annex B countries. This last one should generate Certified Emissions Reductions (CERs). Among them, the CDM proposed by Brazil established itself as the most agreed upon scheme accepted in the weeks preceding the Kyoto Conference of Parties in 1997 (Shishlov et al. 2016).

Technicalities around the CDM project cycle and sophistication of the crediting system evolved from the development of a Project Design Document (PDD), validation by an independent auditor, project registration, monitoring of emissions reductions, independent verification and CER issuance among others (Lucatello 2011). Within the projects' documentation process, an important discussion includes the use of methodologies, measuring sustainable development co-benefits among other issues and more generally, the design process for implementing the CDM (Shishlov and Bellassen 2016).

In this first stage of the evolution of carbon markets, the main arguments from the economics and policy of climate change have been set. Emissions trading and its ability to cap emissions at a desired level may make it possible to achieve abatement of emissions at the lowest overall cost as well as provide the right incentives for firms to innovate in environmentally friendly technologies. On the other, they could generate enough revenues for financing climate actions and promoting sustainable development in developing countries. This first CDM phase offers clearly important lessons for the current process of designing the ETS mechanism in Mexico.

The "Golden Era" of the Carbon Markets: 2005–2011

The period between 2005 and 2011 saw strong growth of international carbon markets, triggered by the European Union decision in 2004 to allow the use of credits from CDM and JI for compliance under the EU Emissions Trading Scheme (ETS). Under these circumstances, the mechanisms gained sudden popularity in the private sector and carbon markets grew much more than originally expected (Michaelowa et al. 2019).

Ten years after the adoption of the KP and five years from its entry into force, CDM became an immense global market, having more than 5,000 projects registered and a value of several billion Euros (DTU/UNEP 2011). In this sense, the CDM has been a great success in developing a new market for GHG emission reduction projects and providing tools for mitigation actions worldwide (World Bank 2012). CDM had been growing in many developing countries up to 2011, but it was concentrated in few geographical areas, such as Asia and Latin America. China dominates the market both in number of CDM and volume of CERs (40%), followed by India (14%), Brazil (8%), Mexico (6%) and South Korea (5%). Thus, 82% of expected CDM emission reductions by 2012 were concentrated in just five countries (UNCTAD 2009). However, the CDM has been strongly criticized for many other reasons, not only due to the difficulties in implementation but also for not delivering on its environmental and sustainable development objectives among other issues (Wara 2008). The CDM has recently come to terms with its future structure (post-Kyoto 2012) and its structural inadequacies: part-time governing bodies, inappropriate division of responsibilities and, among other factors, neglect of due process as well as lack of transparency. Those are also lessons learned for the implementation of an ETS.

In terms of governance, the CDM was also characterized by the involvement of private actors. On one side, private actors have a role in the rule-making process because they can submit proposals for new CDM methodologies and sectors of analysis. Additionally, all private stakeholders of CDM projects may publicly participate in project design documents and its content. Private actors are instrumental for the CDM implementation because a variety of stakeholders like consultancies, certification companies, and project owners themselves, among others, are the ones actually implementing emission reduction measures. This systematic involvement of private actors in governance arrangements has raised many expectations of higher effectiveness and efficiency in the achievement of governance results (Börzel and Risse 2002).

Crucial to an understanding of the need to reform the CDM is that these problems do not result from the lack of efforts from any part of those working within the system but are signs of systemic limitations of flexible mechanisms and the overall climate change environmental architecture, including the KP and the UNFCCC itself. The Conference of the Parties (COP) of the UNFCCC and the Members of the Protocol (MOP)—which regulate and monitor the implementation of the Protocol— have authority over the CDM and its guidelines and decide on recommendations concerning CDM rules. COP/MOP also decides on the designation of the Designated Operational Entities (DOEs), provisionally certified by the Executive Board (EB). It reviews annual EB reports and regional and sub-regional distribution of DOEs and project activities. Finally, it helps obtain funding for CDM project activities (Lucatello 2011).

In this regard, the CDM has been a big success in developing a new market for GHG emission reduction projects in developing countries. It is widely acknowledged as a mechanism that has changed emission trends in some industries and enabled entities in developing countries to participate in the emerging global carbon market. It has also contributed to raising awareness of public and private entities for climate change. Although treated further in the book, those concerns constitute the core of proposals for the ETS scenarios after CDM implementation.

An important legacy of the CDM was that these projects helped developing countries to build technical capacity regarding structuring of emissions reduction projects and carbon accounting (Mehling and Mielke 2012). A common view among experts points to the fact that capacity building for low-carbon transition in developing countries was one of the most important impacts of CDM (Spalding-Fecher 2011).

Especially in large emerging economies like India, China, Mexico and Brazil an important group of experts like consultants and different stakeholders emerged to play different roles in the growing carbon markets. Private financial institutions were actively participating in the carbon markets as intermediaries, enhancing market liquidity (Weber and Darbellay 2011), mostly in bigger development economies. As demonstrated by several studies (Haigh 2011), carbon funds can play a fundamental role in pooling demand for credits. Moreover, carbon funds were instrumental during this phase to enable development banks to support CDM dissemination.

In critical terms, this phase of the evolution of carbon markets was characterized by strong criticism about economic efficiency, environmental integrity and contribution to sustainable development (Lucatello 2011). At some point, the CDM was considered an ineffective instrument with limited results in reducing global GHG emissions and its mayor success was that of being an economic instrument to increase revenues for just a number of restricted developed countries. Although treated further in the book, those concerns must be taken into account when considering ETS scenarios implementation.

The Fall of CDM and Market Fragmentation (2011–2015)

International negotiations around climate change suffered severe blows during COP15 in Copenhagen (2009) when international commitments to emission reductions almost derailed. As a result of that, subsequent COPs like the one in Cancun in 2010 and through to COP21 in Paris (2015) were characterized by important changes around the reforms of the CDM and more general on the KP's flexible mechanism.

The project focus of CDM was soon considered outdated. Some evolutions took place such as the introduction of the concept of NAMA, which escalated the CDM projects. Nationally Appropriate Mitigation Actions (NAMAs) are policies, programmes and projects that developing countries undertake to contribute to GHG emission reduction efforts. They are central instruments in addressing emission reductions in developing economies. Though the CDM is still existing and functioning, NAMAs slowly substituted the system of projects set by the CDM by moving to complex and larger scale projects mainly in renewable energies.

NAMAs do not represent a legal obligation under the UNFCCC since they represent voluntary actions taken by developing countries to reduce GHG emissions to levels below those of "Business as Usual" (BAU) (Bakhtiari et al. 2015). A common characteristic of NAMAs is that they constitute a transformational change for a given sector of the economy and they provide support for such change.

In the case of Mexico, NAMA projects during this phase combined different political agendas of the Mexican Government: combating climate change, fostering sustainable urban development and housing, as well as improving the quality of life of low-income groups in the social housing sector. An important and flagship case was the Mexican housing NAMA that created a project for transforming the housing sector by including diverse groups of actors and institutions. The Housing NAMA included over 40,000 houses in the energy efficiency sector and was backed by the Mexican National Housing Commission (CONAVI) though subsidies from 2013 to 2017. Major achievements included great impact on national emission reductions in line with the National Special Plan for Climate Change (PECC).

A crucial lesson for this particular NAMA and others implemented in Mexico during this phase is that these projects worked with both domestic and international financing, typically through existing lines of credits from national development banks. In a second stage, NAMAs that used domestic sources could leverage international funding, ideally from institutions that are already active in the country such international cooperation agencies. In such cases, the creation of an enabling environment for private/corporate financing were part of the design process from the outset through the end.

During this period known as *"fragmentation"*, volatility and decline of carbon markets due to the falling demand for carbon credits were also crucial issues. Additionally, as a second major issue, voluntary markets started to gain domain within the climate change arena. Concerning the voluntary carbon markets, they have emerged in various jurisdictions, mostly in North America. However, the total volume of credits traded in voluntary markets is only a small percent of the international and national compliance markets (Hamrick and Galant 2017).

During this period, the supply of carbon credits rapidly began saturating aggregate demand—from the EU-ETS and national governments—which was estimated at between 1.6 and 1.9 billion 15tCO2e until 2015. Based on this supply–demand imbalance, CER and ERU started to collapse (Bellassen et al. 2011). This was due mainly to an overall downward trend following the economic recession of 2009, emissions reductions due to other policies (e.g. renewable energy), as well as the inflow of international offsets (Koch et al. 2014). Prices of credits changed significantly and the fall in carbon prices combined with regulatory uncertainty on the future of the CDM in the post-2012 climate regime resulted in a drastic decrease and distrust regarding CDM as a tool for reducing GHG emissions worldwide. Africa was the continent that most suffered the market decline, which threatened the capacity to develop low-carbon projects.

After this drawback, accelerating the CDM reform became an imperative for the UNFCCC. In September 2011, the CDM executive board decided to establish a high-level panel to review the mechanism and prepare it for the post-2012 period. The panel published the final report consisting of 51 recommendations that address not only the 16 CDM EB, but also other stakeholders including national governments, the UNFCCC and project 17 participants (UNFCCC 2012). Key issues addressed in the CDM policy dialogue were (i) streamlining the project cycle; (ii) changing the methods for determining additionality; (iii) modifying the role of the secretariat; (iv) improving the validation and verification model; (v) professionalization of the EB; (vi) implementation of an appeals mechanism and (vii) strengthening the current stakeholder consultation system (Classen et al. 2012).

From the Paris Agreement to the Present and Towards the Global Stocktake (2015–2024)

The year 2015 represented a milestone in taking action on climate change. In Paris, world leaders reached an agreement at the 21st Conference of the Parties (COP 21) to keep the global average temperature increase to well below 2 °C and pursue efforts to hold the increase to 1.5 °C. For the first time, all countries had to make individual, voluntary commitments to contribute to this global goal. Under the Paris Agreement, the vast majority of governments around the globe—189 countries representing 96% of global GHG—have committed to reduce their emissions by submitting the *Intended Nationally Determined Contributions* (INDCs).

Under the INDCs, countries determine their contributions in the context of their national priorities, circumstances and capabilities that should lead to collective actions and transformations toward a zero-carbon future. INDCs represent voluntary guidelines for governments that are intended to communicate the steps they will take to address climate change and resilience in their own countries. Some countries also communicate not only mitigation actions and steps, but how they will adapt to climate change impacts, and what support they need from, or will provide to, other countries to adopt low-carbon pathways and to build climate resilience (WRI 2019).

An important role will be played by carbon pricing in support of the efforts to decarbonize national economies (World Bank 2020). Article 6 of the Paris Agreement provides a basis for facilitating international recognition of cooperative carbon pricing. Since the entry into force of the PA in 2016, a growing number of jurisdictions are implementing or planning to implement a carbon tax or an emission trading system—a total of 57 initiatives compared to 51 in 2018 and this number is set to grow, according to countries' climate pledges (Lucatello, 2017).

As of 2019, 57 carbon pricing initiatives have been implemented. This consists of 28 ETS, spread across national and subnational jurisdictions, and 29 carbon taxes, primarily implemented on a national level. In total, as of 2019, 46 national and 28 subnational jurisdictions are putting a price on carbon. Carbon pricing initiatives implemented and scheduled for implementation cover 11 gigatons of carbon dioxide equivalent (GtCO2 e) or about 20% of GHG emissions (World Bank, Ecofys 2019).

Most of this action has taken place in the Americas, and particularly in Canada where the federal carbon pricing approach has prompted new initiatives at the provincial level. Important developments have also occurred in other parts of the world with new carbon taxes in Singapore and South Africa—the first carbon pricing instrument implemented in Africa—and new initiatives explored in Colombia, Mexico, the Netherlands, Senegal, Ukraine and Vietnam. Countries are committed to using carbon pricing to meet national climate targets. One hundred eighty-five parties have submitted their Nationally Determined Contributions (NDCs) to the Paris Agreement—representing 55% of global GHG emissions—stating that they are planning or considering the use of carbon pricing as a tool to meet their commitments. That is an increase of eight parties from 2018 (World Bank, Ecofys 2019).

Since the entry into force of the PA, the international climate regime has thus changed its character from a top-down approach based on mandatory emissions commitments to a bottom-up system of voluntary government pledges. The combination of existing, emerging and potential carbon market mechanisms can be regarded as an emerging global carbon market landscape based on differing bottom-up market-based approaches (Redmond and Convery 2015).

Currently, the overall picture of world carbon initiatives shows the following distribution: national carbon markets (ETS), offsetting mechanism and carbon taxes. Over the past few years, national and subnational carbon markets, mainly emissions trading systems have proliferated. Both developed and developing nations have introduced emissions trading as a tool to reduce emissions instead of the CDM. Offsetting against carbon taxes has also started to work in the past few years. The World Bank offers a yearly international report of the State and Trends of carbon pricing with a clear and updated mapping of carbon initiatives (WB 2019 and previous versions).

According to this same report (2019), many jurisdictions are deepening their carbon pricing ambition to better align with their climate goals, and many ETS are being created. Governments are increasingly recognizing carbon pricing as a key policy instrument to deliver on climate mitigation targets and are looking to raise carbon pricing ambition—either through price increases, removing exemptions or increased stringency. In some countries—most notably China—the CDM is being transformed into a domestic offsetting mechanism under the newly piloted national carbon trading scheme with more than 2000 projects re-validated for this purpose (Lo and Cong 2017).

In this context, Mexico has established the ETS pilot programme to sum up to the already existing worldwide experiences in carbon pricing and pointing like China to upgrade CDM and NAMAs experiences to an ETS scheme (Table 8.1).

The Global Financial Architecture for Supporting Carbon Markets and ETS

Why is external financial support needed to implement an ETS and who will finance it?

Over the past two decades, many international financial institutions, donor agencies and national development banks have supported the creation of different initiatives to build carbon trading markets under the UNFCCC process and beyond. Carbon finance is vital to achieve low-carbon, climate resilient development and emission reductions. Likewise, the global climate finance architecture is a very complex issue

Table 8.1 Evolution of carbon markets

Evolving periods	Major features
1997–2005: emergence	• Parties negotiate for the definition of the flexible mechanisms and for the definition of their operational rules and procedures • After initial testing through AIJ, the CDM, JI and IET are agreed • Carbon markets created and catalyzed to demonstrate the potential for low cost emission reduction and compliance with Kyoto targets—Environmental integrity and economic efficiency of the mechanisms are studied in detail
2006–2011: "Gold rush"	• After the initial testing period the carbon markets start a phase of great expansion • EU is the main source of demand for CDM credits while China and India dominate their supply • Improvements to the CDM rules, with operationalization of the PoA concept reducing transaction costs of small-scale projects and contributing to a more balanced distribution • Governance and institutional setup, including capacity building needs, emerge as a key element for the carbon market functioning
2012–2014: fragmentation	• Uncertainties on the future climate regime and lack of mitigation ambition of Annex I countries affect the carbon markets negatively • Prices drop quickly reaching all-time low. Investors have less confidence in market mechanisms—NAMAs start to outscale CDM • Governance and institutional setup, including capacity building needs, emerge as a key element for the carbon market functioning registrations and issuances, although with limited numbers • CDM reforms in order to reduce transaction costs for the KP second commitment
2015–2020: post-Paris agreement	• Prices in the carbon markets are still very low. Limited activities in the international carbon markets—The PA brings positive developments regarding market instruments through Article 6. Detailed modalities and procedures for the new mechanisms (i.e. the SDM and CAs) are still to be defined—An increasing number of developed and developing countries implement or plan to implement carbon pricing initiatives, some of which allow use of credits

Source Author's elaboration based on Michaelowa (2019)

to track and monitor, as it is always evolving.[1] Generally speaking, available funds for carbon initiatives flow through multilateral channels and increasingly through bilateral, regional and national climate change channels and funds. The wide range of climate finance mechanisms continues to challenge coordination and follow-up (ODI 2018).

Despite differences in the amount of money disbursed as well as technical procedures for implementing the carbon initiatives, the primary and common function for all these funds is to encourage the development of a global carbon market and support carbon pricing or other instruments like ETS, which are aimed at reducing global GHG.

A brief overview of the international financial architecture for carbon finance can help to clarify why it is important to consider carbon financial funds for launching and sustaining the future Mexican ETS.

Multilateral Development Banks (MDBs) are the first and foremost source of carbon finance initiatives. They play a prominent role in delivering multilateral climate finance. The World Bank Group (WBG) has emerged as a major actor in helping carbon initiatives around the globe. The WBG currently supports 15 individual carbon funds and facilities worldwide and it works like a trustee for those funds. The WBG has been crucial in promoting the *'proof of concept'* of carbon trading schemes or Emissions Trading Systems (ETSs)—by creating the first carbon fund named *"Prototype Carbon Fund"* in 1999. Since then, its activities have expanded considerably: according to a 2017 report, the WBG's carbon finance portfolio reached $4.8 billion (IEG 2017). It's worth bearing in mind that this money was used to fund different projects and initiatives, some of which include support for launching ETS in different parts of the world.

Regional banks such as the European Investment Bank, which administers the EU Global Energy Efficiency and Renewable Energy Fund (GEEREF) as a source of finance for the EU-ETS, are also important actors in funding carbon initiatives. As stated by the Joint Report on Multilateral Development Banks Climate Finance, the African Development Bank (AfDB), the Asian Development Bank (ADB), the European Bank for Reconstruction and Development (EBRD), the European Investment Bank (EIB), the Inter-American Development Bank Group (IDBG) have reported their commitment of almost US$237 billion in climate finance since 2012, in developing and emerging economies (JR-MDB 2018). All of them are helping to boost ETS projects in their specific geographical areas.

Bilateral cooperation agencies, such as the GIZ, the German cooperation agency, are also very active in supporting efforts to build ETS in different parts of the world.

---

[1] Carbon and Climate Finance are used in this article in different ways. Carbon markets are part of the climate finance architecture. The first are economic instruments for effectively managing emissions of GHG in economically efficient ways for the society as a whole. In the absence of specific market-based mechanisms, climate finance is an essential tool to ensure that emission reduction opportunities are successfully implemented through other approaches such as "results-based" proejcts. Climate finance could also be used in the context of the "non-market approaches to sustainable development" mentioned in Article 6 of the Paris Agreement.

GIZ acts mainly as an advisor to relevant government-level actors during the introduction of the emissions trading system. In the specific case of Mexico, as stated in the introduction of this book, GIZ has assisted Mexico's Ministry of Natural Resources in implementing the ETS.

In general, after a revision of carbon finance initiatives to support ETS around the world and specifically in emerging economies, we can say that activities financed by international or bilateral donors offer the following set of solutions:

(a)   Scientific analyses and policy recommendations to inform and support decision-making authorities on the design of the ETS. This is done by previously identifying what sectors will be covered by the system as well as identifying allowed emissions limits.

(b)   Technical expertise on how reinforcing national emissions gas inventories per sector and strengthening methodologies for baseline emissions at national or local level.

(c)   Capacity-building and stakeholder engagement for different actors and institutions both in public and private sectors. Thus, stakeholders can assume their roles and responsibilities in the market.

(d)   Academic support through scientific approaches with local epistemic communities.

(e)   International dialogue and exchange with jurisdictions that already have similar systems in place in order to facilitate learning.

(f)   Communications strategies to different stakeholders in order to promote awareness and the importance of ETS systems as well as their benefits.

(g)   Conduction of workshops, meeting and training sessions in which the emissions trading system is discussed and if necessary, revised.

A crucial point regarding the establishment of an ETS is the delicate balance of supply and demand governance mechanisms established by the government. On the supply side, the distribution of the total supply of emissions units determined depends upon several factors. Supply in particular depends on parameters set by policymakers, for example, by clearly establishing from the beginning the level at which the cap is set, or through the rules set relating to offsets, banking and borrowing, or linking (ICAP 2018).

On the demand side, total demand for emissions units in an ETS depends largely on the behaviour and characteristics of market stakeholders and depends on shocks unrelated to ETS design features, such as the level of emissions under Business as Usual (BAU) scenarios, or the costs of abating emissions within the covered sectors.

A final and very important issue for financing ETS design and implementation has to do with enforcement. Any ETS acts with strict and rigorous rules for market oversight and enforcement. Emissions must be traced clearly and reported consistently. A lack of compliance and oversight may threaten the environmental integrity of the system and the basic functionality of the market, deriving in losses and damages for all stakeholders involved. An important prerequisite for effective compliance must identify all participating entities which are regulated by the system. Government has to be very effective to develop specific processes and features to identify new regulated entities, especially private firms involved in the ETS.

Lessons from and for Mexico: The Creation of the Financial Funds for Technical Assistance and Financing of CDM Projects (FOMECAR)

Following the international evolution of carbon markets, Mexico has paralleled many international phases and steps of the above described carbon history. Particularly, with the "golden era" of carbon markets and under increasing awareness by the Mexican government, both private sector and the Mexican scientific community began to work on climate change mitigation challenges' and their consequences for the country.

An important cornerstone for Mexican mitigation policy was the Clean Development Mechanism (CDM) scheme that was an opportunity for the Mexican private and public sectors both to participate in a regulated market in the emission of CERs and thereby contribute to efforts to reduce GHG emissions. At the same time, it could also serve as an additional source of revenue in the implementation of those projects by cashing CERs derived from international projects.

The CDM was a clear economic instrument to encourage the participation of the public and private sectors in efforts to reduce greenhouse gases and in the implementation of mitigation measures. In institutional terms, at the end of 2006, Mexico established the Mexican Carbon Fund (FOMECAR), which was hosted at the Banco Nacional de Comercio Exterior, S.N.C., (Bancomext) and under the decided support of the Secretaría de Medio Ambiente y Recursos Naturales (SEMARNAT).

FOMECAR was created as a trust within Bancomext through the initial contribution made by the Mario Molina Center for Strategic Energy Studies, and initially received contributions from public and private, national and foreign institutions.

From Bancomext's perspective, FOMECAR involved different features. (1) It constituted a typical development-banking product that, once it has taken advantage of the market niche, would be operated by private banks. (2) It encouraged the development of investment projects with multiplier effects for the Mexican economy; (iii) It supported the export of the Mexican mitigations of greenhouse gases or carbon credits generated through the projects, and it complemented the range of financial products offered by the Mexican government (Crespo-Chiapas 2018).

For SEMARNAT, the Mexican Carbon Fund meant providing assistance and technical and financial resources to promote the development of CDM projects before private initiative and public sector entities. It also helped to fulfil the country's international commitments regarding climate change before the United Nations Framework Convention on Climate Change (UNFCCC) (Crespo-Chiapas 2018).

Through this trust, a selection of eligible projects for support was identified, resources were provided so that participants could comply with the strict stages established in the CDM so that their project would be eligible to receive CERs, providing technical advice and monitoring over the implementation process.

FOMECAR provided non-refundable financial resources by assuming the risk should the project not achieve its certification, which would be reimbursed by the beneficiaries once the project has been developed and the stages have been completed. At the end of this process, the trust received its issuance of CERs from the United Nations Board.

For the purposes of this article, the functions of FOMECAR can be summarized as follows:

- Organizing outreach events, pays to generate a culture of clean technologies in the country
- Supporting preparation of CDM projects with technical assistance and financial resources
- Operating mandates to promote and structure CDM projects (Heredia 2011).

Thus, the Mexican Development Bank became involved with resources and technical assistance to provide to the public and private sectors in the monitoring of projects focused on reducing GHGs emissions, becoming a channel for the distribution of resources for this purpose.

FOMECAR not only provided resources for monitoring the various stages and protocols established in the CDM scheme but it could also provide resources for financing the project itself, that is, for the implementation of projects such as the following: renewable energy projects and efficient use of energy, fuel change projects, waste management projects in landfills and waste from livestock farms, transportation projects and forest projects.

In carrying out its activities, FOMECAR was established as an instrument of technical assistance and training in the subject, both for medium-sized companies as well as transnational corporations and commercial banks, receiving donations from the governments of European countries and counting on the support of multilateral banks.

FOMECAR's results from the period 2006 to 2011 included several outcomes. On one side, it provided technical assistance to 800 CDM project initiatives, from simple proposals for sustainable projects to complex projects with an impact on GHG reductions from PEMEX and CFE. Secondly, it was a major tool for showing Mexico's commitment to mitigation in specialized seminars and exhibitions. Thirdly, it was considered an instrument for financial assistance to five projects with expected reductions of 1.3 million tCO2e, for a total amount of USD 400,000 to support CDM documentation, validation and registration expenses. The investment amount necessary to implement these projects represented USD 123 million (Lokey 2009). Despite these efforts, its scope was limited since none of its projects managed to register the CDM, suffering the enormous bureaucracy established to finally obtain the CERs, along with the 2008–2009 economic recession and the fall of the carbon markets.

The Experience of FOMECAR and the New Emissions Trading System in Mexico

It is noteworthy that within the agreement establishing the preliminary bases of the emissions trading system pilot programme in Mexico, its Chap. 4 provides for the existence and development of a flexible compliance mechanism, which we could summarize as follows:

- The flexible mechanism may consist of a compensation scheme through eligible mitigation projects or activities or the recognition of early actions for mitigation projects or activities that have received external compensation credits before the entry into force of the pilot programme.
- The secretariat will establish a compensation scheme, defining which compensation protocols, national or international, can be used by the interested parties to develop the eligible mitigation projects or activities, and can develop their own protocols.
- The secretariat may issue compensation credits to those activities that reduce or prevent emissions or increase absorption of said gases, in compliance with the protocols provided.
- In addition to being carried out under said protocols, in order to obtain said compensation credits, the activities in question must be carried out in the national territory, be validated and verified by a greenhouse gas emission verification and validation body and be registered in the National Registry of Emissions provided for in the General Law on Climate Change.
- Participants may only offset with offset credits up to 10% of their allowances with delivery obligations during the pilot programme.

That is why, given that the existence of flexible compliance mechanisms is provided for in the emerging emissions trading system in Mexico, it is estimated that the experience of FOMECAR constitutes a benchmark that can be considered in its replication. Either through the involvement of Mexican development banks or international and regional multilateral banks and the contribution of public and private resources together with technical assistance from international cooperation (like GIZ), a new financing scheme can be created for projects that may issue compensation credits that participate in this emissions trading system.

Given the current circumstances of government-imposed economic restrictions as well as the selectivity in channelling financial resources that the COVID-19 pandemic may imply, it is crucial to rethink new financial mechanisms for supporting ETS implementation after the pilot phase. The creation of coordinating entities and efforts both in generating new projects that can reduce GHG emissions and providing technical and financial assistance in their implementation are of utmost relevance for these flexible compliance mechanisms to see the light within the Mexican market.

The very dissemination of the existence of a flexible compliance mechanism within the emissions trading system pilot programme in Mexico and its possibilities and scope as an incentive for the implementation of GHG emission reduction projects can open a window of opportunity. Additional resources could be transferred in for commercialization of compensation through the replication of financial funds such as FOMECAR. Its replication, with corresponding adjustments, seems somewhat relevant when considering implementation of the Mexican ETS.

Conclusions

In this article, the authors offered a first attempt to understand how past experiences for technical support for climate mitigation projects such as the CDM can be used as lesson for the new ETS in Mexico. Building on the experience from FOMECAR and other internationally established ETS, some interesting reflections can be shared before the Mexican ETS will take shape.

Major experiences drawn from the CDM point to the following issues: any mitigation policy instruments must be supported by robust, transparent, and constantly updated information on emissions by sector. This is crucial for Mexico, where major stakeholders are public energy companies such as PEMEX and CFE.

Secondly, in terms of setting cap's stringency, the government and technical advisors financed by international funds need to set clear and affordable reduction targets via the ETS. Further, these targets should be stringent enough to guarantee significant contribution to the achievement of the Mexican INDC. Thirdly, setting the procedures and mechanisms to facilitate, promote and enforce compliance to achieve the ETS objectives should be a priority since it was not the case for the CDM and NAMAs.

Finally, in the process of setting up the ETS, a national fund like FOMECAR or similar should be implemented as an instrument of technical assistance and training in the subject, both for medium-sized companies as well as stakeholders involved in the market generated by the ETS.

References

Bakhtiari F, Gagnon- Lebrun L, Olsen KM, Bizikova L, Harris M, Boodoo Z (2015) Framework for measuring sustainable development in NAMAs. https://unepdtu.org/publications/framework-for-measuring-sustainable-development-in-namas/

Bellassen V, Stephan N, Leguet B (2011) Will there still be a market price for CERs and ERUs in two years time? CDC Climat, Paris

Börzel TA, Risse T (2002) Public-private partnerships: effective and legitimate tools of international governance?

Classen M, Arumugam P, Gillenwater M, Olver C, Lo M, de Chazournes LB, Swanepoel E (2012) CDM policy dialogue research programme research area: governance. CDM Policy Dialogue. UNFCCC, Bonn

Crespo MT (2018) Los Instrumentos de Mercado en la Política de Cambio Climático, el caso de México 2005–2015. Universidad Nacional Autónoma de México, Facultad de Economías, División de Estudios de Posgrado. Tesis de Doctorado. Febrero de 2018

Fan ZOU, Hui YANG, Qin YL (2011) Thinking on the development of "Carbon Finance" in Commercial Banks of China. Energy Procedia 5:1885–1892

Haigh M (2011) Climate policy and financial institutions. Clim Policy 11(6):1367–1385

Hamrick K, Gallant M (2017) Unlocking potential: state of the voluntary carbon markets 2017. Forest Trends' Ecosystem Marketplace, Washington, DC

Heredia M (2011). https://www.pymempresario.com/2011/03/los-bonos-de-carbono-al-alcance-de-inversionistas-mexicanos/

IEG (2017) Approach paper "Cool Markets" for GHG emission reduction in a warming world: evaluation of World Bank Group's support to carbon finance

JR (2018) Joint report on multilateral development banks's climate finance. file:///C:/Users/slucatello/Downloads/2018-joint-report-on-mdbs-climate-finance.pdf

Karani P, Gantsho M (2007) The role of development finance institutions (DFIs) in promoting the clean development mechanism (CDM) in Africa. Environ Dev Sustain 9(3):203–228

Koch N, Fuss S, Grosjean G, Edenhofer O (2014) Causes of the EU ETS price drop: recession, CDM, renewable policies or a bit of everything? New evidence. Energy Policy 73:676–685

Lo AY, Cong R (2017) After CDM: domestic carbon offsetting in China. J Clean Prod 141:1391–1399

Lucatello S (2017) La gobernanza de los mercados de carbono: del Protocolo de Kioto al acuerdo de Paris. In "Gobernanza Climática en México: Aportes para la consolidación estructural de la participación ciudadana en la política climática nacional", vol 2. PINCC-UNAM, México. http://www.pincc.unam.mx/IMG/pdf_gobernanza/vol2.pdf

Lucatello S (2011) The clean development mechanism and sustainable development in Mexico: what contribution? Lambert Publishing, UK

Mehling M, Mielke S (2012) Market-based instruments for greenhouse gas mitigation in Brazil: experiences and prospects. Carbon Clim Law Rev 6(4):365–372

Michaelowa A (2016) Carbon markets or climate finance? Low carbon and adaptation investment choices for the developing world. Ed. Routledge, London

Michaelowa A, Shislov I, Brescia D (2019) Evolution of international carbon markets: lessons for the Paris agreement. Zurich Open Repository and Archive. https://www.zora.uzh.ch/id/eprint/175354/1/ZORA17354.pdf

ODI (2018) Global climate financial architecture. https://www.odi.org/sites/odi.org.uk/files/resource-documents/11850.pdf

Redmond L, Convery F (2015) The global carbon market-mechanism landscape: pre and post 2020 perspectives. Clim Policy 15(5):647–669

Shishlov I, Bellassen V (2016) Review of the experience with monitoring uncertainty requirements in the clean development mechanism. Clim Policy 16(6):703–731

Shishlov I, Morel R, Bellassen V (2016) Compliance of the parties to the Kyoto protocol in the first commitment period. Clim Policy 16(6):768–782

Spalding-Fecher R (2011) What is the carbon emission factor for the South African electricity grid? J Energy South Afr 22(4):8–14

UNCTAD (2009) The state of play of the clean development mechanism review of barriers and potential ways forward. https://unctad.org/en/Docs/ditcbcc20093_en.pdf

UNFCCC (2012) Climate change, carbon markets and the CDM: a call to action. CDM Policy Dialogue. UNFCCC, Bonn

Wara M (2008) Measuring the clean development mechanism's performance and potential. UCLA law review. University of California, Los Angeles. School Law 55(6)

Weber RH, Darbellay A (2011) The role of the financial services industry in the clean development mechanism: involving private institutions in the carbon market. Int J Priv Law 4(1):32–53

World Bank, Ecofys (2012) State and trends of carbon markets 2012. World Bank, Washington, DC

World Bank, Ecofys (2019) State and trends of carbon pricing 2019. World Bank, Washington, DC

Chapter 9
Emission Trading System and Forest: Learning from the Experience of New Zealand

Benjamin Rontard and Humberto Reyes Hernandez

Abstract In the area of international policy to mitigate climate change, the forest has been important in achieving the objectives of liable countries. The Emissions Trading System in New Zealand (NZ ETS) is the only case of an ETS integrating forestry as a mandatory actor. This is the result of prolonged political discussions and the characteristics of New Zealand forestry. Forest landowners are liable to surrender allowances for deforestation and can potentially receive allowances for the level of carbon sequestered. This scheme created new opportunities for forestry activities and impacted the decision-making trade-offs related to land-use changes. In Mexico, the implementation of an Emissions Trading System in 2020 is evidence of the country's commitment to controlling domestic emissions under the Paris Agreement. Nevertheless, for now, the forestry sector is not involved as a liable actor. It is possible to envision the integration of the forest sector because of the extensive forest cover in the country, which provides a livelihood for a large part of the population. Mexico has the experience and institutional framework to integrate forestry into national emission accounting and carbon forest projects in the voluntary market. The potential impacts of this integration are both positive and negative. Environmental impacts are positive because forest areas can help mitigate emissions, but intensive carbon farming disrupts native forests and biodiversity. The economic impacts would be highly favorable for forest landowners if market volatility were controlled, but there is a potential loss of public revenue for the State. Finally, carbon forestry has the potential to cause conflict between economic sectors involved in land use and among participating communities.

B. Rontard (✉)
Programa Multidisciplinario de Posgrado en Ciencias Ambientales (PMPCA), Universidad Autonoma de San Luis Potosi (UASLP), Alvaro Obregon 64, Zona Centro, 78000 San Luis Potosí, SLP, México
e-mail: benjamin.rontard@uaslp.mx

H. R. Hernandez
Facultad de Ciencias Sociales Y Humanidades, Universidad Autónoma de San Luis Potosí, Avenida Industrias #101; Fracc. Talleres, C. P. 78494 San Luis Potosí. S. L. P., México
e-mail: hreyes@uaslp.mx

© The Author(s) 2022   169
S. Lucatello (ed.), *Towards an Emissions Trading System in Mexico: Rationale, Design and Connections With the Global Climate Agenda*, Springer Climate,
https://doi.org/10.1007/978-3-030-82759-5_9

Keywords Forestry · New Zealand · Forest projects · Emissions trading system ·
Land-use change

Introduction

Deforestation and forest degradation have caused 30% of the anthropogenic CO_2
accumulated in the atmosphere. However, forests are an essential component for
achieving climate change targets since they capture one-third of current anthropogenic CO_2 emissions (Federici et al. 2017). The United Nations has denounced
the lack of commitment regarding forestry. Many countries have committed to
achieve net-zero deforestation targets for the next decades, but very few of them
have presented measures for doing so (United Nations Environment Programme
2019).

Article Five of the Paris Agreement presents the necessity to preserve and improve
existing carbon sinks and support the development of a new source of carbon sequestration. Article Six highlights the importance of international cooperation in the
development of mitigation schemes. Both articles are the successors of the treatment
of Land Use, Land-Use Change, and Forestry (LULUCF) in the Kyoto Protocol.
The text established the conditions for generating and trading Certified Emissions
Reductions (CERs) or Emissions Reduction Units (ERUs) under Clean Development
Mechanism (CDM) or Joint Implementation (JI) projects (UNFCCC 2009). Globally, units from forest activities only represent 0.5% of the total traded in carbon
markets, and they are mostly from voluntary activities (Gren and Aklilu 2016).

The concept of Emissions Trading Systems (ETSs) or carbon markets is to
commodify an ecosystem service or avoidance of environmental damage (GHG
emissions). However, integrating carbon units from emissions permits and units for
carbon sequestered into the same market assumes that they are similar products. This
is reasonable because when buyers purchase carbon units in the secondary market,
they correspond to emissions avoided somewhere else or carbon sequestered, which
results in less carbon in the atmosphere.

ETS are the most popular response by governments to meeting their climate
change targets. In 2020, there are 21 ETS in force and nine in the process of implementation (ICAP 2020). Some of them allow emissions to be offset with units from
forest carbon projects or forestry CDM projects. The European Emissions Trading
System (EU-ETS), currently the largest ETS, excludes the international units from
carbon forest activities (European Commission 2020).

The EU-ETS omitted emissions and storage from LULUCF, but the Kyoto
Protocol set the rules for net emissions accounting (Delbeke and Klaassen 2015).
Under the Paris Agreement, the EU included LULUCF in the 2030 target. This
target represents the third pillar of European climate policy after the EU-ETS and
the Effort Sharing Regulation. The measures to support carbon sinks and limit emissions were integrated into the Common Agricultural Policy and seek to improve
agricultural productivity and the provision of ecosystem services (Runge-Metzger

and Wehrheim 2019). The New Zealand Emissions Trading Scheme (NZ ETS) is the only ETS that includes the forestry sector (ICAP 2020). NZ ETS covers 52% of national emissions. The scheme includes the energy sector, industry, domestic aviation, transport, buildings, waste, and forestry.

In the NZ ETS, the different sectors do not respond to the price signal in the same way. Emissions from transport do not respond to carbon pricing as strongly as the other sectors do (Chris Livesey, personal communication 2020). The price sensitivity is not the same in the emitters sector as in forestry. In forestry, price incentive is efficient up to 19 USD to encourage carbon storage (which is close to the current price). However, it is not efficient enough to be an incentive for emissions abatement in industry and energy consumption. 60 USD is the estimated threshold where we could expect to observe a significant emissions reduction (Ollie Bolton, personal communication 2020). This gap in the price incentive demonstrates that units from carbon captured and units from emissions avoided are different economic products, and the relevancy of integrating them into the same scheme is limited.

In 2015, Mexico signed the Paris Agreement and presented its Nationally Determined Contribution (NDC) to the global target (United Nations 2015). Mexico set the target of reducing its greenhouse gas (GHG) emissions by 25% by the year 2030 compared to the business-as-usual scenario (Gobierno de la República 2015). Nevertheless, the Mexican ETS (Sistema de Comercio de Emisiones, ETS) does not integrate forestry as a liable sector. However, the scheme allows entities to offset up to 10% of their surrendering obligations with units from mitigation projects, which can include carbon forest projects (SEMARNAT 2019).

Mexico has plenty of forest resources, and reduction of emissions from LULUCF is essential to achieve international commitment. Nevertheless, the current political measures to increase the national carbon sink are fragile (Ranero 2018). The experience from other countries where ETS and policies to support carbon sequestration have been enforced represents an essential source of learning. Mexico is supported by international organizations in the implementation of the ETS and takes inspiration from other experienced ETS. Given the potential of forest resources and the recent launching of the ETS, observing the functioning of NZ ETS, and learning from it is an important step.

This chapter analyzes the potential for Mexico to establish an ETS with the inclusion of forestry and the likely impacts, using the experience of New Zealand. The first part describes the forestry sector in New Zealand and the functioning of the NZ ETS. The second part presents the forest resources in Mexico and the current political strategy in this sector. The third part tackles Mexico's strengths and weaknesses for the inclusion of forestry in the ETS and the potential impacts.

Forestry and the NZ ETS

Forestry in New Zealand

The forest industry has been an essential factor for economic development in New Zealand. In consequence, native forests were impacted by European settlement. In the middle of the past century, the country started to massively plant exotic species, mostly *Pinus radiata*, to avoid deforestation of native species and to provide for domestic consumption and exportation of forest products (Ministry for Primary Industries 2020a). *Pinus radiata* is the most common exotic tree in New Zealand, covering 90% of forest plantations (New Zealand Forest Owners Association 2019).

Forest covered 80% of the country before Europeans started to arrive. At present, 38% of the land (10.1 million hectares) is covered by forest (8 million of native forest, and 2.1 million of exotic species). The government owns 75% of the native forest and the rest is under private ownership. The private land occupied by exotic species (1.7 million hectares) is available for production (Ministry for Primary Industries 2020a). In 2018, forestry activities produced USD4.7 billion (1.6% of the national gross domestic product). The sector employs 35,000 people and is the third-largest exporting industry in the country after dairy and meat (Ministry for Primary Industries 2020b).

The Forest Act (1949) and the Resource Management Act (1991) set out the forestry rules. The government is not allowed to conduct any productive activity in the forest. Productive forestry plantation and logging activities are carried out on private property. New Zealand forestry is market-oriented and responds to price variation. In the mid-1990s, with a significant increase in the price of timber products, 300,000 ha of new forest area had been planted (Carver et al. 2017). Private forest landowners are mostly individuals who have a small holding. More than 2,000 smallholders are registered with the Farm Forestry Association (FFA), and some 10,000 are not affiliated. The Forest Owners Association (FOA) also includes more than 200 owners of large forest holdings (David Rhodes, personal communication 2020).

The FFA and the FOA represent 70% of harvest volume. The large landowners can be individuals, investors, forestry corporation, or Maori communities (Iwi). Two types of contracts (forest lease and forest right) allow forest landowners to transfer the rights for forest operation. A forestry lease is a leasing contract on a land title. This contract allows the beneficiary to occupy and take resources on the land. The owner of a forestry right has a property right on the trees but not on the land. Generally, this contract is preferred due to its simplicity of enforcement.

The Maori people have a spiritual and intergenerational relationship with the forest. While the occidental approach to forest heritage is on capital bequeath, they consider it a responsibility to past and future generations to preserve what the parents gave and to allow a future generation to live in the same world as the current one (Kingi 2008). Maori forest is held under individual private ownership, but for commercial activities, they join their lands through forestry trusts or corporations that can be either self-organized or under agreement with a forest company. Since land use and forest management is treated by means of a long-term approach in their culture, decision-making is a prolonged process.

NZ ETS and Forest

The NZ ETS was launched in 2008 after the enforcement of the Climate Change Response (Emissions Trading) Amendment Act. Agriculture is the only sector not covered by the program. However, this sector is the principal source of emissions. In 2018, 45% of emissions reported at the national scale were from agriculture, meaning that the ETS covered 55% of national emissions (Environmental Protection Authority 2019a). The NZ ETS works with upstream points of obligation, meaning that the upper part of the supply chain of the source of emissions is subject to forfeit allowances to the government. Forest landowners are points of obligation since deforestation is a direct source of emissions.

Forestry was the first sector in the NZ ETS. Integration of forestry into the NZ ETS seeks to promote carbon sequestration and storage by discouraging deforestation of pre-1990 forest and by encouraging the planting of the new post-1989 forest, replanting of the existing post-1989 forest, and increasing carbon density with longer harvest rotations. The government decided to distinguish pre-1990 and post-1989 forest because of the Kyoto protocol scheme setting 1990 as the baseline year (Ministry for the Environment 2019). For the program, the definition of forest land is an area with at least 30% covered by forest species with at least 30 m of diameter on average (Ministry for Primary Industries 2020c). In 2018, there were 1,412,323 ha of exotic pre-1990 forest and 682,439 ha of post-1989 forest (Environmental Protection Authority 2019b).

Owners of land registered as forest area before 1990 are liable to surrender units for the quantity of carbon released when they deforest more than two hectares in five years. Indigenous forests are excluded from the NZ ETS since they are already protected by the Resource Management Act and the Forest Act (Karpas and Kerr 2011). The obligation is to notify the Ministry for Primary Industries of any deforestation, calculate the emissions, and surrender units. They can pay units directly to the government at a fixed price or buy them on the secondary market. It is also possible to offset deforestation by establishing an equivalent forest elsewhere (Ministry for Primary Industries 2017a).

Participation is voluntary for post-1989 forest owners. They can register their land with the NZ ETS and earn units during each obligation period (between one and five years) according to the net quantity of carbon captured. Until 2013, new participants could claim units according to the level of carbon captured from 2008. Since then, new participants receive units starting from their entrance into the NZ ETS (Carver et al. 2017). According to the evolution of the stock, the participant can earn or surrender units. In the case of total deforestation or if the owner wants to leave the NZ ETS, the participant must repay all units received.

Figure 9.1 illustrates the evolution of carbon storage for a single-age, single-

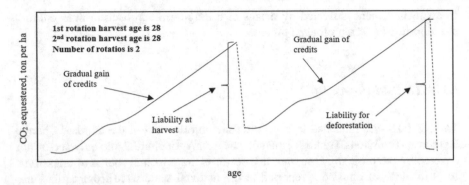

Fig. 9.1 Credits and liabilities over two forestry rotations (Karpas and Kerr 2011)

species forest planted in 2008 or later. It assumes harvest followed by replanting for the first rotation with a partial liability to surrender units, and harvest followed by land-use change for the second rotation with a total liability to surrender units.

The NZ ETS reform (2019) introduced a new accounting methodology based on averaging. Instead of gradually gaining credits until harvesting, participants will gradually gain credits until the average age and then receive a constant level of credits corresponding to the average level of carbon sequestered without liability at harvesting. The average age is the age at which the forest achieves the average level of carbon sequestered. This accounting seeks to simplify and reduce the cost of participation in the NZ ETS. Forest areas registered from January 2021 will apply this methodology, participants with forest registered in 2019 and 2020 will have the option to use it and forest registered before 2019 will continue with the previous methodology of stock change accounting (Ministry for Primary Industries 2020c).

The program uses tools provided by the Ministry for Primary for self-reporting. For pre-1990 deforestation reporting and post-1989 participants below 100 ha, the level of carbon stored is estimated by a look-up table (Ministry for Primary Industries 2017b) which is used to calculate the carbon released by deforestation and the evolution of the carbon stock by comparing to previous years. The calculation is based on information provided by participants (area, species, age, region). For post-1989 forest larger than 100 ha, the participants must use the Field Measurement Approach (FMA) (Ministry for Primary Industries 2018a). This instrument requires GPS data and detailed analysis on a sample plot, gathering data on tree diameters and heights, species, shrub types, crown cover, past and planned silvicultural activities, and adverse events (disease, fire, storm). The Ministry validates reporting processes for primary industries and may carry out audits and monitoring in the field or by remote sensing. Penalties for cheating or omitting information range from a monetary fine to a term of imprisonment up to five years according to the severity of the offense (Ministry for Primary Industries 2015a).

The scheme represented a high cost for pre-1990 forest owners if they planned to carry out land-use change or sell their land. As compensation for the loss of asset value, the government decided to give them a one-off free allocation, which is a fixed amount of allowances (Leining and Kerr 2018). The forest owners can use these units to meet their liability when they decide to deforest or sell them on the secondary market to receive a monetary payment.

Two complementary programs have been implemented by the government to substitute for participation in the NZ ETS: The Permanent Forest Sink Initiative (PFSI) and the Afforestation Grant Scheme (AGS). The PFSI seeks to improve conservation of post-1989 forest. Landowners who enter in the program sign a covenant with the government to definitively conserve their forest (with the possibility of withdrawing after 50 years). Limited harvesting is possible, but a minimum canopy cover must be maintained. Even if the landowner sells the land, the covenant stays in force (Ministry for Primary Industries 2015b). The landowner receives units according to the quantity of carbon stored, using the same reporting instrument. In 2021, the government will enforce the Climate Change Response (Emissions Trading Reform), which includes the end of the PFSI and transfers the participants to the NZ ETS in order to simplify the administrative process (New Zealand Government 2019).

The AGS was a grant distributed by the government to small and medium landowners for planting between five and 300 ha. The landowner receives USD 800 per hectare, and the government continues to receive the carbon units for the first ten years. After that, the forest enters into the NZ ETS (Carver et al. 2017). In 2019, the One Billion Trees Program replaced the AGS. This program operates with the same objective to promote new forest planting and native regeneration. Depending on the type of forest (native or exotic species), the landowner can receive between USD 320 and 2,500 per hectare. Landowners can participate in the NZ ETS unless they grow *Pinus radiata*, in which case the landowner can apply after six years (Ministry for Primary Industries 2018b).

Participation and Impacts

In 2018, 39 pre-1990 forest landowners were liable to surrender units corresponding to 74,363 tCO2e released (0.1% of total emissions covered by the NZ ETS). The same year, 2,106 post-1989 forest participants had registered 324,819 ha (48% of the total). 94% of post-1989 forest participants are owners, 6% are forestry right-holders, and 1% are forestry leaseholders (Environmental Protection Authority 2019b). Most of the post-1989 forest landowners participating in the NZ ETS were already carrying out commercial logging activities, which gave them the possibility of having the capital to finance carbon farming.

Forestry right-holders or leaseholders are investors (individuals or companies) that have a contract with the forest landowners to share profit from carbon units or to pay a fixed rent. They represent a small portion of the number of participants but potentially cover a large part of the registered land. Since most of the forest landowners in New Zealand are smallholders, most NZ ETS participants are registered in small areas. Among the 2,206 participants in 2015, 1,439 listed between one and 49 ha covering

Table 9.1 Distribution of pre-1989 forest participants by size class of forest (Carver et al. 2017)

Size class of forest (ha)	Number of participants	Percentage of participants (%)	Total area registered (ha)	Percentage of land registered (%)
1–49	1,439	65	26,000	9
50–99	353	16	22,000	7
100–499	343	16	66,000	22
500–999	24	1	14,000	4
1,000+	45	2	177,000	58
Total	2,206	100	304,000	100

only 9% of the total, while 45 participants registered more than 1,000 ha, which covered 58% of the total (Table 9.1).

The price increase in timber products prompted forest landowners to massively plant new forest areas between 1993 and 1996. In consequence, even if most of the new pre-1989 forest areas (planted after 2008) are registered in the NZ ETS, the most significant part of this area is forest planted before 2004 (Carver et al. 2017). It is hard to determine whether the NZ ETS had a material impact on pre-1990 deforestation and post-1989 afforestation. Previous empirical observations have shown that carbon pricing had a minimal impact on afforestation (Manley 2016). However, the carbon price is strongly affected by the volatility of the international market. In 2014, the price decreased by USD 3 but in 2015 the price of national units increased and has remained stable between USD 12 and 15 since 2018 (Leining and Kerr 2018).

At first, timber producers' participation in the NZ ETS was driven by a business-as-usual strategy. The recent evolution of the national carbon price encouraged registration of new forest areas (David Rhodes, personal communication 2020). Figure 9.2

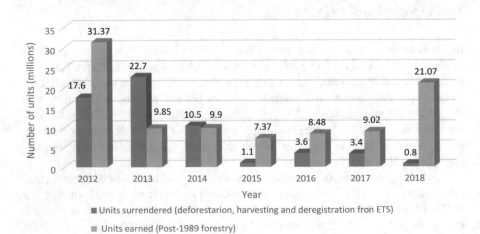

Fig. 9.2 Units earned and surrendered from forestry (millions) (Environmental Protection Authority 2019b)

shows the evolution of units surrendered for deforestation, harvesting, or deregistration of post-1989 areas and units earned by post-1989 forestry. The effect of the carbon price on deforestation and harvesting is strong. However, the effect on removing activities is not immediate, probably due to the phenological stage of registered forest.

Nevertheless, there is an increase in post-1989 forest areas registered from 277,212 ha in 2014 to 324,819 ha in 2018 (Environmental Protection Authority 2015, 2019a, b). Planting of forests in New Zealand is not regular. The average new forest area planted was 100,000 ha per year as of 1990. At present, it is around 15,000 ha (Steven Cox, personal communication 2020). Experts are expecting a wave of harvesting in 2020, followed by a considerable increase in planting.

Forest and Climate Change Policy in Mexico

Forestry in Mexico

Forest area covers 70% of the country (137.9 million hectares), but most of this area is dry forest. Current wooded forest areas in Mexico cover 65.7 million hectares, consisting of 52% of temperate forest, 45.7% tropical rainforest, and 1.4% mangrove (CONAFOR 2019). Deforestation in Mexico has been irregular. In the most recent decades, the deforestation rate decreased from 0.52% (1990–2000) to 0.10% (2010–2015) (Camara de Diputados and CEDRSSA 2019). The target under the Paris Agreement (Nationally Determined Contribution) is to reach a rate of 0% by the year 2030 (Gobierno de la República 2015). Deforestation varies from one type of forest to another. Most of the deforestation occurs in rainforest areas and is caused by land-use change for pastures, while dry forest areas have tended to increase in the most recent decades (Bonilla-Moheno and Aide 2020). At the national scale, land-use cover/change, and forestry (LULUCF) is the source of 4.9% of total GHG emissions (Ranero and Covaleda 2018).

Common property is dominant in the country (51% of Mexican territory) and is the result of numerous agrarian reforms in the twentieth century, which impacted forestry management (Bray et al. 2006). However, most of the forest areas are under private ownership (50%), while 45% are under communal ownership and 5% state-owned (Reyes et al. 2012). Timber production is mostly carried out in common properties. 4.4 million hectares of forest with logging activity is under communal ownership and 1.1 million hectares under private ownership (CONAFOR 2019). The National Forest Commission also supports 220,000 ha of commercial planting in the country.

At the national scale, forestry is a weak activity (0.23% of the gross domestic product in 2016), and imports of timber products are five times higher than exports. However, forestry has an essential social value in the country since the sector produces 166,664 jobs (CONAFOR 2019). Forest areas are an essential resource mainly for a large part of the population who depend on non-timber forest products for their livelihood.

Mexico had a purely productive vision of its forest until 1980 when it started to convert its national forest strategy into a sustainable logging and ecosystem conservation plan. In 2001, the country started a centralized plan for restoration and conservation of forests, support of timber and non-timber products for commercial purposes and livelihood through the creation of the National Forest Commission (CONAFOR 2001). Forest landowners (individuals and communities) can receive training, technical support, and subsidies from CONAFOR either for productive activities or conservation and restoration.

The iconic program is the payment for environmental services (PSA) program. This program was launched in 2003 with the goal of preserving hydrological services from forest funded with a fiscal instrument controlling water consumption, and payments are based on the opportunity cost of land (Muñoz-Piña et al. 2008). Payment can vary according to the region and type of forest. Currently, the maximum amount is around USD50 per hectare per year. The payment must fund technical support, which is compulsory in any CONAFOR program and restoration, reforestation, or conservation. The community can use the rest of the money for collective projects or investments (CONAFOR 2020). The PSA program supported 13,200 participants between 2003 and 2019, with total payments of USD 725 million (CONAFOR 2019). However, one of the limitations is budget availability. Figure 9.3 shows that the budget allocated decreased from USD 242 million in 2016 to USD 101 million in 2019.

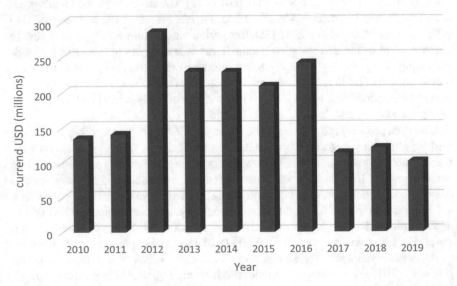

Fig. 9.3 Budget allocated to PSA program from 2010 to 2019 (current US dollars) (CONAFOR 2020)

Forest Carbon Policy in Mexico

Mexico began hosting forest carbon projects soon after the Kyoto Protocol through the voluntary carbon market and REDD + program (Reducing Emissions from Deforestation and forest Degradation). Although Mexico is the second greatest recipient for Clean Development Mechanism (CDM) projects in Latin America and the fifth in the world, the country never hosted any carbon forest project under this scheme (MEXICO2 et al. 2018; Ranero and Covaleda 2018). The limitation in the EU-ETS to accept CDM offsetting only from Least Developed Countries (LDC) and the collapse of the price of Kyoto units after 2012 limited the opportunity for new CDM. Afforestation and reforestation (A/R) projects had not been successful under the CDM scheme because of methodological limits to unit accounting, the long-term process to generate units, and the compliance conditions to preserve carbon storage.

While forest carbon projects are almost absent in the CDM, they covered up to 35% of the credits offset in the voluntary carbon market between 2009 and 2016 (Ranero and Covaleda 2018). In Mexico, there have been carbon forest projects for the voluntary carbon market. Environmental organizations work with local communities and sell carbon units to private companies that want to offset their emissions certified under international standards (Rontard et al. 2020). Table 9.2 shows the projects in Mexico currently in force.

The Scolel'te project implemented in 1997 in the southern state of Chiapas involves 1,200 participants from 90 communities throughout the state covering 7,660 ha. The Cooperativa Ambio organization launched the project and worked directly

Table 9.2 Forest carbon projects in Mexico (Rontard et al. 2020)

Project	State	Area (ha)	Standard	Organization
Sierra Gorda Biodiversity Carbon	Querétaro	21,491	VCS - CCB	Grupo Ecologico Sierra Gorda
San Juan Lachao	Oaxaca	2,388	Climate Action Reserve	ICICO A.C
Sustainable Climate-Friendly Coffee	Oaxaca	292	Verified Carbon Standard	UNECAFE S.C
Carboin	Oaxaca	3,000	NMX-SSA-14064	ICICO A.C
Captura de carbono Santiago Coltzingo	Puebla	3,092	Climate Action Reserve	ICICO A.C
Captura de carbono San Bartolo Amanalco	Estado de Mexico	1,005	Climate Action Reserve	ICICO. AC
Scolel'te	Chiapas	7,660	Plan Vivo	Cooperativa AMBIO
Fresh Breeze Afforestation Project	Tabasco, Nayarit, Chiapas	4,270	Verified Carbon Standard	Proteak UNO S.A.B. de C.V

with the forest landowners. Participants are individuals, members of the communities, even if the forest area is officially communal property. They can harvest trees for logging activities without authorization from the Ministry of Environment (SEMARNAT), which provides them additional income if they respect the 25 years of commitment to Scolel'te.

The organization ICICO A.C. has been very active in setting up forest carbon projects in the last decade. Today, they manage four projects in three different states, and they are the first to develop a forest carbon project with the Climate Action Reserve standard (CAR). Unlike Cooperativa Ambio, they work with communities as a whole.

The government supports actions to encourage carbon storage in the forest through the ENAREDD + political framework. Since 2010, the Mexican government has been working on a strategy to develop a REDD + program across the country (Comisión Intersecretarial de Cambio Climático 2017). ENAREDD + seems to be the principal instrument used to meet Mexico's climate change targets in the forest sector. In REDD + programs, national governments are recipients of payments from other countries. The conditions to be met before starting the program are very strong. After ten years of national strategy design, Mexico has not launched and continues to work on the funding scheme.

Integration of Forestry into the Mexican ETS

The government has not made plans to include the forest sector in the Mexican ETS to date. However, integration of the forest sector has the potential to contribute to climate change targets and support rural communities. Mexico has points in its favor for developing a policy scheme like the NZ ETS. In this section, we present the strengths and weaknesses in the Mexican context to integrate forestry into the ETS and the potential impacts (positive and negative) that such integration would entail.

Strengths in the Mexican Scheme

The NZ ETS has done consistent regulatory work in the Climate Change Response (Emissions Trading) Amendment Act and has efficient public organizations such as the Ministry for Primary Industries, Ministry for the Environment, and Environmental Protection Authority. Mexico has gained experience in forest and ecosystem services policy. Over the years, this experience has enabled institutions and governmental organizations to develop their abilities to develop cross-sectoral policy on the national scale, such as the ETS.

Mexico was the first of the developing countries to have a regulatory framework to achieve climate change targets through the General Climate Change Act (SEMARNAT 2019). This law launched the creation of the National Emissions

Registry (RENE) in 2013. Entities in the energy, industry, transport, agriculture, waste, commerce, or services sector who emit more than 25,000 tCO$_2$e per year are required to report their emissions in the RENE (Ramírez Bautista et al. 2016). The registry is based on self-reporting from entities following guidelines issued by SEMARNAT. Even if forestry is not included in the RENE, the institution would be able to establish a similar reporting instrument in this sector.

Since its creation in 2001, CONAFOR has showed its ability to manage forestry and ecosystem services. The 17 years of PSA and other forestry programs have strengthened CONAFOR functioning. Therefore, this institution should be able to manage the participation of forest landowners in the ETS as Ministry for Primary Industries does in the NZ ETS. Reporting process failures in New Zealand have been detected due to lack of knowledge about the methodology and liabilities. These failures are mostly due to landowners' mistakes and induce monetary sanctions more expensive than the cost of hiring a forestry consultant (David Rhodes, personal communication 2020).

Many landowners work without any technical support because the state limits interactions with landowners to avoid any influence in their decisions (Ollie Bolton, personal communication 2020). The Ministry for Primary Industries must deal with cases of non-compliance caused by lack of knowledge about the reporting process. Landowners also need support to understand market functioning in order to avoid losing money in the sale of carbon units. In Mexico, participants in CONAFOR programs must hire a forest technician to support landowners (CONAFOR 2020). The institution could teach landowners about program functioning. CONAFOR know-how and ability to train forest landowners could be a definite advantage in the integration of forestry into the ETS.

The experience from the forest carbon project in the voluntary carbon market must also be considered. Indeed, although these projects were developed without the participation of government institutions, they showed the ability of Mexican forest landowners to participate in carbon sequestration. These experiences also demonstrate the socioeconomic impact of carbon forestry on local communities. These internationally certified projects have robust technical requirements. The Mexican government could take these standards as a point of reference in implementing a methodology for accounting and monitoring the participation of the forest sector in the ETS. Data from these projects could give an estimate of the level of carbon sequestration and support the development of a national protocol (Ranero and Covaleda 2018).

Large forest areas in many communities are an advantage of Mexican land tenure. Fifteen thousand five hundred eighty-four communities in Mexico have more than 200 ha each. It means an opportunity to avoid the problem that smallholders face in the NZ ETS. When landowners register a small area, it is likely to contain the one unique stand of trees of the same age. Then, at harvesting time, they must surrender units for the totality of their registered area (Fig. 9.1). For them, the balance between harvest and growth is healthy. Large landowners have a different class and age of trees and compensate the surrendered units by units earned in another part of the land where it has not been harvested. In New Zealand, most of the participants with small areas are strongly exposed to this income volatility. Because of this, the government implemented averaging accountability. In Mexico, most of the communities with large forest areas can manage income stability.

Weakness in the Mexican Scheme

In the NZ ETS, most of the forest landowners are farmers who also do commercial logging, even on a small scale. However, in Mexico, commercial logging is not that common, and only a small portion of the communities with forest area have experience in forest management. Although the CONAFOR provides capacity-building and technical support in its programs, the lack of experience in forest management could cause two problems: failures in forest management and difficulty in investing in the first step of participation.

In the NZ ETS, participants have capital when they register their land because they already carry out farming and logging activities. The first income from carbon farming can take time because the land must pass the first report period and the participant must sell the distributed carbon units in the secondary market. It could be difficult for Mexican rural communities to finance the first investment and wait for the first income without any financial support.

Asymmetric information between stakeholders in the contracts is a problem of trust and uncertainty with commitment and price variation. Mexico is quite vulnerable to this issue, which limits the incentive to trade carbon credits from the forest (Van Kooten 2017). Illegal deforestation would represent a severe difficulty in the enforcement of unit surrendering. In Mexico, the Office for Environmental Protection (PROFEPA) identified 108 areas of illegal forest activity (SEMARNAT 2020). Illegal deforestation and forest degradation are mainly due to land-use change without authorization, illegal commercial logging commonly linked to organized crime, and extraction of non-timber products for livelihood or local commerce. It would be a challenging task to achieve compliance of unit surrendering in these zones where people have carried out extraction activities for a long time, and the law has not been enforced.

The NZ ETS experienced ups and downs in the incentive impact of carbon pricing in forest participation according to the variation of the unit price. The incentive impact depends on the cost of carbon farming activity and the potential income from alternative activities. In Mexico, the ETS will start with a total free allocation. The 300 participants will be allocated allowances based on historical emissions, and 5% of the total available units may be sold by auction (MEXICO2 2019). Moreover, no economic sanctions will be enforced in the pilot phase (2020–2022). With this high level of free units and without sanctions for non-compliance, we can assume that the price of carbon units at auctions and in the secondary market will be quite low.

A low price would be insufficient to avoid deforestation and to support afforestation and reforestation at the same time. In current carbon pricing mechanisms, the prices are already low. In the voluntary carbon market, units from forest carbon projects are sold at around USD 10 (Leticia Espinosa, personal communication 2017), and the carbon tax for fossil fuel has a rate of USD $2.4/tCO_2e$. In the NZ ETS, the limit from which the price of carbon units can have a potent incentive effect for forest landowners is at around USD 19 (Ollie Bolton, personal communication 2020).

There are also some political and macroeconomic limits on the integration of forestry into the ETS. In New Zealand, including forestry in the ETS and the international market was an opportunity to support an important sector of the national economy and resulted from political pressure from the forestry sector. In Mexico, forestry has always been an important economic sector, and forest areas were an opportunity to meet the Kyoto Protocol target because it was a positive carbon sink. In international negotiations, Mexico said it would include forestry in the international market. Under the Kyoto Protocol, the forest units were given to the government. However, the forest landowners pressured the government to receive these units. The forestry sector, through the FOA, put intense lobbying pressure on the government to include them in the NZ ETS (David Rhodes, personal communication 2020). In Mexico, the situation is different, and there is no political or economic interest in including forestry in the ETS. The forestry sector does not have any strong political influence in the country because it is not an important economic sector. The share of forest products in the GDP is shallow (0.23% in 2016), and the level of imports in the forestry sector is five times the level of exports (CONAFOR 2019).

Potential Impacts

Environmental Impacts

The motivation to integrate forestry into the ETS is obviously to improve the level of carbon storage at the national scale. This would support achievement of the forestry target in regard to Mexico's climate change commitment. Moreover, the potential increase in forest area would support mitigation of Mexico's emissions and meet the country's NDC. There is severe environmental additionality in the distribution of carbon units for carbon sequestration in new forest areas. In New Zealand, most of the post-1989 forest participants were already carrying out logging activity, which was their incentive to make new plantations. Participation in the NZ ETS is a new source of income, but these forest areas could probably have been planted without it. In Mexico, besides CONAFOR programs, there is no incentive to reforest or plant new forest. Participation in the ETS by bringing income for new forest areas could be a net influence in the trade-off between forest and other land uses.

However, the environmental impact of carbon farming can also be harmful. Landowners participating in this activity are motivated to plant species with a high level of carbon sequestration. In New Zealand, *Pinus radiata* was the most commonly planted species because it is fast-growing and cheap to establish and to log. Additionally, it achieves a high level of carbon storage in a short time compared to native species. Today, 75% of the forest areas registered in the NZ ETS are planted with this species (Carver and Kerr 2017). However, native forests are essential because they are adapted to the local environment. These forests can provide more ecosystem services than carbon sequestration.

Native forests are composed of a high diversity of species and represent an essential habitat for other animal and plant species. Non-timber products from the native forest are also an essential resource for the livelihood of local populations in Mexico (Delgado et al. 2016). Moreover, the biodiversity of native species is vital for soil stability and water quality. Native forests have social and cultural value. For the local population, there is recreational interest in their conservation, and these forests can have spiritual value for indigenous people. The risk of encouraging planting for carbon sequestration is the substitution of native forests by more productive species.

Economic Impact

One of the first motivations for deforestation in Mexico is land-use change because communities do not earn income from their forests. Payment in the PSA program has been set according to the opportunity cost of forest areas (Muñoz-Piña et al. 2008). With time, livelihood support became more of a target than a positive externality of PES programs, and the marginalized population has been more likely to participate (Liu and Kontoleon 2018). The PSA has a small impact on poverty alleviation (Sims and Alix-Garcia 2017) because the payment is quite low, and forest conservation or planting has a high cost. There is also a limited number of recipients. The public budget limits the program, and available funds have been decreasing sharply.

In carbon forest projects, the income from the sale of carbon units would be enough to fund the projects and remunerate people working on them. However, it is still not enough to generate a sustainable income for the whole of participating communities. The advantage of the integration of forestry into the ETS is that the income would not depend on the public budget. As long as the unit sellers find buyers in the market, they will receive an income. In a way, it is a mechanism to collect private funding to support forest conservation and reforestation. In this scheme, the landowners must meet conditions to participate, but there is no limitation on the number of participants or forest area.

The income that participants will receive from the secondary market is exposed to price volatility. In the NZ ETS, unit price was quite unstable before 2015 because of variation in the price of international units (Leining and Kerr 2018). This volatility has been reflected in the participation of post-1989 forest and the deforestation of pre-1990 forest (Carver et al. 2017). The government tackled this problem by delinking domestic ETS from the international carbon market. In Mexico, it would be necessary to have a mechanism to ensure income stability for forestry participants.

In the operation of the ETS, the government receives income when selling units through auctions, so distributing units to forest landowners for carbon sequestration can also have an impact on public revenue. The level of income from auction is proportional to the demand for carbon units. This income also depends on the proportion of free allocation, but this portion is decided by the government. If we assume the integration of forestry with more units distributed for carbon sequestration than units surrendered for carbon released, it will induce a larger supply of units in the secondary market, which will reduce the price and consequently the public revenue from auctions.

Conflicts

Carbon farming is an alternative economic activity and land use. Focusing on this activity means another activity is foregone. In New Zealand, a conflict grew between forestry and farmers. Farming is the largest economic activity in the country, and it also has a strong political lobby. The farmers' organization, 50 Shades of Green, has been struggling against the impact of the NZ ETS on land-use change. Farmers criticized investors buying farmland to convert into intensive forest plantations (Chalmers 2019). Foresters claim they have the right to plant forests to participate in climate change mitigation. However, the farmers highlight the economic loss to agriculture and the ecological impact of intensive forest plantation. The government must find a compromise between the first and third economic sectors.

In the Mexican voluntary carbon market, the carbon forest project created internal conflicts among the participant communities. In the Scolel'té program, conflicts rose between participants and non-participants because of the difference in land-use practices and the benefit they could obtain. In fact, participants can legally harvest their forest for commercial use after the commitment period (25 years). This practice generally requires harvesting authorization. Non-participants felt it unfair that other members in the same community could have more access to natural resources (Osborne 2015). However, this is specific to this program, where participants are individuals instead of whole communities. In communities with collective decision-making about land use and participation in carbon forest projects, this kind of conflict has not been detected (Rontard et al. 2020).

Conclusion

The NZ ETS model is unique because of the importance of the forestry sector in New Zealand's economy. Nevertheless, the scheme is still far from perfect and needs further improvement. Mexico has potential in its forest resources, the experience of existing institutions in forest management, and emissions registration. In other words, the country may be ready for the integration of forestry into the ETS. The carbon forest projects developed under the voluntary carbon market can be considered as a laboratory for their extension to the national scale. However, Mexican forest landowners are mostly communities with fragile technical knowledge about forest management.

The forestry sector does not have the same political influence in the two countries. Unlike New Zealand, forestry is not an essential economic sector in Mexico. New Zealand has developed a standardized system to account for carbon storage. Mexico has not yet achieved this step. This offers an explanation as to why Mexico has not considered the inclusion of forestry in the ETS. The environmental effect will be helpful if the negative impacts of intensive carbon forestry are controlled. In the economic aspect, there is high interest in the communities benefitting from a new source of income, even if this resource would be exposed to market price volatility. Integrating forestry into the ETS induces prevented emissions and comparable carbon sequestering, while units from allowances to the industry and units from carbon forestry are similar economic products. This is arguable, regarding the difference in price incentives between forestry and other sectors in the NZ ETS. Integrating prevented emissions and sequestered carbon into the same market is a political choice, as is the use of an economic instrument to control environmental impacts.

Mexico has the potential to successfully enforce the integration of forestry into the ETS. Nevertheless, this task will require strong technical support for forest landowners from CONAFOR and significant investment for monitoring and verification. Finally, the integration of forestry into the NZ ETS is the result of socio-political discussion among the government, farmers, and forest landowners. In Mexico, it would be essential to establish a democratic process with the forestry sector and discuss its integration into the ETS with forest landowners and forestry organizations in the country.

Acknowledgements The principal author was hosted by Motu Economic and Public Policy Research, Wellington, to carry out this study. We acknowledge their support during this visit, especially the political and technical expertise of Catherine Leining.

References

Bonilla-Moheno M, Aide TM (2020) Beyond deforestation: land cover transitions in Mexico. Agric Syst 178:102734

Bray DB, Antinori C, Torres-Rojo JM (2006) The Mexican model of community forest management: the role of agrarian policy, forest policy and entrepreneurial organization. For Policy Econ 8(4):470–484

Camara de Diputados, CEDRSSA (2019) La Actividad Forestal en Mexico, Estrategias y Acciones contra la Deforestación. Palacio Legislativo San Lázaro, Ciudad de México. Octubre 2019

Carver T, Kerr S (2017) Facilitating carbon offsets from native forests (No. 1124–2019–2385)

Carver T, Dawson P, Kerr S (2017) Including forestry in an emissions trading scheme: Lessons from New Zealand

Chalmers H (2019) Lobby group 50 Shades of Green calls for pause on blanket forestry. Stuff. https://www.stuff.co.nz/business/farming/112878988/lobby-group-50-shades-of-green-calls-for-pause-on-blanket-forestry. Accessed 20 May 2020

Comision Intersecretarial de Cambio Climatico (2017) Estrategia Nacional para REDD+ 2017–2030. ENAREDD+

CONAFOR (2001) Programa Estratégico Forestal para México 2025. SEMARNAT-CONAFOR

CONAFOR (2019) El sector Forestal Mexicano en Cifras 2019. Bosques para el Bienestar Social y Climático. SEMARNAT-CONAFOR

CONAFOR (2020) Reglas de Operación del Programa Apoyos para el Desarrollo Forestal Sustentable 2020. Diario Oficial Jueves 16 de enero de 2020

Delbeke J, Klaassen G (2015) Framing Member States' Policies. Delbeke/Vis (Hrsg.), 92–108

Delgado TS, McCall MK, López-Binqüist C (2016) Recognized but not supported: assessing the incorporation of non-timber forest products into Mexican forest policy. Forest Policy Econ 71:36–42

Environmental Protection Authority (2015) 2014 facts and figures

Environmental Protection Authority (2019a) 2018 New Zealand emissions trading scheme report

Environmental Protection Authority (2019b) 2018 facts and figures.

European Commission (2020) Use of international units. EUROPA. https://ec.europa.eu/clima/pol icies/ets/credits_en. Accessed 15 Apr 2020

Federici S, Lee D, Herold M (2017) Forest mitigation: a permanent contribution to the Paris agreement? Working paper

Gobierno de la República (2015) Intended nationally determined contribution

Gren M, Aklilu AZ (2016) Policy design for forest carbon sequestration: a review of the literature. Forest Policy Econ 70:128–136

ICAP (2020) Emissions trading worldwide: status report 2020. International Carbon Action Partnership, Berlin

Karpas E, Kerr S (2011) Preliminary evidence on responses to the New Zealand forestry emissions trading system. Motu working paper no. 11–09. Motu working paper. Motu Economic and Public Policy Research.

Kingi T (2008) Maori landownership and land management in New Zealand. Making Land Work 2:129–151

Leining C, Kerr S (2018) Guide to the New Zealand emissions trading scheme. Report for the Ministry for the Environment (Motu, 2018)

Liu Z, Kontoleon A (2018) Meta-analysis of livelihood impacts of payments for environmental services programmes in developing countries. Ecol Econ 149:48–61. https://doi.org/10.1016/j. ecolecon.2018.02.008

Manley B (2016) Afforestation responses to carbon price changes and market certainties. University of Canterbury School of Forestry

MEXICO2, Environmental Defense Fund, IETA (2018) Mexico: a market based climate change policy case study

MEXICO2 (2019) México publica el límite de emisiones y asignaciones sectoriales para la fase piloto del ETS. MEXICO2. http://www.mexico2.com.mx/noticia-ma-contenido.php?id=420. Accessed 20 Apr 2020

Ministry for Primary Industries (2015a) Penalties and offences under the emissions trading scheme. Ministry for Primary Industries

Ministry for Primary Industries (2015b) Guide to the permanent forest sink. Ministry for Primary Industries

Ministry for Primary Industries (2017a) Offsetting deforestation of Pre-1990 forest land. MPI. https://www.mpi.govt.nz/growing-and-harvesting/forestry/forestry-in-the-emissions-trading-scheme/offsetting-deforestation-of-pre-1990-forest-land/. Accessed 12 Apr 2020

Ministry for Primary Industries (2017b) A guide to carbon look-up tables for forestry in the emissions trading scheme. Ministry for Primary Industries

Ministry for Primary Industries (2018a) A guide to the field measurement approach for forestry in the emissions trading scheme. Ministry for Primary Industries

Ministry for Primary Industries (2018b) One billion trees programme, our future our billion. New Zealand Government

Ministry for Primary Industries (2020a) New Zealand's forests. MPI. https://www.mpi.govt.nz/ growing-and-harvesting/forestry/new-zealand-forests-and-the-forest-industry/new-zealands-for ests/. Accessed 12 Apr 2020

Ministry for Primary Industries (2020b) New Zealand's forest industry. MPI. https://www.mpi.
govt.nz/growing-and-harvesting/forestry/new-zealand-forests-and-the-forest-industry/new-zea
lands-forest-industry/. Accessed 12 Apr 2020

Ministry for Primary Industries (2020c) Emissions trading scheme improvements. MPI. https://
www.mpi.govt.nz/protection-and-response/environment-and-natural-resources/emissions-tra
ding-scheme/emissions-trading-scheme-improvements/. Accessed 12 Apr 2020

Muñoz-Piña C, Guevara A, Torres JM, Braña J (2008) Paying for the hydrological services of
Mexico's forests: analysis, negotiations and results. Ecol Econ 65(4):725–736

New Zealand Forest Owners Association (2019) New Zealand plantation forest industry facts and
figures. NZFO Association (Ed.), 2018, 19

New Zealand Government (1991) Resource Management Act. New Zealand Government

New Zealand Government (2019) Climate Change Response (Emissions Trading Reform) Amendment Bill: Explanatory note. New Zealand Government

Osborne T (2015) Tradeoffs in carbon commodification: a political ecology of common property
forest governance. Geoforum 67:64–77

Ramírez Bautista AJ, Maldonado Martínez MA, Morales Paredes G (2016) Guía de Usuario Registro
Nacional de Emisiones (RENE), Para el Reporte de Emisiones de Compuestos y Gases de Efecto
Invernadero. SEMARNAT-GIZ

Ranero A, Covaleda S (2018) El financiamiento de los proyectos de carbono forestal: Experiencias
existentes y oportunidades en México. Madera y bosques, 24(spe)

Reyes JA, Gómez JP, Muis RO, Zavala R, Ríos GQ, Villalobos O (2012) Atlas de Propiedad Social y
Servicios Ambientales en México. Instituto Interamericano de Cooperación para la Agricultura.
Cooperación Técnica Registro Agrario Nacional - Instituto Interamericano de Cooperación para
la Agricultura. México. 157 pp

Rontard B, Hernández HR, Robledo MA (2020) Pagos por captura de carbono en el mercado
voluntario en México: diversidad y complejidad de su aplicación en Chiapas y Oaxaca. Sociedad
y Ambiente 22:212–236

Runge-Metzger A, Wehrheim P (2019) Agriculture and forestry in the EU's 2030 climate target.
Towards a climate-neutral Europe. Routledge, pp 165–179

SEMARNAT (2019) Programa de prueba del sistema de comercio de emisiones. 27 de noviembre de
2019. SEMARNAT. https://www.gob.mx/semarnat/acciones-y-programas/programa-de-prueba-
del-sistema-de-comercio-de-emisiones-179414/. Accessed 11 May 2020

SEMARNAT (2020) Ilicitos forestales. SEMARNAT. http://dgeiawf.semarnat.gob.mx:8080/ibi_
apps/WFServlet?IBIF_ex=D4_R_PROFEPA01_03&IBIC_user=dgeia_mce&IBIC_pass=dge
ia_mce. Accessed 11 May 2020

Sims KR, Alix-Garcia JM (2017) Parks versus PES: evaluating direct and incentive-based land
conservation in Mexico. J Environ Econ Manag 86:8–28

United Nations (2015) Paris agreement

United Nations Environment Programme (2019) Emissions gap report 2019. UNEP, Nairobi

UNFCCC, U. N. (2009) Kyoto protocol reference manual on accounting of emissions and assigned
amount. eSocialSciences

van Kooten GC (2017) Forest carbon offsets and carbon emissions trading: Problems of contracting.
Forest Policy Econ 75:83–88

Chapter 10
Non-additionality, Overestimation of Supply, and Double Counting in Offset Programs: Insight for the Mexican Carbon Market

Marcela López-Vallejo

Abstract Mexico utilizes an emissions trading system as one of its carbon pricing instruments. Mexico's planning, like that of other countries, includes flexible mechanisms such as offsets. Offsets allow market participants to compensate for their emissions through mitigation projects. Offsetting via participation in the Clean Development Mechanism and Joint Implementation was fundamental to the Kyoto Protocol. In contrast, the Paris Agreement is ambiguous about its use. Other national or regional offset programs, such as the EU, Australia, New Zealand, Japan, or Korea, work within emission trading systems. Subnationally, the California-Quebec program has been in effect since 2014. As Greenhouse Gases (GHGs) are global, offsetting allows market participants to compensate for their emissions through mitigation projects, whether domestically or abroad. Given their global scope, such programs present a wide variability in quality. This chapter presents an overview of offset programs worldwide and argues that non-additionality, overestimated supply, and double counting are their three most pressing quality problems. This analysis sheds light upon the nascent Mexican system and its offset program.

Keywords Mexico · ETS · Offsets · Supply · Additionality · Double counting

Introduction

Mexico takes part in more than 40 carbon pricing initiatives that utilize Emissions Trading Systems (ETS) (World Bank 2019: 13). A common instrument for extending mitigation options is *offsetting*, which seems to have regained importance after the 2015 Paris Agreements. ETS and offset programs were designed on a global scope in the 1990s by the United Nations through a regime supported by the United States Framework Convention on Climate Change (UNFCCC) and the Kyoto Protocol (KP) (Meckling and Hepburn 2013; Egenhofer 2013).

M. López-Vallejo (✉)
Departamento de Estudios del Pacífico (CUCSH), Centro de Estudios Sobre América del Norte, Universidad de Guadalajara, Zapopan, Jalisco, Mexico
e-mail: marcela.lopezvallejo@academicos.udg.mx

© The Author(s) 2022
S. Lucatello (ed.), *Towards an Emissions Trading System in Mexico: Rationale, Design and Connections With the Global Climate Agenda*, Springer Climate,
https://doi.org/10.1007/978-3-030-82759-5_10

191

The KP allowed for Annex I countries to acquire Certified Emissions Reductions (CER) through three market-based instruments: an emissions trading system and two offset programs, Clean Development Mechanism (CDM) and Joint Implementation (JI) (Kyoto Protocol 1997). In the KP ETS, one country could reach its mitigation goals by transferring part of its assigned emissions to other countries that had fewer and could thus compensate for their pollution. Offset programs promoted the development of climate-mitigation projects in other nations. Under CDM and JI, polluting countries, especially developing nations, paid for these projects (Egenhofer 2013: 359; Meckling and Hepburn 2013: 476).

Although pricing initiatives and ETS are subject to criticism and face many detractors among scholars, practitioners, and NGOs (Swyngedouw 2016; Alcock 2008; Monbiot 2006), they are for others an essential part of the public policy toolkit for addressing climate change (Egenhofer 2013: 359; Ellerman et al. 2010). ETS seem to have some advantages over other pricing strategies, such as taxes. Martínez (2019) and Rabe (2018) make the case for ETS as they are useful in mitigating GHG and meet an amount of country reduction targets without using public resources or politicizing price-setting. ETS are also perceived as a socio-environmental technology-transfer mechanism among countries or market participants (Martínez 2019; Rabe 2018). Rabe (2018: 8) notes that ETS and cap-and-trade systems deliver the exact level of emissions reductions by enforcing non-compliance penalties. This helps offer jurisdictions predictability when it comes to advancement toward their targets. That is, ETS participants are informed of short-term expectations and long-term adjustments they need to make.

Apart from the KP ETS, there are other systems with regional, national, or subnational approaches to using offset credits. The European Union (EU) pioneered the use of ETS and offsets, putting this system in place to comply with its KP commitments. Other countries followed and designed their own systems, such as Australia, New Zealand, Japan, and Korea, among others (Egenhofer 2013). Although the most commonly used mechanism for ETS is cap-and-trade, other schemes are also in effect. Australia uses a baseline-and-credit system and the Canadian province of British Columbia works within a baseline-and-offsets structure.

Subnationally, Quebec and California inaugurated their joint ETS in 2014. This was later the basis for other Canadian provinces to design individual schemes, aiming to join this regional system to comply with federal legislation. In the US, the Regional Greenhouse Gas Initiative has served as an ETS since 2011 and, as in Canada, other states are developing their own ways to join the regional initiative (Rabe 2018; López-Vallejo 2014). China's subnational ETS pilots informed and evolved into a national market (Zhang and Zhou 2020). Some new national initiatives, like the Mexican carbon cap-and-trade system, are starting to pilot ETS schemes. All national and subnational ETS have offsetting programs. Some jurisdictions have even reshaped their ETS in attempts to adapt to their Nationally Determined Contribution as submitted to the Paris Agreement. This is the case of the EU, New Zealand, Kazakhstan, Regional Greenhouse Gas Initiative (RGGI), and California (World Bank 2019). Despite these efforts, carbon prices are still too low to comply with the Paris Agreements. In 2019, prices ranged from \$1USD to \$35USD a tonne of CO_{2eq}

(Broekhoff et al. 2019: 9; World Bank 2019). Experts note that if the world truly aims to comply with the Paris Agreement, offset programs should disappear slowly over time, as ambition requires achieving net-zero GHG emissions in this century (Dufrasne 2018).

Until the world achieves said neutrality, offset programs are useful to complement climate regulation or ETS. As GHGs are global and it does not matter where exactly they are reduced (Broekhoff et al. 2019), offsetting allows market participants to compensate for their emissions through mitigation projects domestically or abroad. These account for quality variability or how the projects preserve environmental integrity (Broekhoff et al. 2019: 18). Experts categorize them in Type A being focused and setting specific policy targets, whereas Type B help other ETS policies as they have broader goals. Type C is the most comprehensive; it influences other climate or clean energy policies (Gillenwater 2012; PMR 2015; Michaelowa 2011). These help countries to reach global GHG targets at lower costs to society than regulation policies do and can function as a technology and funding transfer mechanism (Egenhofer 2013: 359). Mexico's trading system utilizes offsets as a flexibility mechanism.

This chapter contends that offset programs usually present three major quality problems: *non-additionality, overestimated supply*, and *double counting* (Broekhoff et al. 2019; Michaelowa 2011). It further explains these problematic issues and gives recommendations to try to prevent them in the nascent Mexican market. The first two sections present an overview of offset programs worldwide, their characteristics, and scope. The third section analyses why non-additionality, overestimated supply, and double counting are problematic issues for offset programs. In the fourth part, these three issues are discussed along with the current structure of the Mexican market trials. Lessons learned by other programs need to be taken into account for the success of the Mexican initiative. This chapter then concludes with the final section offering recommendations.

Overview of Offset Programs Worldwide

An *offset* is a mechanism compensating for emissions by investing in environmental projects beyond regulated participants or in other market jurisdictions (World Bank 2019; PMR 2015; Egenhofer 2013; Meckling and Hepburn 2013; Fujiwara and Egenhofer 2007). An *offset credit* is a "transferable instrument certified by governments or independent certification bodies to represent an emission reduction of one metric tonne of CO_2, or an equivalent amount of other GHGs" (Broekhoff et al. 2019: 6). Offsets work when ETS participants pay an extra quota to compensate for greenhouse gas emissions from specific projects or standards (Broekhoff et al. 2019; PMR 2015; Egenhofer 2013). They can compensate for individual or companies' entire pollution or for specific sector caps. They function for example when a company makes up for its emissions by financing reforestation, transportation, ecotourism, agriculture, waste, buildings, or clean energy projects elsewhere.

As Broekhoff et al. (2019: 8) suggest, offset programs worldwide have three goals: (1) to develop and approve quality eligibility criteria or standards for offset credits, (2) to develop registries of projects and assess them against these criteria, and (3) to operate the credit transfers. The use of offsets in ETS lowers costs of compliance with socio-environmental policies or regulations by allocating additional funds to specific domestic or international projects (Martínez 2019; Matsuki 2015; Fujiwara and Egenhofer 2007). If international, the range of options is wider but more difficult to standardize and more costly to operate. When offsetting is performed domestically (regionally, within the jurisdiction, or within a sector), compliance costs can be lower, encouraging non-capped participants to move toward decarbonization in a controlled environment (Fujiwara and Egenhofer 2007: 19).

Offset programs range from international intergovernmental to those run by national or subnational governments and to voluntary efforts, generally operated by non-governmental institutions. Of these, some are independent and others are linked to ETS with cap-and-trade systems. The two most important international and intergovernmental offset programs are the Clean Development Mechanism (CDM) and the Joint Implementation (JI) in operation through the Kyoto Protocol. They work globally, and participants may include countries that are official members of the protocol, along with private or voluntary buyers (PMR 2015; Egenhofer 2013; Marcu 2012). As the PMR (2015) reports, the CDM offered Annex I countries offset projects to meet their specific KP targets. The offset credits were allocated to developing countries that had ratified the protocol. The JI linked Annex I countries to help meet their reduction targets.

Together, these programs accounted for the majority of offsetting practices worldwide (PMR 2015). Nonetheless, CDM and JI present serious problems. Research by Cames et al. (2016: 11) suggests that around 85% of offset credits from the CDM up to 2012, and 73% of the 2013–2020 projects may not have led to real emission reductions, especially with industrial gas destruction and other such projects in the energy sector; this may have resulted in an increase of roughly 600 million metric tonnes of emissions through 2015.

Since the Paris Agreement (2015) was ambiguous regarding their use, the permanence of CDM and JI has been debated. Advocates for environmental integrity and ambition tend to suggest discarding them (Carbon Market Watch 2019; Dufrasne 2018). At COP25 in Madrid in 2019, Article 6 of the agreement (addressing voluntary cooperation approaches, such as ETS and offset programs) was still pending. Negotiators could not agree upon several issues. Article 6.8 includes non-market cooperation mechanisms, which are not yet defined. Article 6.2 was more controversial, as it refers to helping reduce NDC emissions through cooperative approaches involving *Internationally Transferred Mitigation Outcomes* (ITMOs). However, the Paris Agreement parties could not reach a consensus for designing a trustworthy global offset accounting system. The main goal of a solid system is to prevent double-counting practices common in the CDM and JI, where emission reduction figures were counted simultaneously by both cooperation partners (Alloisio 2020; Environmental Defense Fund 2019; Schneider and La Hoz Theuer 2018; Gehring and Phillips 2017).

The third source of the debate was Article 6.4, which establishes a Sustainable Development Mechanism (SDM), to be supervised by a body determined by the COP. A third party could guarantee that offset projects met COP criteria or standards (Alloisio 2020; Gehring and Phillips 2017). COP also facilitates direct access for offsetting SDM to the private sector and other social actors.

The second type of offset programs is either national or subnational, generally linked to mitigation policies and cap-and-trade systems. These programs can be found in the European Union, Australia, China, California, Quebec, Nova Scotia, Alberta, Switzerland, and Japan (Broekhoff et al. 2019: 9; PMR 2015; Egenhofer 2013: 365–367). For example, the two North American carbon markets, California-Quebec (called the WCI market, which, since 2019, also includes Nova Scotia) and Regional Greenhouse Gas Initiative (RGGI), utilize offsets to promote projects within their member jurisdictions (Rabe 2018; López-Vallejo 2014). In 2020, the percentage of offsets allowed in these ETS ranged from 8 to 12% and prices from $8USD to $18USD per tonne. Offsets under such programs usually address sectors not covered by other mitigation policies and ETS (PMR 2015).

The third type of offset programs is non-governmental or voluntary, such as Gold Standard (GS), Climate Action Reserve (CAR), or Verified Carbon Standard (VCS). GS was established in 2003 by the World Wildlife Fund and a coalition of other NGOs to promote sustainability in CDM projects (Gehring and Phillips 2017; Meckling and Hepburn 2013). CAR proposes specific protocols for various project types. These may involve coal mine methane, forests, grasslands, nitric acid, nitrogen management, organic waste digestion, rice cultivation, urban forest management, and urban tree planting (Gehring and Phillips 2017: 5). VCS includes a set of parallel standards to generate verified carbon units for emissions reduction (Egenhofer 2013).

These programs complement governmental carbon markets. For example, GS aids CDM and JI voluntary programs; CAR and VCS verify the California-Quebec ETS (PMR 2015). California utilizes voluntary verification, monitoring, and compliance with its offset programs, the American Carbon Registry Standard (ACR) (Gehring and Phillips 2017: 5). Such voluntary programs have two main features: they are verifiable through constant auditing, guaranteeing that offset projects work as they promised. They also require projects to offer social benefits at a local level (PMR 2015). Some voluntary programs are linked to others, such as REDD+, which has caused debate as REDD+ projects are usually difficult to verify.

Characteristics of Offset Programs and Projects

There are significant differences among offset programs. Gillenwater (2012) proposes that they be substantially categorized, with their goals and scope specified. Table 10.1 shows three types of programs and their characteristics.

In general, Type A programs have specific targets and are less ambitious but more focused: KP offset programs are examples of this. Newly developed programs, like those nascent in Mexico, tend to start with this approach. Type B shares Type A

Table 10.1 Types of offset programs according to their goals

	Goal of offset program	Operative characteristics	Pricing	Allocation
Type A	Giving income or funds through offset credits and sending a pricing signal	Resembling a subsidy	Depending on the supply and demand of credits (stakeholders' proposals for projects)	• Where it can make the most significant change of behaviour • Sector-based • Can be standardized or project based
Type B	Promoting other ETS policy instruments directly linked to the offset program	Direct funding plus: • Capacity building, technical, educational, legal, and financial support • Enforcement instruments • Recognition of advances or shaming programs	Uncertainty of subsidies depending on market, but certainty on concrete projects related to other policy instruments Tending to lower transaction costs of offsets and ETS policies	• Financial and non-financial factors are taken into account • Allocation to offset program but with broader ETS policy criteria
Type C	Generating broad market effects	New market developments (other than baselines)	Including non-offset prices (trying to prevent leakage or non-compliance)	Within the offset program and Indirectly outside the offset program (spillover effect)

Source Author's elaboration with information from Gillenwater (2012), PMR (2015), Michaelowa (2011)

features but goals are broader, as programs aim to aid other ETS policies. They offset emissions and develop local renewable markets: CAR and VCS are linked, for example, to Renewable Portfolio Standards as in the California-Quebec offset program. National programs tend to fall within this type. Type C is the most complex as it includes both A and B. It aims to modify participant behaviour and impact ETS, policy instruments, and other social agents. Some programs, for example, require projects to have social co-benefits, as in the EU program or the proposed version of the SDM. Voluntary programs, such as GS, even strive to implement the Sustainable Development Goals.

Within offset programs, quality assessment is fundamental to overcoming problems. The quality of projects is generally measured by how well they prevent practices of non-additionality, overestimated supply, and double counting. As the baseline is generally established by business, as usual, it is critical to assess which projects contribute more to an offset program and environmental integrity. Table 10.2 summarizes the most common types of projects and what experts perceive to be their main strengths or weaknesses.

Table 10.2 Type of projects and quality

Type	Quality[a]	Co-benefits	Risks
Renewable energy (small scale)	High	Reduced air pollution/off-grid electrification	Baseline uncertainty easily addressed by eligibility criteria/big investment and uncertain GHG reduction
Energy efficiency (household)	High	Prompt energy transition/lower costs of energy	Baseline uncertainty easily addressed by eligibility criteria/covered by regulation or industrial standards already
Methane destruction	High	Reduced air pollution (and odors) in localities	Baseline uncertainty easily addressed by eligibility criteria/normally covered by regulation
Energy distribution	Medium	Air quality upgrade/connect off-grid communities	Capital intensive projects/covered by regulation already
Renewable energy (large scale)	Medium	Help consolidate industrial change	Already covered by regulation and baselines/local social and environmental disruption
Methane capture or utilization	Medium	Energy generation and benefits to health	Baseline uncertainty addressed by rules and eligibility criteria/could be seen as supporting polluting industries
Industrial gases avoidance (PFCs and SF$_6$)	Medium	High probability to be covered by regulation	Overproduction to attract credits/no incentive to stop polluting
Energy efficiency (industrial)	Low	Involve industry/slow energy transition	Support of polluting industries/not meeting the Paris Agreement goals
Fossil fuel switching (to gas)	Low	Prevent coal and oil	Slow energy transition/not meeting the Paris Agreement goals
Forestry and land use	Low	Local social benefits (involvement of people)/provide ecosystem services	Long-term results and difficulty to measure in time/risk of reversal or non-permanence of projects
Agriculture	Low	Improvement of technology for local people	Linking to environmental problems (water pollution and scarcity, or deforestation)

(continued)

Table 10.2 (continued)

Type	Quality[a]	Co-benefits	Risks
Biomass	Low	Beneficial use of waste and renewable energy production	Indirect reduction of industry (double counting)/problems assessing land use
Fugitive gas capture	Low	Helps with energy transition	Does not prevent the use of fossil fuels/does not aim for the Paris Agreement goals
Low-carbon transportation	Low	Improve air quality	Mitigation costs above offset prices

Source Author's elaboration with information from Broekhoff et al. (2019), Gehring and Phillips (2017)

[a]Quality is measured by the volume of GHG reductions or removals that are additional, not overestimated, permanent, not claimed by another entity, and not associated with significant social or environmental harms (Broekhoff et al. 2019: 18)

Although general categorizing may be deceptive, it provides a broad view of types that meet benchmarks for quality. Establishing project eligibility criteria needs to be consistent with program goals (Table 10.1): achieving environmental integrity is more likely with less ambitious programs (e.g. Type A) and higher-quality offset projects. Quality projects may foster corporate social responsibility, consolidate community development, or promote the provision of socio-environmental public goods by governments (Broekhoff et al. 2019). Sometimes a buyer must choose lower project quality such as when profitability may be higher, or compliance with environmental criteria greater, and co-benefits less extensive (Broekhoff et al. 2019: 33). Michaelowa (2011: 19) explains that countries or participants must sometimes opt for lower quality projects due to domestic industry pressures affecting international competition: in such cases, the best choice would be technologies in other than directly competitive ways.

Critiquing Offset Programs

Many experts have strong criticism for offset programs. Some argue that they are an excuse for business as usual because they fail to create incentives for changing behaviour toward climate mitigation (Monbiot 2006). In 2019, during COP25 in Madrid, civil society and some NGOs protested against carbon markets and offsets. Their argument was that carbon offsets do not prevent pollution, but simply relocate it. In other words, offsets serve as "greenwashing" mechanisms, locking in high-emitting activities over the long run and discouraging regulation (Broekhoff et al. 2019). Other critical views see them as relying on subjective criteria and methodologies (Broekhoff et al. 2019: 21; Millard-Ball and Ortolano 2010) or as insufficient for encouraging developing countries into decarbonization (Wara and Victor 2008). In this sense, offsets are considered environmental externalities, mitigating in one place while transferring pollution elsewhere.

In 2019, the UN even drafted a harsh critique of traditional offsetting practices. The report by Niklas Hagelberg of the United Nations Environmental Program claimed that offsets generally offer free passes to polluters, at unrealistically low cost. To function as a complementary source of mitigation, "they should change the equivalence of 1MT to 1 offset credit and factor it by the percentage of GHG emissions decrease necessary (45% reduction) to achieve ambition under the Paris Agreement" (Hagelberg 2019). Supporting this argument, Broekhoff et al. (2019: 13) suggest that, although meeting carbon neutrality goals is desirable, there is the risk of "masking" this achievement by relying on carbon offsets as the primary reduction source. Instead, institutions need to view carbon offsets as merely additional measures for achieving neutrality by 2050, as pledged by the Paris Agreement. Offset price increases would set necessary thresholds to further change environmental behaviour (Gillenwater 2012).

A more substantial critique deals with project quality within offset programs. In theory, offsets should guarantee that projects reduce GHG. Well-designed and applied reduction measurement methodologies are fundamental. This argument posits three points regarding evidence to challenge offset programs worldwide. First, they may fail to perform as additional mitigation measures because they are part of existing project designs. If additionality cannot be demonstrated, such criticism by sceptical experts is valid. *Additionality* means that a project needs to demonstrate that it complements other efforts and that it would not have happened without offset funding. Quality also relates to program operation: for projects to succeed, there must be a pool of potential mitigation spaces, technologies, or options. When there is *overestimated supply* and leakage, offset programs tend to fail. Another quality issue is how projects and GHGs are counted. Sometimes, the same project is counted in different offset systems at different prices. *Double counting* creates confusion, interferes with prices which can derail the cap-and-trade system, and undermines environmental integrity.

In sum, there is no one-size-fits-all for offset programs, and there are several conditions that affect their design: examples are the scope, market segment, regulatory framework, institutional setting, and technical capacities to operate them (Matsuki 2015). The next section discusses in detail the three quality issues challenging offset programs: non-additionality, overestimation of supply, and double counting.

Problematic Issues: Non-additionality, Overestimating Supply, and Double Counting

The thorniest issues for offset programs arise from project quality. Unclear definitions of how additional they are, over- or underestimation of supply, and double-counting practices are discussed in this section. Gillenwater (2012) and Meckling and Hepburn (2013) suggest that we need better models and understanding to heighten the benefits offset systems offer over alternative pricing policies (e.g. taxes). If offsets do not meet quality criteria, they will be of no use in the context of the Paris Agreement (Dufrasne 2018).

Non-additionality

Additionality means that an offset project differs from its GHG baseline and is "additional" to expected emission reductions from any regulation or ETS cap (Gillenwater 2012: 26; Michaelowa 2011: 18–19). In other words, a project is additional when it is prompted by the offset program, not by policies or other factors (e.g. technological advances, incoming investment and projects from external agreements, or new governmental approaches). The projects most often considered additional are those which are not expected to attract investors or governmental funds, are difficult to finance due to technical reasons, are innovative and not considered common practice, have financing sources, face implementation gaps, and that are not mandated by regulatory entities (Broekhoff et al. 2019: 21; Gehring and Phillips 2017: 4). Broekhoff et al. (2019: 19) warn of the common mistake of categorizing a project as additional if it reduces GHG emissions beyond what they would have been without the project. Aside from emissions reduction, the main criteria for using offsets as additional to regulation and ETS is that without the credit a project could not be undertaken at all.

Assessing the additionality of projects can be problematic in two senses. First, offset programs need to demonstrate that they can cause a change in GHG mitigation behaviour via awarded projects. Several CDM offset projects initially failed to provide evidence of additionality, which delegitimized the offset approach (Wara and Victor 2008; Schneider 2007).

There are different approaches for defining additionality worldwide. Michaelowa (2011: 17–18) compares how the EU, the US, and some developing countries confront this. The EU, he explains, has strict definitions using investment tests and ambitious technology benchmarks. In contrast, the US industry favours robust general tests and flexible technical approaches. Least developed countries and islands foster strict additionality which may be effectively reflected in mitigation. Heavily industrialized developing countries (e.g. China and India) adopt a flexible concept of additionality to keep profiting from offsetting.

What is additional and what is not? Broad scope offset types (e.g. Type B and especially Type C, Table 10.1) would need to reject projects which overlap with other policies. This is difficult in the context of an NDC pledge. In contrast, as Type A programs are more specific, they tend to limit the allocation of large-volume offset credits and clearly define GHG reduction calculations (PMR 2015: 6). This proves useful for market participants as it lowers the risk.

Assessing additionality also varies by whether the program is project-specific or standard-focused. Project-specific approaches contextualize additionality, making the definition of objective criteria difficult. In contrast, approaches relying on standards evaluate smaller sets of projects under pre-determined eligibility criteria, reducing subjectivity. CAR utilizes a standardized approach and has developed 20 protocols, contrasting with the VCS and Gold Standard which use over 200 project-specific methodologies and protocols (Broekhoff et al. 2019: 21).

The second problematic issue regarding additionality in project-by-project approaches is how to establish a *baseline*. A baseline predicts the quantity of emissions that would have happened in the absence of the offset credit, holding all other factors constant (Broekhoff et al. 2019: 23; Gillenwater 2012: 26). In other words, "a project's GHG reductions are quantified by comparing the actual emissions that occur after the project is implemented to its predicted baseline emissions" (Broekhoff et al. 2019: 23). Baselines are produced by negotiation between stakeholders (Michaelowa 2011: 18), which generates uncertainty among participants of carbon markets and can transform one additional project into non-additional with time or contextual changes.

Subjectivity project assessment creates multiple methodologies. Michaelowa (2011: 19), for example, notes that countries with environmental high standards might ask for datasets to define baselines, while host countries of participants working under less stringent environmental rulings will prefer simpler requirements. Although contextualizing projects might be a good practice for developing local capacities, it can be challenging for meeting quality standards, especially when the offset market grows (Gillenwater 2012: 14).

Another fundamental element to assess projects is time and the duration of crediting criteria. Time and permanence of projects matter, as they directly impact the volume of offset delivery (Michaelowa 2011: 19). Additionally, the timing of credit releases may influence project quality. Releasing credits for sale ahead of actual emission reductions or baselines may harm the project (Gehring and Phillips 2017: 4). As Table 10.2 suggests, different projects represent different risks and benefits. For example, reforestation generally works under long-term schemes (100 years or more) to effectively deliver the absorption volume from an offset program. Normally, offset programs have pre-defined timelines where policy intervention remains additional in time (Broekhoff et al. 2019; Gillenwater 2012). The peril of long-term projects is that they may lock in policy and technological innovation, making it difficult to verify and prevent leakage. This is why the EU banned forestry and land-use credits (World Bank 2019; PMR 2015). Facing new technologies, unexpected events (political or environmental), and new policies (global, national, or local) may require periodically reassessing the baseline to arrive at a flexible timeline or, as in Japan, none at all (World Bank 2019).

Offset providers rely upon certain strategies to prevent non-additionality. The first one is a robust methodological approach, most commonly dealing with (1) the documentation of alternative scenarios to the proposed project, (2) assessment of the financial obstacles projects encounter and how offset credits may help overcome them, and (3) a pre-determined catalog of projects (PMR 2015: 6). The second strategy is to implement crediting systems by sector, which may help prevent subjectivity and diverse methodologies when assessing offset project additionality, especially for jurisdictions not able to cap the entire economy, but aiming to evolve into more comprehensive trading systems. This strategy credits emissions reductions from a covered sector against a threshold, where credits are granted to projects initiated below a certain level (Egenhofer 2013: 366; Michaelowa 2011: 32). This is aided by establishing a threshold that can be expressed in absolute emissions, carbon intensity, or technological transfer (Fujiwara 2009: 44). The third strategy is for offset providers to guarantee their emission savings over time. If the project becomes non-additional, the provider promises to compensate by developing another project. Clark (2009: 47) showed that "as the offset market grows, some offset companies have enough capital to invest in projects speculatively: they fund an offset project and then sell the carbon savings once the cuts have actually been made".

Overestimating Supply

There are three ways in which the supply of projects can be problematic: (1) overestimating the GHG emissions reduction, (2) not having quality projects which grant socio-environmental co-benefits or even impact negatively in localities (Broekhoff et al. 2019: 23–24), and (3) not having enough projects for the offset programs. Overestimation occurs when the baseline is miscalculated and it establishes more potential reductions than they really are. Overestimation is also present when a project fails to account for leakage. Leakage occurs when taking care of one forest implies that agricultural activities just moved to some other area that will be deforested. A study by Haya (2019) suggests that 82% of the CARB offsets for forestry (36 projects) might present some sort of leakage and have been over-credited. Another source of leakage was N_2O, which relocated its production from the EU to developing countries, where CDM credits were more profitable (Michaelowa 2011: 29).

Preventing socio-environmental disadvantages is crucial for quality project development. For example, some need prior social consultation or even insistence upon co-benefits to localities (Broekhoff et al. 2019: 30). To help projects meet such qualifications and guarantee implementation requires participatory processes with indigenous peoples, local communities, international experts, and civil society. As Gehring and Phillips (2017: 4) note, "consultation is a key factor for nearly all top-level certification schemes". Further, quality offsets may even promote network cooperation among various participants. Martínez (2019: 252–253) asserts that collective work can address this purpose, where business, social entities, small-communitarian associations, or indigenous peoples cooperate in developing wind and solar energy projects, i.e., sharing property, management, and benefits. For example, the organization VERRA certifies that offset projects meet the *Climate, Community, and Diversity Standard* (*CCB*) by assessing land management projects which benefit climate-change mitigation along with local community development and biodiversity. It also helps CARB projects follow the right sustainable protocols (VERRA 2020).

Oversupply of projects is also dangerous. When an offset program allocates excess credits, prices tend to fall. This can be harmful when the system requires a fixed percentage of offsets: it may derail the program, especially in economic or global crises when emissions naturally tend to be reduced (Rabe 2018; Öko-Institut e.V. 2018). Flexibility may solve this issue by adjusting the emissions mitigation percentage the program covers. In other words, it might be necessary to temporarily or permanently withdraw a certain number of project offerings, as the EU ETS does, to keep the price as stable as possible (Michaelowa 2011: 30).

In contrast, under-supply deals with a poor quality offer of projects. Offset participants and jurisdictions tend to seek the most profitable, sometimes low-quality projects. As shown in Table 10.2 of the previous section, there are very few project types meeting quality criteria. Because of low quality, the EU ETS reduced its share of CDM as 2012–2020 offset options and banned credits in 2013 for certain mitigating activities, including capturing and destroying GHG emitted by landfills and feeding farm animals, or offsets from industrial pollutants (Meckling and Hepburn 2013: 482; Michaelowa 2011: 16). In a race to the bottom, companies participating in the EU offset program overproduced large amounts of HFC-23 and asked for credits to develop HFC-22, an only slightly less polluting gas (Schneider and La Hoz Theuer 2018).

Supply can also be adjusted when global crises emerge. For example, this is the case of the aviation industry during the COVID-19 pandemic. As Lang (2020: 1) notes, "it was expected that airlines bought offsets above a baseline set by the average of the aviation industry's emissions in 2019 and 2020. Because of the coronavirus pandemic, global air travel has fallen dramatically. As a result, the baseline is far lower than predicted, and airlines will have to buy far more offsets than anticipated". There is a debate over the need to include offsets from REDD+ into aviation (Carbon Pulse 2019; Egenhofer 2013; Yuvaraj 2011).

During the KP, there was an oversupply of projects; it was not difficult to find low-quality carbon offsets. The potential supply of GHG reductions was huge because there were so many GHG emission sources with no legal or economic incentives to change. With each country pledging to the Paris Agreement, however, quality project supply became complicated. The need to reach NDC goals or the existence of various offset programs worldwide may lead to the third problem, which is double counting.

Double Counting

When country or market participant "A" claims a certain emission reduction volume, it cannot count upon NDC mitigation registries or the cap of country or market participant "B". If it does, double-counting problems emerge: these are expressly prohibited by Article 4.13 of the Paris Agreement and clarified by its Article 6 (Broekhoff et al. 2019: 15–16; Obergassel and Asche 2017). Double counting, then, happens when two or more offset participants claim GHG reductions from the same project or when, through fraudulent practices or legitimate mistakes, accurate registration fails. This is the case in renewable energy projects, for example, when both consumer and producer, or the project and a power plant, claim the GHG reduction for the same clean electricity, and the offset program grants separate credits (Broekhoff et al. 2019: 25).

Defining the amount of GHG reduction per property clarifies credit ownership rights and may prevent double counting. There is a clear link between avoiding double-counting practices and project temporality (or permanence), where establishing a baseline and lifetime for the credit is fundamental. When credits are awarded, they should be retired from the market, or in the words of the Paris Agreement, the corresponding adjustments should be applied (Broekhoff et al. 2019: 16; Gehring and Phillips 2017: 4). If not, countries with NDCs would be tempted to reduce emissions through domestic policy while also selling a credit to a more polluting country (if technology came from there) for the same reduction. Robust counting methods and verification by third parties are needed to prevent "cheating" by splitting a credit in two. In other words, double counting implies reducing by half, which is said to have been mitigated (Carbon Market Watch 2019). Double counting can thus deter NDCs, as it did with the JI. JI had double-counting problems where both participants were committed under KP and claimed the same credit as their own (Elsworth et al. 2012; Elsworth and Worthington 2010). Among other reasons, this is why several voices are raised against using old KP mechanisms such as CDM, within the context of the Paris Agreement (Carbon Market Watch 2019; Environmental Defense Fund 2019; Dufrasne 2018).

Clark (2009) warns of another more subtle type of double counting. It occurs when different offset companies quote different prices for the same credit. This happens, he explains, because of overestimating GHC supply, as was discussed above. It may happen due to the project nature and type. If it is immersed in a Type C offset program, costs may be higher due to co-benefits. In this scenario, apart from external double counting (host country or sector with receiving partner), it can happen cross-sector, *within* the same ETS. Double counting can also happen within offset programs themselves when policies or standards are credited simultaneously with projects (Michaelowa 2011: 16).

One iconic example of the perils of double counting is the aviation sector. The International Civil Aviation Organization (ICAO) is working together with the Paris Agreement institutions to develop its Carbon Offsetting and Reduction Scheme for International Aviation (CORSIA). Starting in 2021, this industry is committed to compensating for any increase in its GHG emissions under a 2020 baseline (Broekhoff et al. 2019: 16; Dufrasne). CORSIA has an "open architecture" design where airlines can purchase and retire offset credits issued by ICAO. Verification will be performed by various certification programs working for either ICAO or project developers. Given the different methodologies and participants, this could create a double-counting problem. Who records mitigation? The airline, company purchasing the ticket, or individual passenger? To avoid such fragmentation, unified double-counting rules must be created to meet environmental integrity (VERRA

2017; Michaelowa 2011: 17–18). The fear of double counting for lack of coordinated methodologies was intensely discussed at COP25, especially when assessing the proposed inclusion of REDD+ projects into CORSIA.

To prevent double counting as established by Article 6 of the Paris Agreement, the Environmental Defense Fund (2019) proposed several exchanges among NDC countries with non-NDC and voluntary schemes for aviation (e.g. CORSIA) and the use of CDM. As COP25 in 2019 did not result in an agreement as to the functioning of carbon markets and how to deal with KP mechanisms, those protocols and rules are still pending. Robust double-counting prevention methods need to be put in place. Experts and international organizations (see Broekhoff et al. 2019; Environmental Defense Fund 2019; PMR 2015; Schneider et al. 2014) recommend preventing double counting with restrictive eligibility criteria, inventory-based accounting, emission balances (spare emissions vs. removals covered by conditional NDC), international accounting rules, tracking systems, and/or third-party verification.

Aside from recommendations under the Paris Agreement, there are other offset programs seemingly better suited to prevent double counting. For example, entities covered by the California and Quebec joint ETS can use 8% of their cap as offsets. This market prohibits issuing offset credits in sectors covered by ETS or those already under regulation in both jurisdictions (PMR 2015: 8).

To sum up, unclear additionality, problems with supply, and double counting are core issues that can derail any offset program and impact any ETS. New ETS and offset programs need to take into account these three issues before setting up operations. This applies to the nascent Mexican ETS, which includes an offset program. The next section explains how the pilot ETS has worked and how its offset program endeavours to face these three challenging issues.

MexiCO2 ETS and Offset Program

Mexico has participated in the UNFCCC since it was designed in 1992, ratifying all instruments and, though categorized as a Non-Annex country in the KP, setting pricing strategies for reducing emissions (e.g. taxes). The General Law on Climate Change (LGCC) enacted in 2012 offered a legal basis for developing policies, programs, and instruments to reach mitigation and adaptation goals. At COP21 in Paris, 2015, Mexico presented its NDC, which included two innovations. One disaggregated its goals into non-conditional (realistic goals the country could meet if policies were implemented) and conditional (needing external sources of funding to be put into place). The other innovation included black carbon in the covered commitments. The specific non-conditional commitments accounted for a reduction of GHG and short-term pollutants of 25% below business as usual by 2030 (22% of GHG and 51% of black carbon). Conditional goals account for 36% GHG and 70% of black carbon reduction by 2050 (with a baseline at the year 2000). Peak emissions are estimated for 2026 (MexiCO2 2019b).

In 2015, as part of a comprehensive energy and fiscal reform, Congress passed the Law for Energy Transition (LET) and other related laws. The LGCC and LET legally supported deploying carbon pricing instruments for reaching NDC goals; such as Clean Energy Certificates (CEC), carbon taxes, and ETS to start operating in 2023. Clean Energy Certificates (CEC) account for a certain amount of electricity generated from clean energy sources since 2014. This means that if a power plant generated clean energy before that year, they would not be able to obtain these certificates. Once clean energy is produced, its generators put CEC on the market; companies from consuming sectors needing to mitigate pollution could buy CEC to meet their obligations. CECs are auctioned by the National Center of Energy Control (CENACE) or may be traded on the spot market or through bilateral contracting.

Worth noting is that the federal government in power since 2018 made three decisions impacting the CEC market and the path to decarbonization. First, it cancelled any auctioning for electricity projects. In previous auctions, all awarded projects were for renewable energy deployment. Second, in October 2019, it changed the rules of CECs to include all Federal Electricity Commission (CFE) power plants (generating clean energy before and after 2014) to grant these certificates and put them on the market (DOF 2019b). Having been a monopoly for 70 years, CFE is thus guaranteed to receive the most CECs. Having CFE in the CEC market will result in oversupply, which will lower prices (García 2019). In 2016, CEC started at $25USD per MW; by 2018, the price dropped to $18USD and future estimates are not optimistic. The third decision came in April 2020. It stated that, due to the 2020 COVID pandemic and for energy-security reasons, all renewable private providers would need to suspend activities; only CFE would stand (DOF 2020). Renewable-energy companies immediately filed for judicial protection against abuses of public authority (under the legal figure called "amparo"). Companies such as Mexsolar I, Dolores Wind, among others, won a provisional suspension in the courts against the government's ruling in May 2020 (Elceo 2020). One month later, Greenpeace Mexico and Centro Mexicano de Derecho Ambiental (CEMDA) won a definitive suspension (CEMDA 2020). As expected, these decisions had several consequences, among which were derailment of the CEC market and arrested renewable energy deployment in the country.

Apart from CECs, Mexico has three carbon taxes as pricing mechanisms. Two fall under the umbrella of Special Taxes Upon Services and Production (Impuesto Especial sobre Producción y Servicios [IEPS]). *IEPS oil* taxes fossil fuel imports; *ISAN* covers new car purchases. Most interesting is *IEPS carbon,* which sets GHG prices for different types of fuel.[1] This is the only tax that can be paid through CDM offset credits. This model resembles that of South Africa, where 5 to 8% of carbon taxes can be covered by offset programs (Mehling and Dimanchev 2017: 24). Since 2018, the Mexican government has accepted CDM offsets to cover 20% of the tax payment, under certain conditions: they must be developed in Mexico and not emitted before 2014, they ought to be sold on the European Emissions Market, and they need to address post-Kyoto goals. Because of low prices (¢0.30USD per tonne), as of 2019, Mexican fiscal authorities had not received tax payments via offsets (MexiCO2 2019a).

The first phase of the carbon market started in 2017 with a simulation program with no obligatory participant information disclosure, real data, or economic impact; it was a "role playing" exercise. The LGCC was reformed in 2018 to set a timeline for the second phase and future implementation of the ETS. The second phase directs participants into market trials divided into two periods. From January 2020 to December 2021, there is a pilot testing; from January 2022 to December, there will be a transitional testing program. The goal of this two-phase period is to prepare participants and familiarize them with market dynamics. Having a long-term testing period (36 months) may evidence the pressure of the Mexican industry resisting committing to real GHG reductions in a carbon market in the short-term, arguing loss of competitiveness (Arteaga 2019; Flores 2019). The Mexican ETS utilizes a most-polluting-sector approach; including facilities with in situ CO_2 emissions of 100,000 tonnes for a given year after 2016. The participating sectors are as follows: fossil-fuel energy (deployment, production, distribution of oils; generation, transmission, distribution of electricity), large industries (automakers, cement, chemical, food and beverages, glass, mining, petrochemical, paper, iron, and steel), and other industries which emit from static sources. These sectors represent 45% of reported national emissions (DOF 2019a). The logic behind the decision is that energy companies and energy-intensive manufacturing regard ETS as imposing lower burdens than other pricing strategies (e.g. taxes) (Meckling and Hepburn 2013: 479).

The Mexican ETS includes two flexible mechanisms, offsets and early action. As Mehling and Dimanchev (2017: 30) note, having an offset program can help adjust the ETS cap and face short-term fluctuations and incoming policies. In the case of Mexico during ETS simulations, the offset program established that credits can cover up to 10% of emissions and cannot be linked to CECs (see results in Table 10.3). When the market officially starts operations in 2023, the inclusion of CECs will depend on their performance (e.g. in terms of prices).

Table 10.3 shows that the results of market simulations are very different, with the number of offsets having increased dramatically. This may be explained by price increases, which made offsetting a very competitive practice. When ETS implementation arrives in 2023, early action and offsets acquired during the trials may still be valid if no more than six months old. This is important because some early-action projects are already operating in Mexico through CDM or voluntary markets. As of 2016, there were 13 projects ranging from methane capture from landfills, to energy efficiency and wind and solar generation, to reforestation and sustainable forestry (see

Table 10.3 Offset results of the three-year market simulation

	Simulation 1 (2017)	Simulation 2 (2018)	Simulation 3 (2019)
Offsets awarded to government (tonnes)	6,899,943	13,676,755	37,000,000
Prices of offsets (Mexican pesos)	54	83	198

Source Author's elaboration with information from SEMARNAT (2018a, b), SEMARNAT (2017)

Annex 1). These projects have been developed through CDM and voluntary markets such as GS, VCS, and Plan Vivo (MexiCO2 2016). As of 2020, for example, CAR reported one landfill and more than 25 forestry projects (Climate Action Reserve 2020).

To operate, Chapter IV of the Mexican ETS basis document (DOF 2019a) mandates the Ministry of Environment and Natural Resources (SEMARNAT) as its governing authority, for designing offset protocols and grant credits. Protocols will set proceedings, requirements, and methodologies to quantify emissions reduction, absorption, and prevention in eligible projects. During ETS simulations, participants could utilize previous projects as offsets as transaction instruments. As of 2018, the market simulation had a $3–$12USD price range.

In sum, Mexican stakeholders can utilize offset credits to pay for the IEPS carbon tax through CDM projects, enter into international voluntary offset programs, and use offset credits to improve reductions under ETS. As elsewhere, offset programs need to prevent non-additionality, overestimated supply, and double counting.

Mexican Offset Program and Problematic Issues

It seems unfair to judge the Mexican ETS offset program while it is still under construction: at the time of this article, offset protocols are still being drafted. Early-action projects and voluntary markets, and ETS simulations, even at this early stage, may suggest that the Mexican ETS offset program falls under Type A. It still has not addressed links with other climate policies and instruments such as CECs, GHG mitigation mechanisms, taxes, or other offset programs operating in Mexico. It certainly has no broad spillover effect into such other policy domains as poverty, education, or health. In this limited context, the Mexican offset program aims to address quality issues by mandating that projects be real, quantifiable, permanent, verifiable, and operative. Projects awarded during the ETS simulations have mainly been of medium quality, focusing on methane capture and use from landfills, followed by low-quality projects in forestry and large-scale renewable energy; with only three projects ranked as high quality due to co-benefits.

Non-additionality

To guarantee additionality, the Mexican ETS has been careful to prohibit offset projects which directly mitigate CO_2. Offsets will need to fulfil almost the same criteria as that of CDM: developed on Mexican soil and verified by a third party. Still, of the projects registered under voluntary schemes, additionality may be questioned in the cases of projects for wind energy deployment from wind parks in Oaxaca and solar power in Baja California. As Table 10.2 suggests, additionality in the case of large-scale clean and renewable energy might not be considered additional. This can also apply to forestry projects, where community and local regulation in some rural (mostly indigenous) areas have often promoted local development through pluri-harvesting and improved agricultural management. These practices are traditional and may not strictly add up to emissions reduction, but they do directly address co-benefits (e.g. funding, greater income, market promotion, or biological diversification).

In contrast, version 2.0 of Mexico's forestry management offset protocol, approved by the California Action Reserve (2019), includes additionality tests. They guarantee that no forestry projects are developed without the offset program. Projects need to prove that they are not mandated by any law or regulation, and they must pass a performance test demonstrating that forests are in danger of changing their land use or losing their CO_2 balance. While these tests analyse several formal elements (land property rights, local, state, and national regulation, environmental integrity, and co-benefits), they do not include implementation gaps. For example, projects may fail when there is the little governmental capacity to control rezoning practices, in either federal or public areas. A project may become non-additional in these situations, especially when the protocol allows for aggregating different economic activities into one project.

Another risk is that sustainable agroforestry in Mexico commonly faces security issues. Agroforestry areas in certain regions are also used by criminal organizations, which determines how territorial planning will be structured. Although this situation is off-limits for an ETS market, it threatens the viability of Mexican agroforestry projects. Additionality problems might also be present when accepting forests under short-term protection[1] as eligible for credits: baseline and temporality may need to be adjusted over time. However, the protocol correctly assesses the difficulty of proving additionality for forest degradation, declaring these cases non-eligible.

Oversupply

In terms of supply, the offset program design still needs to address problems of leakage and to better promote high-quality projects offering broad co-benefits. Of the 13 projects reported by MexiCO2 (up through 2016), only three may be categorized as high-quality according to Table 10.2 (see Annex 1). Overestimation can also be problematic as Mexico's potential for offering offsets could rely on forestry and low-scale projects. Recalling the EU ETS experience, including REDD+ in the ETS offset can risk promoting low-quality projects. The main challenges to REDD+ projects in Mexico may be constantly renewed territorial planning. Leakage will be difficult to prevent in such long-term projects as forest conservation to take place over 100 years.

Another challenging issue deals with low demand and offers of offset credits and projects. Although it was just a market simulation, there has been low demand for

---

[1] Short-term protection implies that territories do not fall under the categories of national parks and natural protected areas. They are projects lasting about 3-5 years.

offset credits by participants during this first phase of the program. In this context, the costs of certification for offset projects remain high (Santos 2019). Provision of clean and renewable projects can also affect offset demand and offer. As explained above, political decisions made by the current federal government are changing the structure of clean and renewable energy generation. As in the case of CECs, there will likely be an oversupply of CFE clean energy projects if, after 2023, the Mexican ETS accepts CECs as offset instruments.

Double Counting

To avoid double counting, the Mexican ETS base document establishes that offset projects must be listed in the National Emissions Registry (RENE)[2] and to be immediately cancelled once credits are awarded. RENE registers projects directly linked to ETS as well as those issuing from CDM and voluntary programs. As explained earlier, offset projects are already functioning in Mexico. Of the projects mentioned above (see Annex 1), two of the projects reported by MexiCO2 (2016) also appear in CAR (2020) records (methane destruction in Yucatán and forest carbon capture in Santiago Tlacotepec, Estado de Mexico). This is inconsequential unless the two systems, Mexican and California ETS, want to count these projects as their own.

Another risk for double counting might be bilateral offset contracts. Such is the case of the Memorandum of Understanding signed between the California ETS and the Mexican State of Chiapas for forest offsetting (Government of California 2016). Although this initiative has not made any progress, the Mexican ETS should take care to prevent double counting of subnational bilateral offsetting that may not be reported to RENE.

Conclusions and Recommendations

Various ETS and offset schemes have proliferated over the last twenty years. CDM and JI were the first offset programs developed after the global climate governance structure began in 1992. To date, several failures are widely apparent, and the programs have sometimes been adverse to the very environmental goals they claimed to support. CDM and JI co-existed with other national and subnational offset programs, frequently unlinked to either of those flexible mechanisms. There are also voluntary markets promoting projects and standards.

Such fragmentation is detrimental for reaching a global common range of offset prices. Without global or regional pricing, climate policies (ETS, taxes, and offset programs) may suffer from a lack of common ground and an inability to review interconnected pricing systems (Michaelowa 2011: 17–18). A lack of interrelated

---

[2] Since 2016, the Mexican Emissions Registry (RENE) mandates all industries with more than 25,000CO2eq tonnes to submit annual reports on emissions.

pricing or failure to develop a centralized system undermine the notion that mitigation must be a global effort. In offset programs, fragmented pricing impacts project quality and standards, as diverse methodologies are put in place. The cases of CDM and JI prove that quality is necessary for legitimizing the use of market approaches to addressing climate change. Both flexible mechanisms had quality problems, which made some experts discredit them: as of now, it is unclear whether CDM and JI will continue to function under the Paris Agreement. A report by Dufrasne (2018) notes that lower expectations for NDCs (goals they can easily achieve), providing continuity in pollution offset, will result in "hot air" credits. In other words, they would just be a justification to emit and the end result would be an increase in global emissions; to date, there are about 20 gigatonnes of hot air in the NDCs.

The goal of this chapter was to present an overview of offset programs world-wide and discuss three principal issues they face: non-additionality, overestimated supply, and double counting. Although monitoring these issues generates substantial transaction costs (Michaleowa 2011), it is necessary to promote environmental integrity and guarantee that offsets are an attractive instrument for generating more mitigation than has been heretofore scheduled. If offset programs are developed with a broader scope (e.g. Types B and C), transaction costs may rise. However, if these types are combined with offset projects aiming for co-benefits and high quality, offset programs may offer solutions to support ETS or regulatory climate policies.

To address additionality problems in the Mexican ETS, it may be helpful to reduce subjectivity in eligibility criteria. Utilizing fewer methodologies or harmonizing them with verifiers (e.g. VCS or GS) is the first step. The second step might be to draft protocols that add flexibility to the eligibility criteria. For example, there needs to be a range of threshold percentages for when a project might prove additional over time (Gillenwater 2012). This flexibility needs to cover a diverse set of activities, some global, others occurring locally, with distinction also made between metropolitan and regional criteria.

Regarding supply issues, the promotion of co-benefits through prior social consultation seems appropriate for Mexico. To address low prices, decisions must be made regarding CFE's status as a potential main offset provider. As mentioned before, when the Mexican ETS starts operations in 2023, CECs might be linked to the offset program. Having a big clean-energy power producer as CFE may oversupply the nascent offset program (as it has done with the CEC market). Due to an excessive offer of credits by CFE, prices might be impacted with no interest from other market participants to account for these credits. It may also, with so little competition, prevent developing other sources of clean and renewable energy (e.g. distributed generation). Nonetheless, this may provide a window of opportunity for assuring quality (e.g. additionality and co-benefits), by undertaking alternative projects to benefit localities; it may be easier to prove that they could not be developed under the CFE offset credit monopoly. To promote this policy option within the current political context, it may be necessary to establish a two-threshold baseline, considering high (non-CFE) and low (CFE) project quality. This alternative route could also encourage local providers to continue developing projects with real, and broader, social co-benefits. Another option might be to offer support to early investors who promote new technology, allowing them to set their own offset percentages within the established range.

Preventing double counting will only be solved through transparency and a clear, current, and open RENE database. The Mexican offset program needs to acknowledge other forms of intentional and unintentional double counting, such as overlapping policies, regulations, or instruments. The easiest way to prevent external double counting is by linking the offset program to others worldwide.

Although the Mexican trading scheme is still in its first stage, it must address specific issues with its offset program. First, it needs to establish a clear relationship between other climate policy instruments such as international credits coming from CDM or national instruments like IEPS carbon or CECs. It may be that the Mexican ETS is transformed into a hybrid pricing strategy that includes ETS taxes and other regulatory policies (Michaelowa 2011). A clear link between all instruments prevents "regulatory cherry-picking" by market participants and helps stabilize prices.

Second, the Mexican offset program needs to establish project quality and standard protocols. MVR mechanisms, independent of ETS regulators, are fundamental to assessing offset projects (Meckling and Hepburn 2013). GS and VCS are key agents for evaluating projects within the Mexican offset program, which could, for example, work on a "double additionality" basis, where eligible projects must comply with specified co-benefits. In other words, co-benefits must be conceived as additional to the GHG mitigation additionality requirement. Third, ETS price stabilization is crucial. Experts suggest that in the future, setting price floors and ceilings is the best strategy (Michaelowa 2011: 30). This may also apply to the price range for offset programs, where prices must constantly rise to make the program attractive to market participants and project holders.

The use of offset programs worldwide has flourished. The variety of programs, however, impacts credit prices and indicates diverse methodologies and quality standards. Achieving quality is challenging, as projects often fall under non-additionality practices, GHG overestimation and leakage, and double counting. The Mexican ETS is on schedule to address these issues by combining strict eligibility criteria with high-quality projects that include socio-environmental co-benefits, flexible use of credits to pay for other policy instruments, open data, and transparency.

Annex 1

See Tables 10.4 and 10.5.

Table 10.4 Mexico's voluntary offset projects

Project	Goal	Locality	Voluntary program	Co-benefits	Quality
Methane capture and use in 27 farms	Generation of 32,670 MW/h for self-use	Jalisco	CDM	Job creation Income increase Odor management Water conservation	Medium
Methane capture and use in landfill	Reduction of 100,000 tonnes of CO_2 per year	Guanajuato	CDM	Job creation Health positive effects Odor management Local government savings on electricity (possibility to re-invest these in public services)	Medium
Methane capture and use in landfill	Generation of 1.95 MW	Estado de Mexico	CDM	Job creation Health positive effects Odor management Local government savings on electricity (possibility to re-invest these in public services)	Medium
Methane capture and use in landfill	Generation of 2 MW Reduction of 833,396 tonnes of CO_2 in 10 years	Durango	CDM	Job creation Health positive effects Odor management Local government savings on electricity (possibility to re-invest these in public services)	Medium
Methane capture and use in landfill	Generation of 2 to 4 MW Reduction of 1,625,926 tonnes of CO_2 in 10 years	Aguascalientes	CDM	Job creation Health positive effects Odor management Local government savings on electricity (possibility to re-invest these in public services)	Medium

(continued)

Table 10.4 (continued)

Project	Goal	Locality	Voluntary program	Co-benefits	Quality
Methane capture and use in landfill	Generation of 1.95 MW Reduction of 1,625,926 tonnes of CO_2 in 10 years	Chihuahua	CDM	Job creation Health positive effects Odor management Local government savings on electricity (possibility to re-invest these in public services)	Medium
Reforestation "Scolel'te"	Capture CO_2 in 8,958 HA Mixed reforestation	Chiapas	Plan Vivo	Increase agricultural production via mixed reforestation (trees plus corn, beans, fruits, coffee) Reaching 92 indigenous communities Help with territorial planning	Medium
Sustainable forestry	Increase of 1,270HA of forested area in 40 years Reduction of 7,758.129 tonnes of CO_2 in 40 years	Nayarit, Tabasco, Chiapas	VCS	Poverty alleviation Increase in the commercialization of "teca" wood Job creation (fair labour) Support to rural schools Biodiversity conservation	High
Renewable energy generation: 3 wind parks	Generation of 360 MW Prevent 750,000 tonnes of CO_2	Oaxaca	CDM VCS	Job creation	Medium
Energy efficiency: Change of light bulbs into fluorescent lamps	Reaching 57 million families Reduction of 2.78 million tonnes of CO_2 per year (in 10 years of the project)	Countrywide	CDM	Savings of 14,900 million pesos in electricity bills Reduction on federal subsidy to electricity	Medium

(continued)

Table 10.4 (continued)

Project	Goal	Locality	Voluntary program	Co-benefits	Quality
Solar plant "Aura Solar"	Generate 82 GW/h per year	Baja California	CDM VCS	Job creation	Medium
Efficient kitchens "Utsil Naj"	40,000 Kitchens 100,000 people	Baja California, Sonora, San Luis Potosí, Jalisco, Guanajuato, Estado de México, Michoacán, Veracruz, Oaxaca	GS	Poverty alleviation Avoided deforestation for cooking of about 40–60%	High
Forest conservation in Santiago Tlacotepec	CO_2 capture of approx. 25,000 tonnes in 6 years	Estado de México	CAR	Forest conservation Water conservation Biodiversity conservation	Low
Methane destruction in swine farm	Destroy Methane	Yucatán	CAR	Residue used as fertilizer Odor management Water conservation	High

Source MexiCO2 (2016)

Table 10.5 CAR projects

Project	Goal	Locality	Credits granted	Credits retired
Los Bancos Forest	Forest carbon capture	Durango	0	0
San Lucas Amanalco Forest	Forest carbon capture	Estado de Mexico	0	0
Santiago Tlacotepec Forest	Forest carbon capture	Estado de Mexico	0	0
San Bartolo Forest	Forest carbon capture	Estado de Mexico	0	0
San Nicolás Tototlapan Forest	Forest carbon capture	Mexico City	4,304	3,909
San Jerónimo Zacapexco Forest	Forest carbon capture	Estado de Mexico	0	0
San Rafael Ixtapalucan Forest	Forest carbon capture	Puebla	0	0
Santiago Coltzingo Forest	Forest carbon capture	Puebla	15,324	7,354
Tecocomulco ASRTulancingo	Improved forest management	Hidalgo	0	0
La Estancia ASRTulancingo	Improved forest management	Hidalgo	0	0
Ixtula y Sembo ASRTulancingo	Improved forest management	Hidalgo	0	0
Puentecillas y Anexox ASRTulancingo	Improved forest management	Hidalgo	0	0
San Pedro Huixotitla ASRTulancingo	Improved forest management	Hidalgo	0	0
El Ocote ASRTulancingo	Improved forest management	Hidalgo	0	0
El Nopalillo ASRTulancingo	Improved forest management	Hidalgo	0	0
Sangre de Cristo ASRTulancingo	Improved forest management	Hidalgo	0	0
Los Romeros ASRTulancingo	Improved forest management	Hidalgo	0	0
Sabanetas ASRTulancingo	Improved forest management	Hidalgo	0	0
Emiliano Zapata ASRTulancingo	Improved forest management	Hidalgo	0	0

(continued)

Table 10.5 (continued)

Project	Goal	Locality	Credits granted	Credits retired
El Ventorrillo ASRTulancingo	Improved forest management	Hidalgo	0	0
San Lorenzo Sayula ASRTulancingo	Improved forest management	Hidalgo	0	0
Alhuajoyucan ASRTulancingo	Improved forest management	Hidalgo	0	0
Hueyapan ASRTulancingo	Improved forest management	Hidalgo	0	0
Cima de Togo ASRTulancingo	Improved forest management	Hidalgo	0	0
Xahuayalulco ASRTulancingo	Improved forest management	Hidalgo	0	0
Chacalapa ASRTulancingo	Improved forest management	Hidalgo	0	0
Las Puentes y Anexos ASRTulancingo	Improved forest management	Hidalgo	0	0
Arturo Gomez Canales ASRTulancingo	Improved forest management	Hidalgo	0	0
Juan Lachao Forest	Forest carbon capture	Oaxaca	31,470	23,787
Reforesting the Usumacinta River	Improved forest management	Campeche	0	0
Methane recovery in swine farm	Methane recovery	Yucatán	575	0
La Perseverancia Biogas Plant	Energy	Morelos	34,267	0

Source Climate Action Reserve (2020)

References

Alcock F (2008) Conflict and Coalitions within and across the ENGO Community. Global Environmental Politics 8(4):66–91

Alloisio I (2020) Article 6 of the Paris agreement: is no deal better than a bad deal? European University Institute (January 21). https://fsr.eui.eu/article-6-of-the-paris-agreement-is-no-deal-better-than-a-bad-deal/

Arteaga JR (2019) Semarnat pone a prueba el ADN de las empresas. Forbes Mexico (December 16th). https://www.forbes.com.mx/semarnat-pone-a-prueba-el-adn-verde-de-las-empresas/

Broekhoff D, Gillenwater M, Colbert-Sangree T, Cage P (2019) Securing climate benefits: a guide to using carbon offset. Stockholm Environment Institute & Greenhouse Gas Management Institute, Stockholm

Cames M, Harthan RO, Fussler J, Lazarus M, Lee CM, Erickson P, Spalding-Fecher R (2016) How additional is the clean development mechanism? Analysis of the application of current tools and proposed alternatives. Oko-Institut, INFRAS, Stockholm Environment Institute, Berlin

Carbon Market Watch (2019) COP25: No deal on UN carbon markets as a number of countries reject loopholes. Madrid (December 15th). https://carbonmarketwatch.org/2019/12/15/cop25-no-deal-on-un-carbon-markets-as-a-number-of-countries-reject-loopholes/

CEMDA (2020) Conceden suspensión definitiva a amparo contra ley y política de generación de energía eléctrica. Mexico City: CEMDA. https://www.cemda.org.mx/conceden-suspension-definitiva-a-amparo-contra-ley-y-politica-de-generacion-de-energia-electrica/

Clark D (2009) The rough guide to green living. Penguin Books, London

Climate Action Reserve (CAR) (2019) Revisión del Protocolo Forestal para México. CAR, Mexico City. https://www.climateactionreserve.org/how/protocols/mexico-forest/revision-del-protocolo-forestal-para-mexico/

Climate Action Reserve (CAR) (2020) Projects. Climate Action Reserve, Los Angeles, CA. https://www.climateactionreserve.org/how/projects/

DOF (Diario Oficial de la Federación) (2019a) Acuerdo por el que se establecen las bases preliminares del Programa de Prueba del Sistema de Comercio de Emisiones. DOF, Mexico City. http://www.dof.gob.mx/nota_detalle.php?codigo=5573934&fecha=01/10/2019

DOF (Diario Oficial de la Federación) (2019b) Acuerdo por el que se modifican los lineamientos que establecen los criterios para el otorgamiento de certificados de energías limpias y los requisitos para su adquisición, publicados el 31 de Octubre de 2014. DOF, Mexico City. https://www.dof.gob.mx/nota_detalle.php?codigo=5576691&fecha=28/10/2019

DOF (Diario Oficial de la Federación) (2020) Acuerdo por el que se emite la política de confiabilidad, seguridad, continuidad y calidad del sistema eléctrico nacional. DOF, Mexico City. https://dof.gob.mx/nota_detalle.php?codigo=5593425&fecha=15/05/2020

Dufrasne G (2018) The clean development mechanism: local impacts of a global system. Carbon Market Watch, Brussels

Egenhofer C (2013) The growing importance of carbon pricing in energy markets. In: Golthau A (ed) The handbook of global energy policy. Wiley-Blackwell, Chichester, UK

Elceo (2020) Seis empresas obtienen amparo contra freno de pruebas de energías limpias. Elceo, México City. https://elceo.com/negocios/tres-empresas-obtienen-amparo-contra-freno-de-pruebas-de-energias-limpias/

Ellerman AD, Convery F, De Perthius C (2010) Pricing carbon: the European union emissions trading scheme. Cambridge University Press, Cambridge

Elsworth R, Worthington B (2010) E. R. Who? Joint implementation and the EU emissions trading system. Sandbag, London

Elsworth R, Worthington B, Morris D (2012) Help or hindrance? Offsetting in the EU ETS. Sandbag, London

Environmental Defense Fund (2019) Meeting the climate change goals of the Paris agreement. How to avoid double counting of emissions reductions. Environmental Defense Fund, New York

Flores A (2019) El mercado mexicano de carbono, un componente crucial para enfrentar el cambio climático e impulsar la justicia social en México. World Resources Institute Mexico (January 28th). https://wrimexico.org/bloga/el-mercado-mexicano-de-carbono-un-componente-crucial-para-enfrentar-el-cambio-clim%C3%A1tico-e

Fujiwara N, Egenhofer C (2007) Shaping the global arena: preparing the EU emissions trading scheme for the post-2012 period. Centre for European Policy Studies, Brussels

Fujiwara N (2009) Flexible mechanisms in support of a new climate change regime: the CDM and beyond. CEPS Task Force Report, CEPS, Brussels, November

García JL (2019) ¿Qué son los CELs y por qué México podría dar la espalda a la lucha contra el cambio climático? (December 6th). Economía Hoy. https://www.economiahoy.mx/energia-mexico/noticias/10241087/12/19/Que-son-los-CELs-y-por-que-Mexico-podria-dar-la-espalda-a-la-lucha-contra-el-cambio-climatico.html

Gehring M, Phillips FK (2017) Intersections of the Paris agreement and carbon offsetting: legal and functional considerations (Paper No. 42). University of Cambridge, Cambridge

Gillenwater M (2012) What is additionality? Part 2: a framework for more precise definitions and standardized approaches, discussion Paper No. 002 (January). Silver Spring, MD: Greenhouse Gas Management Institute

Government of California (2016) California-Mexico Memorandum of Understanding on Climate Change & the Environment. 2016 Progress Report. Government of California, California.

Hagelberg N (2019) Carbon offsets are not our get-to-jail free card. UN Environment, Nairobi. https://www.unenvironment.org/news-and-stories/story/carbon-offsets-are-not-our-get-out-jail-free-card

Haya B (2019) ARB's U.S. Forest Projects offset protocol underestimates leakage—preliminary results. University of California, Berkeley

Lang C (2020) The aviation industry's denial of the climate crisis. REDD Monitor, April 10th. https://redd-monitor.org/2020/04/10/the-aviation-industrys-denial-of-the-climate-crisis/

López-Vallejo M (2014) Reconfiguring global climate governance in North America: a transregional approach. Routledge, London

Marcu A (2012) Expanding carbon markets through new market-based mechanisms. Centre for European Policy Studies, Brussels

Martínez N (2019) Imaginarios sociotécnicos y los futuros posibles de la transición energética en México. In: Tornel C (ed) Alternativas para limitar el calentamiento global en 1.5° Más allá de la economía verde. Henrich Boll Stiftung, Mexico City, pp 238–267

Matsuki T (2015) Project-Based Carbon Offset Program, Carbon Pricing Training Workshop. World Bank, Vietnam

Meckling J, Hepburn C (2013) Economic instruments for climate change. In: Falkner R (ed) The handbook of global climate and environmental policy. Chichester UK, Wiley-Blackwell, pp 465–485

Secretaría de Medio Ambiente y Recursos Naturales (SEMARNAT) (2017) Ejercicio de Simulación del mercado de Carbono en México. Reporte de Resultados de la Segunda Simulación. SEMARNAT-MexiCO2, Mexico City

Secretaría de Medio Ambiente y Recursos Naturales (SEMARNAT) (2018a) Ejercicio de Simulación del mercado de Carbono en México. Reporte de Resultados de la Tercera Simulación. SEMARNAT-MexiCO2, Mexico City

Secretaría de Medio Ambiente y Recursos Naturales (SEMARNAT) (2018b) Ejercicio de Simulación del mercado de Carbono en México. Reporte de Resultados de la Segunda Fase. SEMARNAT-MexiCO2, Mexico City

Mehling M, Dimanchev E (2017) Achieving the Mexican mitigation targets: options for an effective carbon pricing policy mix. SEMARNAT-GIZ, Mexico City

MexiCO2 (2016) Proyectos—Mercado Voluntario de Carbono. MexiCO2, Mexico City. http://www.mexico2.com.mx/proyectos-de-carbono.php

MexiCO2 (2019a) Nota Técnica. Impuesto al Carbono en México. MexiCO2, Mexico City

MexiCO2 (2019b) Nota Técnica. Sistema de Comercio de Emisiones en México. MexiCO2, Mexico City

Michaelowa A (2011) Fragmentation of international climate policy—doom or boon for carbon markets? In: UNEP Riso Centre (ed) Progressing towards post-2012 carbon markets. UNEP Riso Centre, Denmark, pp 13–23

Millard-Ball A, Ortolano L (2010) Constructing carbon offsets: the obstacles to quantifying emission reductions. Energy Policy 38(1):533–546

Monbiot G (2006) Selling indulgences. The guardian (19 October), London. https://www.monbiot.com/2006/10/19/selling-indulgences/

Obergassel W, Asche F (2017) Shaping the Paris mechanisms part III: an update on submissions on Article 6 of the Paris agreement. Wuppertal Institute for Climate, Environment and Energy, Wuppertal, GR

Öko-Insitut e.V. (2018) Designing an emissions trading system in Mexico: options for setting an emissions cap. Öko-Institute e.V., GIZ, SEMARNAT, Freiburg

PMR (Partnership for Market Readiness) (2015) Overview of carbon offset programs. Similarities and differences. World Bank, Washington DC

Carbon Pulse (2019) Shades of REDD+: Should forests offsets be eligible for CORSIA? Carbon Pulse (December 3rd). https://carbon-pulse.com/87946/

Rabe BG (2018) Can we price carbon? MIT, Cambridge

Santos P (2019) Experiencias de México con el mix de políticas de precio al carbono. MexiCO2, Mexico City

Schneider L, Kollmuss A, Lazarus M (2014) Addressing the risk of double-counting emission reductions under the UNFCCC. SEI working paper No. 2014-02. Stockholm Environment Institute, Seattle, WA

Schneider L, La Hoz Theuer S (2018) Environmental integrity of international carbon market mechanisms under the Paris agreement. Clim Policy 19(3):386–400. https://doi.org/10.1080/146 93062.2018.1521332

Schneider L (2007) Is the CDM fulfilling its environmental and sustainable development objectives? An evaluation of the CDM and options for improvement. Oeko-Institut and WWF, Berlin

Swyngedouw E (2016) CO2 as neoliberal fetish: the love of crisis and the depoliticized immuno-biopolitics of climate change governance. In: Cahill D, Cooper M, Konings M, Primrose D (eds) The SAGE handbook of neoliberalism. Sage Handbooks, London, UK, pp 238–276

UNFCCC (United Nations Framework Convention on Climate Change) (1997) Kyoto protocol. UNFCCC

UNFCCC (United Nations Framework Convention on Climate Change) (2015) Paris agreement

Verra (2017) Double counting in ICAO's CORSIA: issues and solutions. https://verra.org/double-counting-in-icaos-corsia-issues-and-solutions/

Verra (2020) Climate, community and biodiversity standards. https://verra.org/project/ccb-program/

Wara MW, Victor DG (2008) A realistic policy on international carbon offsets, Working Paper #74. Stanford University, Stanford

World Bank (2019) State and trends in carbon pricing. World Bank, Washington DC

Yuvaraj NDB (2011) Voluntary market—future perspective. In: UNEP Riso Centre (ed) Progressing towards post-2012 carbon markets. UNEP Riso Centre, Denmark, pp101–110

Zhang X, Zhou L (2020) Subnational climate-change policies in China. In: Harvard project on climate agreements, overview and framing: climate-change policy in China. Subnational climate-change policy in China. Harvard Project on Climate Agreements, Harvard, pp 25–27

Chapter 11
Capacity Development Associated with the Implementation of Emissions Trading System in Mexico

María Concepción Martínez Rodríguez, Catherine Nieto Moreno, and Mariana Marcelino Aranda

Abstract The creation of an emissions trading system in Mexico as response to international policy on climate change forces the government and corporations to create new activities and responsibilities to address this issue. It is also important to know who will be the decision-maker and who is in charge of the institutional work (representation and negotiation). The main objective of this chapter is to point out who the stakeholders involved in the design, implementation, evaluation and transparency of the system are, or should be, according to the national regulatory framework and international summons. We shall also analyze the mechanics and information provided by the system and how it helps to make environmental policy, which helps to reduce emissions. Finally, we will also analyze whether it also helps to establish strategic alliances and international agreements toward common objectives and priorities. The chapter approaches the topic based on capacity development theory, which focuses on improving governance among different levels and stakeholders: government, companies, civil organizations, and scientists. We emphasize the potential of training spaces as a place for transformation and developing a learning framework whose own relevance relies on the focus of emergent strategies, which ensure the environmental integrity and conditions for the country's competence in the international context. This chapter contributes to existing literature about the understanding of executing such a system, the stakeholders involved at the national level, and their potential to create international networks.

M. C. M. Rodríguez (✉) · C. N. Moreno
Centro Interdisciplinario de Investigaciones y Estudios sobre Medio Ambiente y Desarrollo, Instituto Politécnico Nacional, Calle 30 de Junio de 1520, Barrio la Laguna Ticomán, Alcaldía Gustavo A. Madero, C.P. 07340 Ciudad de México, México
e-mail: mcmartinezr@ipn.mx

C. N. Moreno
e-mail: cnietom2000@alumno.ipn.mx

M. M. Aranda
SEPI-UPIICSA, Instituto Politécnico Nacional. Av. Té 950 Granjas México, Iztacalco, 08400 Ciudad de México, México
e-mail: mmarcelino@ipn.mx

© The Author(s) 2022
S. Lucatello (ed.), *Towards an Emissions Trading System in Mexico: Rationale, Design and Connections With the Global Climate Agenda*, Springer Climate,
https://doi.org/10.1007/978-3-030-82759-5_11

Keywords Emissions trading system · Capacity development · México · Climate change · Governance · Corporate governance

Introduction

This chapter is divided into six parts. The first, the introduction, raises awareness of the topic and of the approach used in this analysis. The second talks about international environmental governance, offering a general picture of how international organizations work and how they influence national policy. The section concludes with an example regarding climate change. The third section includes a review of national environmental governance and how it is addressed in the regulatory framework of climate change, using the emissions trading system as an example. The fourth discusses corporate governance and how it helps to plan, guide, execute, and control businesses in order to hold them accountable to environmental policy, specifically the emissions trading system. The fifth section contains a proposal of methodology. We focus on capacity development in order to draw a results analysis and conclusions where we discuss the basic conditions and priorities that stakeholders ought to take into account in order to successfully execute each phase in the emissions trading system.

There is no agreement among researchers about what governance means. There are several different definitions. Generally speaking, we say that governance is the decision-making process undergone by governmental and non-governmental actors in solving a specific problem (Aguilar 2009; Cerrillo Martínez 2005; Denhardt and Denhardt 2007; Georgiadou and Reckien 2018; Kooiman et al. 2008; Martínez Rodríguez 2015; Marsh 2008; Pierre 2000; Treviño Cantú 2011; Whittingham 2010; Zurbriggen 2011). In the international context, the alliance of different countries and their interests in regard to the issue of climate change interact with each other, providing a frame of reference in which each country voluntarily adopts or rejects the decision, in an ostensibly diplomatic approach.

National governance is how the government includes the different actors involved in decision-making, policy design, implementation, and assessment, and their instruments. Meanwhile, corporate governance is the interaction between companies and their environment, the articulation of their organizations and the new regulatory frameworks, seeking above all to ensure profitability and competitiveness, under a framework of transparency and respect for the environment.

Capacity Building (CB) was a core concept in development, justice, and welfare policies through the 1990s. This approach intended to develop or increase knowledge, output rate, management, skills, and other capabilities from the ground up according to a pre-imposed design. However, it did not acknowledge pre-existing capacities among people or institutions. Inspired by the launch of the Millennium Development Goals (MDGs) and the Paris Declaration on Aid Effectiveness in 2005, there was a shift to a new concept, that of capacity development. This concept is a broader in its approach. It acknowledges pre-existing capacities and knowledge. Operators make a diagnosis in order to identify which areas need their capacities strengthened. Thus, the aid is designed according to the actual requirements using an endogenous process. This approach later became the trend (Greijn et al. 2015; OCDE 2008; Robeyns 2016; Zamfir 2017).

The capacity development approach goes further than a social aid policy. It has evolved to other areas of sustainable development such as industry, energy, climate change, and health and security. It also drives endogenous innovation as it uses global knowledge to find proper solutions for specific local contexts (Greijn et al. 2015; OCDE 2008; Zamfir 2017).

The capabilities approach postulated by Amartya Sen and Martha Nussbaum, among others, adds further insight into capacity development. According to Sen, life is a set of interrelated beings and activities, where capabilities are functional vectors that reflect the freedom to live one type of life or another (1993). In economic terms, this is an alternative approach to the welfare economy with a wider perspective where the society's local empowerment is an unnegotiable strategy for sustainable development and social transformation. This is what the world needs in order to face climate change.

Since the capabilities approach focuses on the extent of real opportunity that people have to accomplish what they value through agency (existence of institutions, democracy, literacy, low levels of poverty, governance) and the set of capabilities at their disposal, it provides an ethical framework for evaluating the achievement of capacity development and how it is obtained.

In this analysis, we use Porter's (2007) definition of capacity development as "the emergent combination of attributes, capabilities and relationships that enables a system to exist, adapt and perform in a manner which expands the real freedoms that people enjoy" (p. 19). This approach is useful for weighing opportunities, challenges, and strategies for the emission trading system in Mexico.

The emissions trading system is a market-based instrument with international compliance to reduce greenhouse gas emissions which cause climate change, a problem that impacts the whole of humanity. Mexico is taking the first steps in order to adopt this trading system by collecting international knowledge and experience.

International Environmental Governance

The norms, rules, and procedures for international cooperation take shape according to ecological interdependencies that are clearly cross-border. When we talk about environmental issues, we must be aware that they are not contained by political boundaries. Environmental issues are borderless and international organizations attempt to address them. Climate change is undoubtedly the most typical of global environmental problems, where there is no solution without international cooperation. (Delbeke and Vis 2019).

International organizations help to process the sum total of knowledge and information. Additionally, they help by providing and setting up institutions for the progressive establishment and implementation of norms and rules within national regimes (Bauer et al. 2006).

How can international organizations influence national governments? International bodies are supposed to influence the behavior of national political actors by changing their knowledge and belief system. They also attempt to influence political processes by creating, supporting, and establishing norm-building processes, rules and procedures for specific international cooperation problems. These organizations are crucial actors in developing inter- and transnational negotiations and discussions on specific issues (Bauer et al. 2006).

However, the influence of international organizations depends on the strength of the national government. In a first instance, being part of these bodies is completely voluntary and the guidelines that these organizations issue are non-binding recommendations. Though, in some cases this depends on the type of regulatory framework developed by international organizations as well as common practice and other sources of international law.

In addition to the above, within these bodies there is a focus on decision-making carried out by consensus of the participating countries (known as the parties). When it comes to environmental issues, strength lies in the technical base. The climate change phenomenon implies taking divergent actions, which have a great impact on traditional economies. The recommendations issued by international organizations are complex and politically controversial.

One of the international organizations created by the United Nations is the United Nations Framework Convention on Climate Change (UNFCCC). The UNFCCC provides a global response to the threat of climate change. Mexico has been party to the UNFCCC since 1992. In 1993, after the ratification act, Mexico stated at the international level its consent to be bound by the guidelines established in this instrument.

There are two UNFCCC meetings that marked the path for international climate change policy. One is the Third Conference of the Parties (CoP3) in 1997, where the Kyoto Protocol was adopted. The Kyoto Protocol was an instrument with specific quantitative goals for reducing Greenhouse Gases (GHGs). It was ratified by Mexico in 2000, and officially came into force on 16 February 2005. The other important meeting is CoP21 in Paris, France, where the Paris Agreement on climate change was approved. This agreement entered into force on 4 November 2016 and has been applicable since 2020. The agreement aims to limit the global average temperature rise to below 2 °C, relative to pre-industrial levels and to continue efforts to limit this increase to 1.5 °C in order to reduce risks and effects of climate change while at the same time strengthening countries' capacity to face the impacts. This is, without a doubt, the most important voluntary commitment that has been made in recent years in the field of climate change (National Institute of Ecology and Climate Change 2018).

The fundamental difference between the Paris Agreement and the Kyoto Protocol is that the former embraces "voluntary" commitments from developed and developing countries, replacing the top-down scheme with the bottom-up one. Hence, a legal regime based on the "principle of common but differentiated responsibilities" is not enough. Something deeper is required: structural changes to economic systems tied to behavioral and value changes (Ibarra 2019).

National Environmental Governance

Below we mention the instruments at the national level that have been developed for the building an emissions trading system in Mexico. This represents an essential step in the process of updating national climate change policy. Mexico's climate change policy reflects the commitment of the Mexican government to reducing greenhouse gases and compound emissions by promoting sustainable development based on a competitive and low-emission economy, while also adapting new strategies. This requires a responsible, coordinated, and continuous effort at the three levels of government, as well as the activation of citizen participation mechanisms and of the sectors involved in the implementation and evaluation of the performance of mitigation and adaptation actions.

Mexico's Regulatory Framework on Climate Change is

- The Political Constitution of the United Mexican States (CPEUM): Article 4 establishes that Everyone has the right to a healthy environment for their development and well-being. The State will guarantee respect for this right.
- General Law of Ecological Balance and Protection of the Environment (LGEEPA): Aims to promote sustainable development and establish bases for guaranteeing Mexicans the right to live in a healthy environment. Article 5 establishes that the federal government is empowered to formulate and execute mitigation and adaptation actions in regards to climate change.
- General Law on Climate Change (LGCC): Aims to regulate compounds and emissions of greenhouse gases in order to stabilize their concentrations in the atmosphere to a level that prevents dangerous anthropogenic interference in the climate system; regulate mitigation actions; as well as promote a transition toward a competitive, sustainable and low-carbon economy.

Policy Instruments

The following two policy instruments are derived from the General Law on Climate Change (LGCC):

- National Registry of Emissions (and its Regulations)
- Emissions Trading System.

National Registry of Emissions

The General Law on Climate Change (LGCC) was published on 6 June 2012 and entered into force in October of the same year. This made Mexico the first developing country to issue a law on the matter. The LGCC dictates the creation of multiple public policy instruments, including the National Emissions Registry (RENE) and its Regulations (RLGCCMRENE, 28 Oct 2014). Both make it possible to compile the necessary information on the emission of compounds and greenhouse gases (CyGEI), such as carbon dioxide (CO_2), methane (CH_4), nitrous oxide (N_2O), black carbon (CN), and various fluorinated compounds from the different productive sectors across the country (established by Art. 87 and 88 of the LGCC, as well as the 6th and 9th articles of LGCC Regulations regarding RENE).

This regulation establishes the creation of agreements that will define the technical aspects for the registry's operation. One of these agreements, the Agreement on the Grouping of Gases and Greenhouse Compounds and their Global Warming Potential, identifies each one of the chemical substances according to an internationally accepted norm defined by associations specializing in the field. Also, it considers the formula and chemical family to which the substance belongs, as well as its global warming potential. This is consistent with what was published in the Fifth Assessment Report of the Intergovernmental Panel on Climate Change (IPCC).

The National Emissions Registry (RENE) is a policy instrument that will make it possible to compile the necessary information regarding CyGEI emissions from the different productive sectors across the country so they are traceable, and trends can be evaluated in order to make national emission reduction strategies. Keeping an emissions registry will allow companies and industries to identify their emission sources with the aim of reducing their carbon footprint, while generating opportunities for business and competitiveness along the way.

RENE has two main objectives:

1. To collect information on:

   (a) Direct emissions derived from the combustion of fixed and mobile sources;
   (b) Emissions derived from reactions in industrial processes;
   (c) Indirect emissions derived from the consumption of purchased electrical or thermal energy.

2. To integrate the registry about CyGEI emission reductions and mitigation projects implemented in national territory and promoted by either individuals or companies.

Emissions Trading System

The General Law on Climate Change was amended on 13 July 2018 to incorporate the international commitments acquired by Mexico in the Paris Agreement and the

Intended Nationally Determined Contributions, which commits Mexico to unconditionally reduce its greenhouse gas emissions by 22% and black carbon emissions by 51% from the baseline by 2030.

The Agreement establishing the preliminary bases of the Emissions Trading System Program was published in the Official Gazette of the Federation (DOF) dated 01/10/2019. This is a market instrument designed to reduce greenhouse gas emissions based on the cap-and-trade principle, where the government imposes a ceiling or cap on total emissions from one or more sectors of the economy. Its legal basis is found at the national level, in Article 7, section IX of the General Law on Climate Change, which attributes the power to create, authorize, and regulate emissions trading to the Federation. In accordance with the provisions of the fourth paragraph of article 92 of the same law, this constitutes an economic market instrument. Article 94 of the General Law itself authorizes the Ministry of Environment and Natural Resources (SEMARNAT), in participation and consensus with the Commission, the Council and representation of the participating sectors, to establish the system progressively and gradually.

The aim of the Emissions Trading System is to promote emission reductions that can be carried out at the lowest possible cost, in a measurable, reportable and verifiable manner, without compromising the competitiveness of the participating sectors in international markets.

The ETS is comprised of two phases, an initial phase to let the actors involved learn how an emissions market behaves. For this purpose, the General Law on Climate Change calls for the development of a pilot program. The transition phase and the proper operational phase of the system will come into effect at the end of the transition stage of the pilot program (DOF 01/10/2019).

In accordance with Article Six of the Agreement (DOF 01/10/2019), the ETS pilot program pursues the following objectives: to make progress in the reduction of emissions, promote them, test the operation of the ETS, identify areas of opportunity, generate information, become familiar with its operation, and develop capacities in terms of emissions trading.

In accordance with the Agreement (DOF 01/10/2019), the ETS pilot program consists of the information reported in RENE by the participants, the allowance ceiling, the monitoring system, market transactions, and the flexible compliance mechanisms.

Article Six of the Agreement (DOF 01/10/2019) also stipulates that the pilot program will last thirty-six months, starting on 1 January 2020 and ending on 31 January 2022.

The pilot program is divided into two periods:

I.   The period from 1 January 2020 to 31 December 2021 is part of the pilot phase of the pilot program, and
II. The period from 1 January 2022 to 31 December of the same year will constitute the transition phase of the pilot program to the ETS' operational phase.

The pilot program will have no economic effects, which means that there will be no monetary penalties and the allocation of allowances will be free in a proportion equivalent to the participants' emissions, regardless of the allowances that are destined to the corresponding reserves.

It could be said that the RENE contains the ETS. The RENE holds the national registry of emissions of all gases provided for in the General Law on Climate Change from all industries. Therefore, it is necessary to have a reliable institution such as the RENE and the commitment of the industries with compulsory reporting, which must have adequate knowledge for participation and proper declaration of their emissions. Both the RENE and ETS come under the jurisdiction of SEMARNAT.

Corporate Governance

Climate change poses many challenges for companies, among them is the fulfillment of a series of increasingly rigorous requirements by environmental authorities. One example of this is the implementation of the Emissions Trading System (ETS). It is necessary to evaluate the capabilities of companies to meet the requirements so that an ETS may be implemented (Durant s/a).

The ETS is a market-based public policy instrument to control pollution by providing an economic incentive to reduce CO_2 emissions. To be successful in managing such a cap-and-trade system, companies need strategic, technical and financial skills.

Competitive advantage, reputation, image, the ability to attract and retain markets, and the different relationships companies have with their environment, depend on environmental compliance. Through carbon trading and management and new, GHG emission-reducing technologies, the new markets are based on reducing environmental impact.

Using data from the National Emissions Registry (RENE), we see that facilities in the energy and industry sectors reporting annual direct emission of 100,000 tons or more of carbon dioxide constitute the primary source of direct Greenhouse Gas (GHG) emissions. Thus, they will be the first included in the ETS.

Only carbon dioxide emissions will be considered in the ETS's pilot phase, since it is the most emitted greenhouse gas nationwide, making it an effective indicator before moving into the system's operational phase.

The different sectors and subsectors that must compulsorily report their direct and indirect emissions of compounds or greenhouse gases from all their facilities when they exceed 25,000 tCO_{2e} (tons of CO_2 equivalent) are indicated in Table 11.1.

SEMARNAT has an electronic reporting platform called COA Web for reporting and compiling emissions registered by RENE. This platform provides for multiple fields required for recording direct and indirect emissions based on the activities carried out by the business. Facilities subject to emissions reporting must therefore have the necessary data on hand in order to fill in the corresponding fields.

Compulsory reporters:

Table 11.1 Sectors and subsectors with compulsory reporting according to RENE regulations

Sectors	Subsectors
Energy	Generation, electricity transmission, and distribution Exploitation, production, transport, and distribution of hydrocarbons
Industry	Chemical industry Iron and steel Metallurgy Metal-mechanic Mining Automotive Cellulose and paper Graphic arts Petrochemicals Cement and lime Glass Electronics Electricity Food and beverage Lumber Textiles
Transportation	Air Rail Marine Ground
Agricultural and cattle raising	Agriculture Cattle raising
Waste	Sewage water Urban solid waste and special handling waste, including final disposal
Services and commerce	Construction, commerce, educational services, recreational and entertainment activities, tourism, medical services, government and financial services

1. Identify emission sources (fixed and mobile)
2. Collect the necessary data in order to apply the corresponding calculation methodologies
3. Measure, calculate, or estimate their direct and indirect emissions according to each activity carried out at the facility, applying the corresponding calculation methodology
4. Add direct and indirect emissions
5. Verify the information reported through the corresponding verification bodies
6. Report emissions annually using the annual operating certificate (COA).

The verification report is the document issued by an accrediting body that verifies the relevance, integrity, consistency, transparency, and accuracy of the information contained in the emission reports that facilities in industries with compulsory reporting must supply to RENE.

Another issue that companies have to work on is registering mitigation projects. Mitigation projects are validated by international organizations such as Climate Action Reserve, Verified Carbon Standard, Gold Standard, Plan Vivo, American Carbon Registry, The Climate, Community and Biodiversity Standards, and by Mexican national organizations (SEMARNAT) in order to guarantee transparency.

In the European Union's experience, companies commented that starting slowly gave both public and private actors the opportunity to learn how to approach issues in practice (Delbeke and Vis 2019).

Capacity Development for the Emissions Trading System in Mexico

The capabilities approach, used in this paper, is a relatively young theory in the field of social sciences. It is based on identifying and using the endogenous resources and potentials of a country or community. According to this perspective, development is defined as a process of transformation of the society, aimed at overcoming existing difficulties and challenges. It also states the importance of social capital and cooperation links with external agents to attract human, monetary and technical resources that contribute to development, in this case, a socioeconomic development limited by greenhouse gas emissions.

Thus, we will use the capabilities approach as a methodological framework for the design of governmental and non-governmental policies for the development of the country (Robeyns 2016).

The concept of "capacity building" through the UNDP document: Rethinking Technical Cooperation (1993), the OCDE report: Shaping the 21st Century: The Contribution of Development Co-operation (1996) and the UNDP report: Capacity Assessment and Development in a Systems and Strategic Management Context (1998) evolved and was criticized because its results were allegedly insufficient, did not strengthen local capacities and did not make a long-lived change. Then, the concept of "capacity development" emerged as a trend after the adoption, in 2000, of the UN Millennium Development Goals and the 2005 Paris Declaration on Aid Effectiveness. Additionally, the 2008 Accra Agenda for Action provided guidelines for systematically identifying areas where it is necessary to strengthen capacities. The document The Challenge of Capacity Development. Working Towards Good Practice (OECD 2008) defined the most accepted and used concept of "capacity development"; Zamfir (2017) mentions the different international organizations that also define the concept of "capacity development" (UNDP, World Bank, UNECA, USAID, FAO).

The current international development agendas (Sustainable Development Goals 2015) have been expanded to regionally and internationally address: climate change, health, and safety. Taken holistically, "capacity development" is considered the key driver of sustainable development, providing the global flow of knowledge and experience through financial support, non-financial support such as cooperation, and public–private partnership strategies (Greijn et al. 2015).

The first step is to carry out a capacity assessment where the internal (personal) and external conditions (for example, regulatory frameworks) are identified as well as prioritize the actions that must be taken to develop these capacities. Which policies are necessary, and which can be implemented at local, state, federal, and international levels? Based on the analysis of the available documentary information.

To acknowledge this, we came up with a table divided into diagnosis or baseline and the opportunity areas where we need to build capacities in the government and private sectors Table 11.2.

The academic sector also participates as a guest on the ETS Advisory Committee, based on Article forty-seventh of the Agreement (01/10/2019). However, researchers are needed to work on the Emissions Trading Systems, emissions markets and what they imply, researchers are also needed in green finance, international environmental trade. It should be noted that researchers are not only needed to work in these areas, but also in the academic sector so human capital can be developed along these lines of research to cover all the needs of both the government and private sectors.

The drafting of curricula for careers should address climate change in general, but with specialties available in emissions finance, new technologies for emission mitigation, and government innovation for policymaking of instruments to incentivize the domestic market and make it competitive. In general, more specialized human resources are required.

Civil organizations along with academia participate in the integration of the Committee, but are required to demonstrate knowledge about Emissions Trading Systems. However, it is also important to add knowledge of how emissions impact society and how an ETS would improve the well-being of society, particularly its health. Not to mention how it contributes to creating or eliminating jobs and how the obsolete technology is going to be disposed. Further, how companies are going to carry out their reconversion and what it is going to be the impact of that conversion on populations. We agree the knowledge about ETSs is important, however, again we refer to participation of citizens with an interdisciplinary focus and training on both the technical and social basis. They must be committed to the interest of national society and not only the interests of their own respective subsidiaries.

Analysis of Results and Conclusions

In order to comply with international and national agreements on climate change, governance in its broad sense is required: one that goes beyond a limited perspective focused on the allocation of power and the management and administration of resources. Environmental governance implies the representation of multiple and diverse actors with different interests, who, together, both vertically (global, regional, national, and local) and horizontally (government, society, companies, organizations,

Table 11.2 Development of capacities

Diagnosis	Government sector's necessities	Private sector's necessities
Mexico's participation in international environmental organizations	Trained personnel, both technically and diplomatically at SEMARNAT, capable of influencing technical negotiations at international level, considering the national context	Designated staff to collaborate with SEMARNAT and provide information about their sector to assist in international negotiations
Implementation of international protocols and agreements. There is a robust regulatory framework and policy instruments	Designate staff to monitor, evaluate, update, and interact with stakeholders. That observes both international and national movements, to avoid the obsolescence of the regulatory framework and its instruments Implementation of a continuously improving system Work on the involvement of all stakeholders on the public policy process	Personnel trained in design, implementation and evaluation of public policies and their instruments, both technically and in government activities, so they participate actively Representatives of sector's chambers that are sensitive to international and national changes Work on the collection of information that serves as input for the design of public policies
There is a National Registry of Emissions and a COA-WEB	Ensure the robustness of the system, with high-capacity equipment and competent personnel in charge of its information management Update the system based on user needs Staff and organizations specialized in emissions control, validation and verification, in emission calculation methodologies, RENE surveillance and that possess in-depth knowledge of COA-WEB management Trainers for the users with the ability to detect their needs	Computer-trained staff Have the infrastructure that allows the proper access to government platforms Personnel with great knowledge of environmental management and specialists in emissions registration Interdisciplinary teams (finance, operations, administration, etc.) Personnel with extensive knowledge on filling out the COA-WEB, methodologies for calculating emissions and using the RENE
ETS pilot program	Trainers for the sectors involved with the ETS Personnel with knowledge about the emissions market, allowances that are issued in each of the three phases of the ETS, based on the companies' emissions history and international commitments Personnel capable of setting the Emission Cap for the ETS, based on the country's emission reduction goals and international mitigation commitments	Personnel trained in: • Emission inventory, • Emission methodology • Emission market (caps, transactions, efficiency, business opportunities, allowances, offset instruments, auctions, etc.) • The operation of the ETS • Emission mitigation projects • Identify opportunities for emission reductions within companies • New technologies to reduce emissions • Innovation of processes in order to reduce emissions, • Innovation of mitigation projects • Emission finance Companies need to check if they have enough qualified personnel to carry out their emissions measurements and mitigation projects Check the organization's chart for who can absorb new responsibilities or create new departments or, if needed, hire another company or create a new organization

(continued)

Table 11.2 (continued)

Diagnosis	Government sector's necessities	Private sector's necessities
ETS operation	Personnel with extensive knowledge about the unilateral, bilateral, multilateral and indirect links tied to ETS-Mexico Personnel trained in: • Emission control and emissions audits • Methodologies for calculating emissions • Emissions market behavior (price volatility and vulnerability to fraud, speculation, stocks, risks, transactions, auctions, identification of new markets, marketing opportunities) • Set up penalties for noncompliance, financial fines and instruments to complete the regulatory obligation such as compensations for offenses • Design ETS surveillance policies • Research and development of new technologies for emission mitigation and pollution control • Supervise national emission validation and verification organisms • Supervise international organizations in the issuance of certificates to participate in mitigation projects • Evaluate trends with ETS results • Develop the traceability of the ETS • Manage ETS operation • Governmental innovation in the environmental sector (regulatory instruments) • Accountability to guarantee transparency and governance of the ETS	Personnel trained in: • Emission control and emissions audits • Methodologies for calculating emissions • Emissions market behavior (price volatility and vulnerability to fraud, speculation, stocks, risks, transactions, auctions, identification of new markets, marketing opportunities) • Research and development of new technologies for emission mitigation and pollution control • How the ETS operates • How the regulations work and the design of the CO_2 management strategy • Technical knowledge about their industrial facilities in terms of emissions and its future potential reduction • Compensation systems • Drafting of monitoring plans • Calculation methodologies considering measurement instruments, sampling or analysis methodologies that could improve the accuracy in determining emissions

etc.), must resolve all the demands or needs so that they are best served individually and collectively toward a common good, within the framework of a common objective (Ibarra 2019).

Without a doubt, the ETS represents a national challenge. Let us begin with the government infrastructure, where we must have technological and computational strength to house the systems that make up the ETS. There must also be human resources prepared for the different national and international positions that we have described. So far in Mexico, these people have been training on the go, so it is necessary to formally prepare them in areas that we have called interdisciplinary cadres that group social sciences and technical sciences, and diplomacy and solid technical knowledge. This way, they will be capable of influencing international organizations and defend the national context and not only be spectators who accept that developed countries impose policies based on their own interests. This has an impact on the competitiveness of companies, for example, by lowering or raising the level of a certain pollutant. We need to train and empower human resources for four sectors: the international sector, and the national public, private and social sectors.

The ETS is an adaptation of an international policy instrument and since the experience of other countries is available, Mexico can develop a better ETS. Mexico will be the first country in Central and South America to implement it. Therefore, we will be an example to our fellow countries. However, despite Mexico being the most advanced, it must also operate, change its bureaucracy through government innovation that allows us not only to issue new regulatory frameworks, but to overcome the international and national context with tailored responses and mechanisms for transparency, validation, verification, trust, operability, and flexibility. The ETS needs to be constantly innovated so in order to be compatible with the ETSs already implemented around the world. Further, it must also take advantage of the regional and geographical assets that we have and avoid obsolescence. On this matter, government innovation must have its priorities. It must create public policy instruments that help the private sector in Mexico to participate, too. If we consider the actual panorama where most companies in Mexico require development of their capacities so that they may access the ETS, starting with emissions accounting and organizational studies, a question emerges: Who will perform this new task? The profile of companies in Mexico, based on the picture of the obligated subjects, is not very encouraging. It will require a strong investment for its transformation and possible participation in the ETS. Because of this, the government and private sector have to go through this change together. The focus of the government on private initiative will be very valuable toward new regulations. As can be observed in the capacity-building chart, companies must develop interdisciplinary cadres with strategic, technical, and financial skills in emissions trading. They must also have a climate strategy with projects to execute and evaluate them administratively and financially. They, too, have to analyze national and international possibilities for linkages. The ETS requires companies to grow or lose out, since the market will generate requirements where the environmental factor, GHG emissions, certifications, will be necessary for supply chains. The self-knowledge of the companies paired up with a deep knowledge of the regulations will provide great help in detecting areas of opportunity for investment, transformation, adaptation, and innovation. The creation of inter-secretary and transversal programs between government agencies such as the Ministry of the Economy, the Ministry of Finance and Public Credit, the Ministry of Energy, and the National Institute of Ecology and Climate Change must be a priority so that the government can actually and synchronously face the challenges that environmental policy poses.

Mexico not only faces the climate change phenomenon in a physical, territorial way: the rising of sea levels and changing temperatures requires specific infrastructure, the development of new materials and the construction of a health system for new diseases brought on by climate change. Furthermore, Mexico must also face the challenge of incorporating environmental public policies into its economy, since it will be seriously affected if it does not comply with international environmental certifications for the commercialization of its products. Climate change also creates new markets, a new green economy, the carbon economy had already been discussed as a benchmark, as a trial, but now with the implementation of the ETS, it is a necessity for countries and companies to begin to participate in that market.

The ETS brings new ways of negotiating, the creation of new markets, new rules on auctions, transactions, and compensation instruments. Further the creation of new careers, new jobs as national verification organizations rise, and maintenance of computer systems, international markets, green prosecutors, new methodologies, modeling of systems, projections of mathematical models of emissions, and market behavior are needed—everything that the "air economy" implies.

Both the government and companies must invest in infrastructure and training to prepare themselves during this first phase. Additionally, this requires overwhelming participation from academia, while government must step forward to create research incentives in this field and create new careers.

We are going through a radical change where the usual subordination of the environment to economy is going down in history. Now, the environmental end, the reduction of GHG emissions, the reduction of the impact on nature, will be what allows companies to be acknowledged in the market. The environmental certifications that companies must obtain by demonstrating innovation in their processes will be a priority. Production with the least environmental impact will be most valuable. International markets, and especially those that have already implemented ETSs, have made great advances in environmental certifications. Both their supply chains and the final product have environmental guarantees. Here in Mexico, we must work hard, the economic sector has enjoyed great liberties and little regulation and monitoring of national compliance and that has vitiated it. Now, the challenge escalates to an international level, because if they want to continue participating in the market, they must get the work done. The government is complying with the issuance of policies and instruments in compliance with the agreements. However, there is still a great gap between designing them and implementing them, because right there is where it is necessary to team up with companies so they are aware of these guidelines and can comply with them. The ETS pilot program allows us to address this topic and know the situation in our country in practice, so government may prioritize work in this area.

Acknowledgements Thanks to: Instituto Politécnico Nacional, SIP: 20200801.

References

Aguilar L (2009) Gobernanza y gestión pública. Ciudad de México: Fondo de Cultura Económica. Primera edición electrónica 2015. ISBN 978-607-16-3364-4

Bauer S, Busch PO, Siebenhüner B (2006) Administering international governance: what role for treaty secretariats? Global Governance Working Paper No 29. Amsterdam et al.: The Global Governance. Project. www.glogov.org. All rights remain with the authors

Cerrillo Martínez A (2005) La gobernanza Hoy: Introducción, pag. 13, en La Gobernanza hoy: 10 textos de referencia, Cerrillo Martínez, Agusti (coord.), Instituto Nacional de Administración Pública, Madrid

Delbeke J, Vis P (2019) Towards a climate-neutral Europe: curbing the trend. Routledge

Denhardt JV, Denhardt RB (2007) The new public service. Serving, not steering, Expanded Edition, M. E. Sharpe, USA

Durant M (s/a) Gobernanza empresarial y organización sobre el comercio de emisiones en el Manual de Capacitación sobre la preparación para el Mercado de Carbono, IETA, Climate Challenges Market Solutiones

Georgiadou Y, Reckien D (2018) Geo-information tools, governance and wicked policy problems problems. Int J Geo-Inf 1–10.https://doi.org/10.3390/ijgi7010023

Greijn H, Hauck V, Land A, Ubels J (2015) Capacity development & technical cooperation: capacity development beyond aid, May 2015. Fecha de acceso: 19de agosto 2020. https://www.die-gdi. de/uploads/media/CAPACITY_BOOKLET_ENG_WEB.pdf

IbarraSarlat R (2019) Cambio climático y gobernanza. Una visión transdisciplinaria. Universidad Nacional Autónoma de México, Instituto de Investigaciones Jurídicas

Instituto Nacional de Ecología y Cambio Climático (2018) Contexto Internacional en materia de Cambio Climático. Fecha de acceso: 25 May 2020 https://www.gob.mx/inecc/acciones-y-progra mas/contexto-internacional-17057

Kooiman J, Bavinck M, Chuenpagdee R, Mahon R, Pullin R (2008) Interactive governance and governability: an introduction. J Transdiscipl Environ Stud 7(1):1–11

Marsh D (2008) Understanding British government: analyzing competing models. J Polit Int Relat 10(1):251–268

Martínez Rodríguez MC (2015) Gobernanza ambiental: Orígenes y Estudios de caso. Plaza y Valdés

OECD (1996) DAC. Development Assistance Committee. Shaping the 21st century: the contribution of development co-operation. https://www.oecd.org/dac/2508761.pdf

OECD (2008) The challenge of capacity development: working towards good practice. OECD J Dev 8/3. https://doi.org/10.1787/journal_dev-v8-art40-en

Pierre J, Peters G (2000) Governance, politics and the state. Basingstoke, Macmillan

Porter S (2007) Who will guard the guardians themselves?: contributions of the capability approach to capacity development evaluation frameworks. Thesis, University of Cape Town, Faculty of Humanities, Department of Political Studies. http://hdl.handle.net/11427/3716

Robeyns I (2016) The capability approach, The Stanford Encyclopedia of Philosophy (Winter 2016 Edition), Zalta EN (ed.) Fecha de acceso: 15 may 2020 https://plato.stanford.edu/archives/win 2016/entries/capability-enfoque/

Sen A (1993) Capability and well-being. In Nussbaum, Sen A (eds) Quality of life. Clarendon Press, Oxford

Sustainable Development Goals (2015) https://www.un.org/sustainabledevelopment/

Treviño Cantú JA (2011) Gobernanza en la administración pública. Revisión teórica y propuesta conceptual Contaduría y Administración, núm. 233, enero-abril, 2011, pp 121–147. Universidad Nacional Autónoma de México, Distrito Federal, México

UNDP (1993) Rethinking technical cooperation—reforms for capacity building in Africa. UNDP Regional Bureau for Africa, Development Alternatives Inc., Elliot J. Berg, Coordinator

Whittingham MV (2010) ¿Qué es la gobernanza y para qué sirve? Revista Análisis Internacional 2:219–235

Zamfir I (2017) Understanding capacity-building/capacity development: a core concept of development policy. EPRS|European Parliamentary Research Service, Fecha de acceso 20 de Agosto de 2020. https://www.europarl.europa.eu/thinktank/en/document.html?reference= EPRS_BRI(2017)599411

Zurbriggen C (2011) Gobernanza: una mirada desde América Latina. Perfiles Latinoamericanos 39–64

International Regulatory Framework

Convención Marco de Naciones Unidas sobre Cambio Climático. https://unfccc.int/es
Intergovernmental Panel on Climate Change, IPCC. https://www.ipcc.ch/

National Regulatory Framework

CPEUM, Constitución Política de los Estados Unidos Mexicanos. http://www.diputados.gob.mx/LeyesBiblio/ref/cpeum.htm
DOF, Diario Oficial de la Federación con fecha 01/10/2019. https://www.dof.gob.mx/index.php?year=2019&month=10&day=01
LGEEPA, Ley General del Equilibrio Ecológico y la Protección al Ambiente. http://www.semarnat.gob.mx/gobmx/biblioteca/index.html
LGCC, Ley General de Cambio Climático. http://www.semarnat.gob.mx/gobmx/biblioteca/index.html
RLGCCMRENE, Reglamento de la Ley General de Cambio Climático en Materia del Registro Nacional de Emisiones. http://www.semarnat.gob.mx/gobmx/biblioteca/index.html

Institutions

Registro Nacional de Emisiones (RENE). https://www.gob.mx/semarnat/acciones-yprogramas/registro-nacional-de-emisiones-rene
Secretaría de Energía (SE). https://www.gob.mx/sener
Secretaría de Medio Ambiente y Recursos Naturales (SEMARNAT). https://www.gob.mx/semarnat

Part III
Mexican ETS Connected Issues with the Broader Climate Agenda

Chapter 12
The Environmental Justice Dimension of the Mexican Emissions Trading System

Danae Hernandez-Cortes and Erick Rosas-López

Abstract Emissions trading systems have the potential of increasing air quality given that GHG emissions are often co-produced with local pollutants such as NO_x, SO_x, and Particulate Matter (PM). Can emissions trading systems exacerbate or alleviate environmental justice concerns in emerging economies? According to the U.S. Environmental Protection Agency, Environmental Justice is achieved when no group is disproportionately affected by an environmental policy or phenomenon. The main objective of this chapter is to estimate the pollution burden faced by marginalized neighbourhoods in Mexico. This is relevant for Mexico given the beginning of the pilot program of the Mexican Emissions Trading System (ETS) and the country's history of income inequality and poverty. Using linear regression and two-way fixed effects methods, we found that the highest emitters regulated under the ETS are located near poor populations. We estimated a 5% CO_2 emissions-reduction scenario corresponding to national targets and associated NO_2 emissions to that scenario. We find that this scenario is consistent with a decrease in the exposure of NO_2 pollution for the most marginalized neighbourhoods. This chapter also discusses other potential sources of environmental injustice that could result after the beginning of the ETS and the potential to address them.

Keywords Emissions trading · Environmental justice · Co-pollutants

D. Hernandez-Cortes (✉)
School for the Future of Innovation in Society, School of Sustainability, Arizona State University, PO Box 875603, Tempe, AZ 85287-5603, USA
e-mail: Danae.Hernandez-Cortes@asu.edu

E. Rosas-López
Coordination for Climate Change Mitigation, Instituto Nacional de Ecología y Cambio Climático, Boulevard Adolfo Ruiz Cortines 4209, Jardines en La Montaña, Tlalpan, Mexcio City, Mexico
e-mail: erick.rosas@inecc.gob.mx

S. Lucatello (ed.), *Towards an Emissions Trading System in Mexico: Rationale, Design and Connections With the Global Climate Agenda*, Springer Climate,
https://doi.org/10.1007/978-3-030-82759-5_12

Introduction

In many countries, Emissions Trading Systems (ETS) as well as other market-based approaches have been considered as instruments to achieve Greenhouse Gas (GHG) emissions reductions. However, there have been concerns regarding emissions trading systems increasing existing gaps in pollution exposure across space. Much of the opposition towards emission trading systems and carbon taxes stem from environmental justice concerns. Environmental justice concerns arise if there are differences in environmental quality across income levels. More specifically, some of these environmental justice concerns are centred around vulnerable people experiencing higher levels of exposure to local pollutants. Local pollutants like particulate matter, SOx, and NOx are short-lived and their damages to nearby populations depend on where they are released, while GHG emissions cause long-lasting global damages.

Studies have analysed the distributional impacts of emissions trading systems from an empirical perspective (Fowlie et al. 2012). Whether they will be effective in reducing disparities in air pollution exposure depends on the answers to three questions. First, what is the current spatial relationship between CO_2 emissions and air pollution exposure? Second, are polluting facilities located near vulnerable populations? Third, what is the spatial relationship between regulated facilities and air pollution exposure after the policy is implemented? The objective of this chapter is to analyse the current spatial relationship between regulated emissions and air pollution exposure as well as the characteristics of the populations near regulated facilities. Understanding these questions will provide background for future work that analyses whether Mexico's GHG emissions trading system can improve environmental justice among Mexican communities.

Mexico started the first year of the pilot of its GHG emissions trading system in January 2020. The program regulates heavily polluting industries that emit more than 100,000 metric tons of CO_2 of total annual direct emissions coming from stationary sources. This program is one of the pillars of an ambitious climate policy in Mexico and, according to authorities, already covers around 40% of national GHG emissions in the first year of its pilot phase (SEMARNAT 2019b). If the target is achieved, besides reductions in CO_2 emissions, Mexico could expect co-benefits in air quality as is the case with other GHG emissions trading systems around the world (Hernandez-Cortes and Meng 2020; Walch 2018). However, despite Mexico having high rates of poverty and inequality, the current policy does not explicitly include a special emphasis on involving vulnerable populations as possible stakeholders in the climate policy. Other cap and trade systems in the world have had a special interest in helping vulnerable communities either through revenue recycling or investments in climate-friendly projects in these communities.[1]

---

[1] For instance, California auction proceeds are used to fund projects in disadvantaged communities (as legally defined by the state) such as air quality monitoring stations and projects with air quality co-benefits.

This chapter will discuss possible air pollution co-benefits from Mexico's emissions trading system and analyse whether vulnerable communities could benefit from decreases in GHG emissions. Given that emissions coming from local pollutants are often co-generated with GHG emissions, by reducing GHG emissions, there could be potential gains in local pollution reductions. A large body of academic literature has documented the impacts of pollution on health, showing that pollution can be especially detrimental for vulnerable populations such as the elderly, children, and low-income groups (Deryugina et al. 2019; Arceo et al. 2016; Gutierrez 2015). Therefore, this chapter will analyse whether regulated facilities under Mexico's ETS are located near vulnerable populations and their pollution exposure. We find that the highest CO_2 and NO_2 emitters are located close to disadvantaged communities, mainly in urban areas. These emitters are electricity generators and oil and cement producers. We find that the Mexican ETS is likely to account for nearly 90% of CO_2 emissions and 40% of NO_2 emissions coming from all stationary sources in our dataset. We simulate CO_2 emissions for the first year of the pilot program following the expectations of the Mexican government and find that disadvantaged communities might experience a large decline of NO_2 emissions compared to baseline levels. Moreover, this decrease is higher for vulnerable communities than other communities.

The rest of the chapter is structured as follows. Section 12.1 defines environmental justice and the role of emissions trading systems in exacerbating or reducing existing gaps in environmental justice. Section 12.2 describes the context of regulated facilities under Mexico's ETS and their characteristics. Section 12.3 describes the data sources used in this chapter. Section 12.4 discusses the empirical framework as well as the results. Finally, Sect. 12.5 concludes and puts forward discussion questions for future research in the context of Mexico's ETS.

Environmental Justice and Emissions Trading Systems

The US Environmental Protection Agency defines environmental justice as the "fair treatment and meaningful involvement of all people regardless of race, colour, national origin, or income with respect to the development, implementation, and enforcement of environmental laws, regulations, and policies". Regarding fair treatment, this definition refers to a situation where "no group of people should bear a disproportionate share of the negative environmental consequences resulting from industrial, governmental, and commercial operations facilities" (EPA 2018). Social justice concerns about environmental policy have a longstanding history. Some of these concerns are focused on how the burden of environmental phenomena like air or water pollution falls on poor and minority communities.

Environmental justice concerns can be further divided into exposure and policy incidence. Environmental justice in exposure can be understood as vulnerable communities being systematically located near more polluting areas due to external reasons such as land prices (Banzhaf et al. 2019). Environmental justice in policy incidence can be understood as vulnerable communities being disproportionately affected by environmental policies such as relocation of facilities (Liu 2013).

Environmental justice concerns have received attention from policymakers while trying to elaborate climate policy. In the case of emissions trading systems, studies have found that poor and minority communities are located near disadvantaged communities (Cushing et al. 2018). Other studies have suggested that emissions trading systems might decrease the amount of pollution exposure that low-income and minority communities face (Grainger and Ruangmas 2018; Hernandez-Cortes and Meng 2020). In the case of the United States, current discussions about climate policy mention justice to all communities but with a special focus on low-income communities, indigenous peoples, and communities of colour.[2]

Existing studies suggest that low-income communities are located near heavily polluted areas (Currie et al. 2011). In the case of Mexico, Chakraborti and Margolis (2017) found that poorer communities in Mexico are located near the firms that release higher toxic pollution. The authors find that plants near vulnerable communities (measured by the Urban Marginalization Index) emit 87% more cyanide and 72% more arsenic and chromium than the average community. Other subnational studies have found a similar result: more polluting firms are located near poorer communities (Lara-Valencia et al. 2009; Grinesky and Collins 2008). These studies have found that vulnerable communities are located close to the highest polluting facilities in Mexico. This means that there might be inequality in environmental exposure to air and water quality in Mexico. Other studies that have approached environmental justice concerns in Mexico are Mahady et al. (2020) and Lome-Hurtado et al. (2019). However, there are no estimates on the incidence of environmental policy on environmental justice. To our knowledge, this is the first effort to analyse the potential environmental justice benefits of a GHG policy in the context of Mexico.

This chapter aims to provide descriptive evidence of the characteristics of the populations living near the facilities regulated by the Mexican ETS. Although the Mexican ETS is regulating greenhouse gases that have no direct impact on local air quality, there could be co-benefits associated with the reduction of CO_2 emissions. By setting a cap on the amount of CO_2 emissions, facilities might emit lower emissions coming from stationary sources such as SO_2 and NO_2. At the same time, these pollutants reduce the amount of harmful secondary pollutants such as $PM_{2.5}$, which have been found to have severe consequences to long-term health in affected populations (Deryugina et al. 2019).

Mexico ETS Context and Inequality

The Mexican ETS started its pilot phase in January 2020. This phase will last three years with a transition period in the third year to the fully operating system in 2022.

---

[2] See the Climate Equity Act (Harris 2019).

The rules currently regulate facilities in the energy and industrial[3] sectors that have emitted at least 100,000 metric tons of CO_2 in any year during the 2016–2019 period. Throughout the pilot phase, emissions allowances will begin to be allocated via grandfathering to the facilities, according to historical emissions, the NDC target, and sectoral targets stated in the General Law of Climate Change (SEMARNAT 2019a), with the possibility of auctioning allowances after the first pilot year. The inclusion of other greenhouse gases, sectors, and allowance allocation processes will be evaluated before the start of the operating phase. The cap was announced in late November of 2019: 271.3 and 273.1 million allowances will be available in 2020 and 2021, respectively. A backup reserve of 20% additional allowances is also in place and rules for offsets will be developed during the initial phase.

Mexico faces inequality not only across income and wealth but also gender, ethnicity, access to public services, and, more generally, opportunity (Altamirano and Flamand 2018). There is also stark geographic inequality: some regions within the country are developing fast, while others have been lagging for years (Esquivel 1999; Dávila et al. 2002). The birthplace of a person may play an important role in determining the opportunities they can access, thereby limiting or enhancing the capabilities they can develop (Altamirano and Flamand 2018, p. 28). For example, according to the Social Mobility Report in Mexico 2019 (CEEY 2019), 86% of Mexicans born in poverty[4] in the south of Mexico remain in poverty during their adult life while the same is the case for 54% of people born in poverty in the northern region.[5] Another dimension of inequality, which may compound and intertwine with the others, is the environmental burden. While some pollutants that affect air quality may be more severe in more prosperous urban places, due to other institutional factors, high-polluting facilities may locate in relatively less developed regions, burdening the less advantaged communities inhabiting therein.

Several studies have analysed the possible sources of environmental injustice due to local climate policy. In the case of cap and trade programs, Kaswan (2008) explains some of the environmental justice concerns due to GHG cap and trade systems. The advantage of cap and trade programs is the low cost of regulation compared to other regulations like command and control. As Kaswan (2008) explains it, facilities can align their emissions to the number of allowances either by reducing emissions to the allowance levels, reducing emissions to less than the given allowances and selling the rest, or buying allowances to compensate their excess emissions. Thus, by buying and selling allowances, the spatial distribution of emissions is expected to change. Given that GHGs are produced together with other co-pollutants such as NO_x, SO_x, particulate matter, and toxic substances, a cap on GHG might decrease pollutant emissions.

---

[3] The energy sector includes electricity generation and oil production whereas the industrial sector includes automotive, cement and lime, chemicals, food and beverages, glass, steel, metal, mining, petrochemical, and paper and cellulose.

[4] Measured in this case as the bottom 40% of a wealth index using household characteristics from a social mobility survey.

[5] The southern states are Campeche, Chiapas, Guerrero, Oaxaca, Quintana Roo, Tabasco, Veracruz, and Yucatán whereas the northern states are Baja California, Chihuahua, Coahuila, Nuevo León, and Sonora y Tamaulipas.

These pollutants and toxins are known to have several health consequences to the populations living nearby the polluting facilities. Therefore, by imposing a cap on GHG, there could be improved health outcomes for those living near regulated facilities. This relationship depends on whether GHG and local pollutant emissions are complements or substitutes in the production activity. Holland (2012) finds evidence supporting the idea that GHG and local pollutants are complements in the production process.[6]

Whether a GHG cap and trade will decrease the production of co-pollutants depends on facilities' technology and abatement options. Fowlie et al. (2012) explain that cap and trade programs could exacerbate pollution in historically disadvantaged communities. The authors mention that large polluting facilities can purchase allowances and produce higher pollution, which could increase pollution exposure near these areas. However, under a cap and trade program, we would expect a higher reduction of pollution coming from low-abatement cost facilities than other facilities (Fowlie et al. 2012; Burtraw et al. 2005). If these low-abatement cost facilities are located near disadvantaged communities, there could be environmental justice co-benefits from a GHG emissions trading system.

Other GHG emissions trading programs have tried to address existing environmental inequalities and future differences in emissions by implementing policies that monitor pollution from regulated facilities in vulnerable communities. For instance, California's emissions trading program under AB 32 has an explicit target to help disadvantaged communities by funding public investments that could improve air quality among these communities using the auction proceeds from the cap and trade system. Moreover, California's AB 32 has implemented environmental justice committees in disadvantaged communities where community leaders can propose new programs to improve air quality in their communities. Other jurisdictions, such as the EU-ETS, address fairness issues at the Member State level, reflecting income disparities between countries: Distributing 10% of allowances for growth and solidarity reasons, financially supporting the modernization of the energy sector, and offering partial free allocation to the power sector in exchange for low-carbon investments (Meadows et al. 2020). Revenues from auctioning allocations are mostly used on further reducing GHG emissions in other sectors, R&D, and supporting lower- and middle-income households in order to address social aspects (Borghesi et al. 2016).

In the case of Mexico's ETS, environmental justice concerns have not yet been fully operationalized as a policy target. However, in supporting documents to the design of the Mexican ETS, the Ministry of Environment (SEMARNAT) has considered the potential of directing revenue collected through the auctioning of emissions allowances to decarbonization projects or to mitigate unwanted distributional effects (SEMARNAT and GIZ 2018). Additionally, the Ministry of Environment will be

---

[6] The main effect found by the author is driven by changes in the amount of output produced. This could be different if there are other abatement strategies such as a change in fuel. For instance, in the case of vehicles, diesel tends to be more fuel-efficient than gasoline but produces more nitrogen dioxide.

able to conduct auctions from the second pilot year on, to gain experience in this regard and further discuss the use of these revenues (SEMARNAT 2019b).

Data Sources

This chapter analyses whether vulnerable communities are located close to the higher polluting facilities in Mexico and simulate likely CO_2 reductions under the ETS. In order to do so, we use data on vulnerability measures as well as emissions data from all polluting facilities in Mexico.

Vulnerability data: We used two main data sources to classify the vulnerability level in communities across Mexico. For the urban areas, we used the 2010 Index of Urban Marginalization calculated by CONAPO at the urban AGEB level. AGEBs are the smallest spatial unit in Mexico used by INEGI.[7] Similar to census tract information, AGEBs are small spatial units comprised of less than 50 street blocks. CONAPO calculates the urban marginalization index for all urban AGEBs by using AGEB-level census data on a set of poverty and income indicators.[8] CONAPO then divides the AGEBs into five different categories regarding their marginalization index at the national level: "very low", "low", "medium", "high", and "very high". We should expect higher income communities to be in the "low" and "very low" marginalization groups and poorer communities to be in the "high" and "very high" categories. To account for rural areas, we also included the 2010 locality index of marginalization at the rural locality level. Rural localities are smaller in extension and population than their urban counterparts; therefore, they are more comparable in extension to urban AGEBs than urban localities. Analogous to the urban marginalization index, the locality marginalization divides localities as "very low", "low", "medium", "high", and "very high" with the same indicators.

Emissions data: Emissions data at the facility level come from the *Registro de Emisiones y Transferencia de Contaminantes* (RETC) compiled by SEMARNAT. This registry contains all toxic emissions as well as CO_2 and NO_2 at the plant level for regulated entities by NOM-165-SEMARNAT-2013. This dataset contains year-level pollution emissions emitted by regulated stationary sources.[9] We consider the toxic, local pollution (NO_2) and CO_2 emissions of all these facilities. We focus on

---

[7] CONAPO (the National Population Council) analyses demographic information. INEGI (the Statistics and Geography Institute) is in charge of compiling and collecting nationally relevant information in Mexico.

[8] The variables are percent of children that do not attend school, adult population without elementary education, population without access to health services, and percent of infant mortality. Other variables included are percent of households without running water, households without sewage, households without a bathroom, households with firm floor, households with a high number of inhabitants, and households without refrigerators.

[9] The dataset contains geographic coordinates in a variety of formats. These were transformed to decimal degrees when possible. Some other important facilities in terms of emissions were manually added when coordinates were incorrect/unavailable. Other errors were manually corrected.

NO_2 given that is the only local pollutant consistently reported in RETC throughout the years we analysed. NO_2 has harmful health effects such as lung damage and is an important precursor of $PM_{2.5}$ and ground ozone, both of which are associated with other health effects such as asthma and chronic bronchitis among others. Although RETC has a very complete record of pollution emissions, it is not the main GHG emissions registry used for Mexico's emissions trading systems. The system used for regulating and monitoring these emissions is the *Registro Nacional de Emisiones* (RENE). However, this system is confidential and the data are not available. This chapter considers the emissions in RETC a good proxy of GHG and pollution emissions coming from stationary sources. RETC only considers stationary sources whereas RENE considers additional mobile sources and indirect emissions coming from electricity use. Differences between RENE and RETC are expected to arise from mobile emissions and electricity use. Therefore, since RETC only has information on point sources, the emissions in RETC are a proxy for overall CO_2 emissions and a lower bound for the emissions in RENE. RETC contains data for the 2004–2018 period. However, we restrict the data to the 2016–2018 period given that these are the relevant years for inclusion into the emissions trading program.

Linking vulnerability level to emissions data: We linked the emissions data to each area's vulnerability level by using the coordinates of the RETC facilities to link each facility to its corresponding urban AGEB. Given the scattered distribution of rural localities, we calculated a buffer of 1 km^2 surrounding the locality and associated the RETC facilities within this buffer. In a few cases, there is an overlap between the rural locality buffer and the urban AGEBs. We kept two records for these facilities to account for both communities.

Analysis

GHG Emissions in Mexico and Covered Entities

Figure 12.1 shows average emissions by industry using data from RETC. Electricity generation produces the largest CO_2 emissions, followed by cement and oil producers. In the case of NO_2, the largest emissions come from electricity generation. Figure 12.2 Panel (a) shows the location of the 2018 RETC facilities by the marginalization level of the locality/AGEB that contains it. As an example of the spatial distribution of firms, Fig. 12.2 Panel (b) shows their location in Greater Mexico City.

While RETC facilities are located across the country, some areas have a higher point density, like the central region of the country—Greater Mexico City and the *Bajío* region—as well as some industrial areas along or near the northern border. No clear pattern of marginalization emerges although it appears that facilities with higher marginalization levels are concentrated in centre-to-south Mexico. In the case of Greater Mexico City, Panel (b) shows that there can be a juxtaposition of different marginalization levels in facilities close by, although a larger trend can be observed in

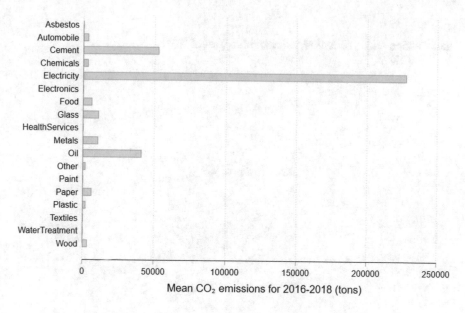

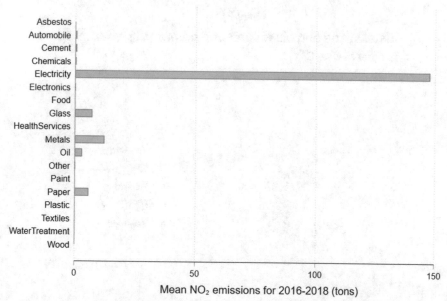

Fig. 12.1 Average CO_2 and NO_2 emissions by industry *Notes* Authors' estimations using publicly available data from RETC. These figures show the average CO_2 and NO_2 emissions for the 2016–2018 period classified by sector. NO_2 and CO_2 emissions are expressed in tons.

Panel a)
Installations and disadvantaged community level, 2018

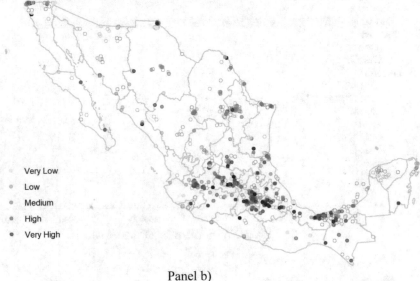

Panel b)
Installations and disadvantaged community level, 2018
Greater Mexico City

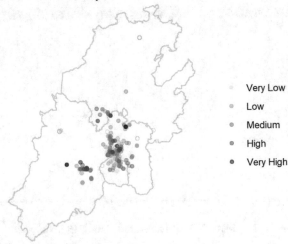

Fig. 12.2 Map of facilities under RETC and marginalization levels *Note* Maps created by the authors using data from RETC installations and CONAPO's urban AGEB and rural locality marginalization index. Panel (a) shows the geographic location of 2018 RETC installations in colours according to the marginalization level of the AGEB or locality that contains it. Panel (b) zooms in on the Greater Mexico City region. Transparent datapoints show facilities without matching AGEBs or localities.

which better-off AGEBs are located within the centre of Mexico City, and some other facilities are concentrated in peripheral areas with higher levels of marginalization.

Mexico's ETS is planned to cover over 40% of total GHG emissions in the country and inclusion in the program depends on whether CO_2 emissions crossed the 100,000 tons threshold in any of the years of the 2016–2019 period. To analyse possible gains on GHG reductions due to the ETS, we considered the regulation threshold for the ETS. In Fig. 12.3, Panels a) and b) show the proportion of CO_2 and NO_2 emissions covered by the ETS as a function of the threshold above which facilities are automatically enrolled in the program. To calculate the coverage, we computed the annual average for the period 2016–2018 considering facilities above the CO_2 emissions regulation threshold. This number was then divided by the yearly average total emissions in our dataset—either CO_2 or NO_2—for the same period. Panel c) shows the number of facilities that participate in the emissions trading system, also as a function of the threshold. The coverage for CO_2 is close to 90% and above 40% of NO_2. In contrast, facilities vary more with the threshold level, showing that regulating 287 facilities (under the current threshold in our dataset) compared to 305 under a more stringent cap may have monitoring and implementation costs and not as much gains in environmental coverage.

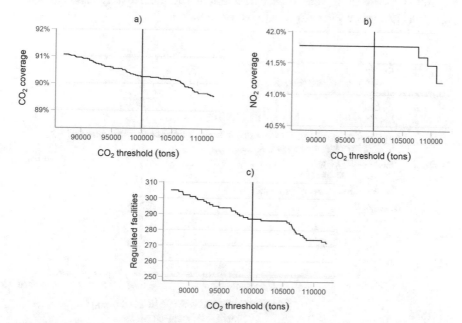

Fig. 12.3 Emissions threshold, coverage, and regulated facilities *Note* Authors' calculations using data from RETC for the 2016–2018 period. Panels a) and b) show the percentage of CO_2 and NO_2 emissions covered by the ETS as a function of the CO_2 threshold expressed in tons, with respect to the total average emissions in our dataset during the relevant period. Panel c) shows the number of regulated facilities with the CO_2 threshold expressed in tons.

Characterization of GHG Emissions and Environmental Justice

Figure 12.4 shows average pollution emissions by marginalization level for the three largest CO_2 emitting sectors for both rural and urban areas. This figure shows that on average, electricity generators with the highest emission levels are located in urban areas with "very high" marginalization levels. However, rural areas with "high" marginalization levels also face high levels of emissions coming from electricity generation. In the case of cement production, both urban and rural communities with "high" marginalization levels have facilities that release the highest levels of CO_2 emissions.

Table 12.1 shows the descriptive statistics of facilities regulated by the ETS. As expected, regulated facilities show higher CO_2 emissions as well as other pollutants emissions. Furthermore, regulated facilities are in neighbourhoods with higher levels of marginalization than non-regulated facilities.

In order to further explore these differences, we use a two-way fixed effect regression where we account for year and municipality fixed effects in order to control for emissions driven by year-to-year fluctuations and municipality characteristics. Equation (12.1) shows our empirical specification.

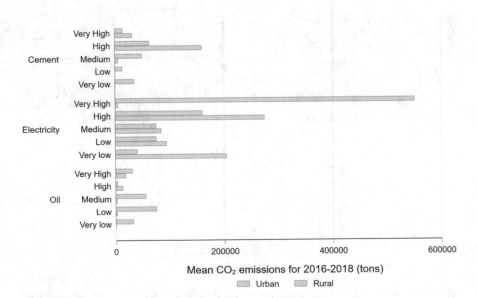

Fig. 12.4 Emissions by marginalization level for the most polluting sectors *Note* Authors' estimations using data from RETC (CO_2 and NO_2 emissions) and the CONAPO's urban AGEB and rural locality marginalization index. The figure shows the average CO_2 emissions during the 2016–2018 period by marginalization level for rural and urban areas. CO_2 emissions are expressed in tons.

Table 12.1 Facilities' descriptive statistics

	(1)	(2)
	Facilities with emissions lower than 100,000 tons of CO_2	Facilities with emissions higher than 100,000 tons of CO_2
CO_2 emissions (tons)	(11,890.756) 2,088.015	(809,918.050) 462,367.479
NO_2 emissions (tons)	(82.808) 1.392	(1577.329) 235.962
Lead emissions (tons)	(238.078) 3.774	(334.841) 31.397
Cadmium emissions (tons)	(4.384) 0.088	(48.080) 4.669
Very low marginalization (share)	(0.423) 0.233	(0.347) 0.140
Low marginalization (share)	(0.388) 0.185	(0.325) 0.120
Medium marginalization (share)	(0.433) 0.250	(0.446) 0.273
High marginalization (share)	(0.380) 0.175	(0.419) 0.227
Very high marginalization (share)	(0.174) 0.031	(0.250) 0.067
Rural (share)	(0.376) 0.170	(0.443) 0.267
Observations (unique facilities)	6,398	150

Note column (1) shows the descriptive statistics for non-regulated facilities under the Mexican emissions trading system and column (2) shows the descriptive statistics for regulated facilities under the Mexican emissions trading system. The emissions are reported in tons. The marginalization levels refer to the populations near these facilities. The regulation threshold is average annual CO_2 emissions higher than 100,000 tons in the 2016–2018 period. This table is restricted to facilities located in populated census tracts

$$Y_{it} = \alpha + \sum_{i=1}^{5} \beta_i 1\{MarginalizationLevel_i\} + X_i + \gamma_t + \mu_m + \epsilon_{it} \qquad (12.1)$$

where Y_{it} is CO_2 emissions at locality/AGEB i in year t during the period 2016–2018. $1\{MarginalizationLevel_i\}$ are indicator variables that equal one for each marginalization level. γ_t are year fixed effects, μ_m are municipality fixed effects, and X_i is an indicator of whether the AGEB is rural or urban. ϵ_{it} is the standard error clustered at the rural locality/AGEB level. Each specific β_i shows the marginal difference in emissions compared to a base category, which in our case will be the "very low" marginalization level. Estimating this regression allows us to control for municipality-specific time-invariant effects during the 2016–2018 period. Examples of these variables are municipality-specific environmental programs, infrastructure, or municipality government characteristics, among others. Adding year-specific effects allows us to control for emission changes affecting all localities that are specific to one year. For example, a drop in emissions due to slower economic conditions during a specific year.

Panel (a) of Fig. 12.5 shows the coefficients estimated from Eq. (12.1) along with their confidence intervals for CO_2, NO_2, lead, and cadmium emissions. These results imply that AGEBs/rural localities with "high" marginalization levels are on average exposed to 7,600 additional tons of CO_2 from the facilities located nearby compared to communities with "very low" marginalization levels. To the extent that these CO_2 emissions are produced with co-pollutants, communities with "high" marginalization levels could be exposed to higher local pollution emissions than communities with "low" marginalization levels. This implies that an emissions trading program that targets facilities with high CO_2 emissions could potentially benefit communities with "high" marginalization levels, conditional on existing co-benefits between CO_2 emissions reductions and other local pollutants. As an illustrative comparison, we estimated Eq. (12.1) using other pollutants (NO_2) and toxic emissions (Cadmium and Lead) in subpanels (b)–(d) of Fig. 12.5. Compared to the results for CO_2, we cannot conclude that the emissions are significantly different across different income groups. However, this does not conclusively prove that there is no detectable difference in emissions by marginalization group, as we may simply lack the precision to estimate it. Future work could look at other pollution data such as air quality monitoring data near these facilities in order to further characterize this relationship. Panel (b) of Fig. 12.5 shows the corresponding coefficients for Fig. 12.5 where the "very low" marginalization level is the base category.

Simulation of Mexico's ETS and Environmental Justice

The Mexican emissions trading program has the potential to create co-benefits in air quality improvements while reducing CO_2 through cap and trade. As explained before, this will be determined by the correlation between CO_2 emissions and co-pollutants. In order to explore the potential improvements in air quality as a result of the emissions trading program, we simulate an emissions-reduction scenario, by means of decreasing CO_2 and NO_2 emissions by 5% in the first year of the program with respect to the 2016–2018 average for regulated facilities. This is consistent with the Mexican emissions reduction target for the industrial sector, as indicated in the General Law of Climate Change.[10] It should be noted that other feasible scenarios include non-uniform reductions within sectors, which also would be consistent with

---

[10] The Law (Cámara de Diputados 2018) establishes targets to be met in 2030 with respect to a baseline of the following sectors: Transport (-18%), Electricity generation (-31%), Residential and commercial (-18%), Petroleum and gas (-14%), Industry (-5%), Agriculture and livestock (-8%), and Waste (-28%). Although the time frame of our simulation is different, we assume that

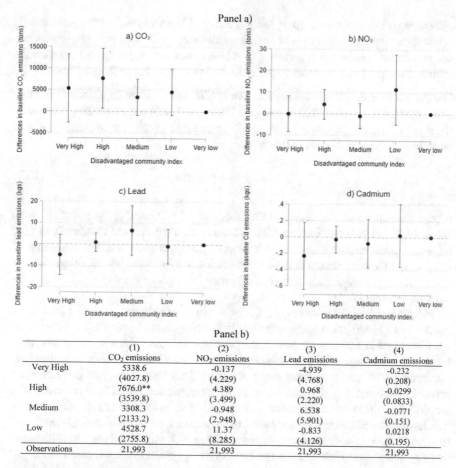

Panel a)

Fig. 12.5 Differences in emissions by marginalization *Notes* Authors' estimations using data from RETC (CO_2, NO_2, Lead, and Cd emissions) and the CONAPO's AGEB and rural locality marginalization index. Panel a): Subpanels a)–d) show a different estimation of Eq. (12.1) where the dependent variable changes for each specific pollutant. The x-axis denotes the marginalization level of the exposed communities. The y-axis denotes the difference in baseline emissions of the respective pollutant for each marginalization level with respect to the "very low" marginalization level. Panels (a)–(d) are using 21,993 observations which represent a yearly observation per plant-locality pair for the 2016–2018 period. The points are the point estimates of Eq. (12.1) with 95% confidence intervals using clustered standard errors at the locality level. Panel (b): Regression results associated with panel (a) results. Standard errors clustered at the locality level in parenthesis.

Panel b)

	(1) CO_2 emissions	(2) NO_2 emissions	(3) Lead emissions	(4) Cadmium emissions
Very High	5338.6	-0.137	-4.939	-0.232
	(4027.8)	(4.229)	(4.768)	(0.208)
High	7676.0**	4.389	0.968	-0.0299
	(3539.8)	(3.499)	(2.220)	(0.0833)
Medium	3308.3	-0.948	6.538	-0.0771
	(2133.2)	(2.948)	(5.901)	(0.151)
Low	4528.7	11.37	-0.833	0.0218
	(2755.8)	(8.285)	(4.126)	(0.195)
Observations	21,993	21,993	21,993	21,993

the Mexican emissions reduction target. This is the case for the electricity sector: it typically has a lower abatement cost than other sectors such as cement production and

---

a linear emissions path that accomplishes the 5% reduction in 2030 would probably also accomplish it in 2019 or 2020.

oil refining (Friedmann et al. 2019; INECC 2018).[11] We could expect that installations in this sector become net sellers of emissions allowances, thereby reducing emissions and associated co-pollutants locally. Our scenario is, therefore, a lower bound for the spatial and equity consequences of emissions reductions due to the Mexican ETS.

For the 5% uniform decrease scenario, we predict the average emissions in the first year of the pilot program (2020) using the 2016–2018 data and estimating a two-way fixed effect regression given in Eq. (12.2) in order to obtain the average predicted emissions in the period after 2016–2018.

$$Y_{it} = \alpha_0 + m_s + r_t + u_{it} \tag{12.2}$$

where the dependent variable is either the tons of CO_2 and NO_2 emissions with sector (m_s) and year (r_t) fixed effects. We obtained the predicted values of CO_2 and NO_2 and simulated a 5% decrease scenario with respect to the average emissions of CO_2. Using these predicted emission reductions, we followed a similar approach to Eq. (12.1) and obtained the percent difference in emissions compared to the "very low" base category.[12]

Panel a) of Fig. 12.6 shows the findings of our simulation. Panel a) shows the results for CO_2 and panel b shows the results for NO_2. In the case of CO_2, we find that a 5% decrease in emissions results in the previous differences across marginalization levels disappearing. Whereas baseline emissions indicate that "high" marginalization areas had on average more emissions than "very low" ones, and this reduction scenario levels the situation by making differences indiscernible. In the case of NO_2, we find that there are differences in the predicted emissions across marginalization levels. Compared to the "very low" base category, communities with "medium" marginalization levels are expected to have higher NO_2 emissions with a 5% decrease in CO_2 emissions. However, in the case of NO_2, communities under the "high" marginalization level do not have higher predicted emissions compared to the "very low" marginalization communities. Therefore, we find a decrease in the exposure of NO_2 pollution for the most vulnerable areas but increases to other communities in the "medium" and "high" marginalization levels. However, for the "high" marginalization communities, the increase is not significant. Panel b) of Fig. 12.6 shows the coefficient results associated with Panel a) where "very low" is the base category.

There are potential limitations to our methods. For instance, we do not consider the fate and transport of pollution in the environment, which could potentially change the conclusions of our NO_2 analysis. Furthermore, we assumed that technology remains constant, which means that facilities do not invest in other technologies that

---

[11] The need for industrial heat in heavy industries (e.g. petrochemical, cement, and steel) limits the options that installations in these industries can invest in. Fuel switching, electric-arc furnaces, shear-burning, and in the future hydrogen and carbon capture and storage are frequently more expensive options per ton of emissions avoided. An important number of electric utilities can switch fuel oil to natural gas or convert from conventional thermal power stations to combined cycle power plants.

[12] We used robust standard errors instead of clustered standard errors given the small number of clusters.

Panel a)

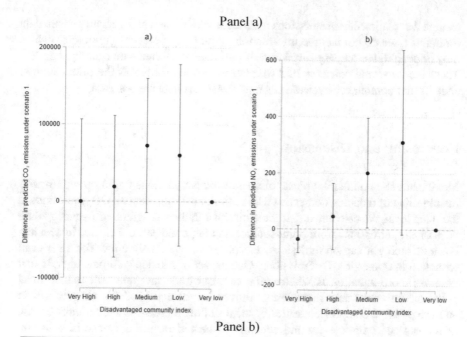

Panel b)

	(1) CO$_2$ (predicted emissions)	(2) NO$_2$ (predicted emissions)
Very High	-116.2	-35.87
	(53937.4)	(24.51)
High	19147.7	46.01
	(46606.6)	(59.68)
Medium	72548.1	201.2**
	(43907.1)	(98.50)
Low	60023.7	310.3*
	(59558.2)	(166.2)
Observations	124	124

Fig. 12.6 Simulation of CO$_2$ and NO$_2$ emissions under scenario 1 *Notes* Authors' estimations using data from RETC (CO$_2$ and NO$_2$ emissions) and the CONAPO's AGEB and rural locality marginalization index. Panel a): Subpanels a) and b) show the estimates for different dependent variables for the simulation where the dependent variable is the simulated emissions of each specific pollutant during the first year of the program. The x-axis denotes the marginalization level of the exposed communities. The y-axis denotes the percent difference in CO$_2$ and NO$_2$ emissions under scenario 1 for each marginalization level with respect to the "very low" marginalization level. Panel b) Regression results associated with panel a) results. Standard errors clustered at the locality level in parenthesis. Panels a) and b) are using 124 observations which represent one observation per regulated facility. The points are the point estimates of Eq. (12.1) with 95% confidence intervals using robust standard errors.

change the relationship in emissions releases from CO_2 and NO_2. Finally, given data limitations, we do not include information about other potential pollutants such as SO_2 or potential secondary formation of pollutants that create emissions of $PM_{2.5}$. These are two valid concerns that we plan to explore further as the pilot program ends its first compliance cycle and as new emissions data are released.

Conclusion and Discussion

Mexico has started the pilot phase of an ambitious emissions trading program with the objective of reducing domestic GHG emissions. Using data on CO_2 emissions at the plant level, we calculated that the emissions trading program will cover around 90% of CO_2 emissions from point sources and large industrial facilities in Mexico. By introducing a cap on emissions, the program will likely allow for an overall reduction in domestic CO_2 emissions. One aspect of Mexico's climate agenda that deserves more attention is whether the cap and trade program will reduce local pollution emissions near regulated facilities. The objective of this chapter was to analyse possible complementarities of local pollutant emission reductions because of the cap and trade system and who would benefit from a decrease in pollution as a result of the program. More importantly, this chapter also examined whether low-income communities would benefit from reductions in local pollution emissions and toxins due to the GHG emissions trading system.

Consistent with other studies, we found that the electricity sector has the highest CO_2 and NO_2 emissions in Mexico. The other two highest emitting sectors are cement production and oil refining. These three sectors are likely to have the highest number of regulated facilities under the GHG cap and trade program. We also analysed the distribution of emissions across communities with different levels of marginalization. We found large disparities between urban and rural areas: high emitting facilities are generally located in urban areas with "very high" marginalization levels, as defined by the Mexican government. However, rural areas with "high" marginalization levels also face high CO_2 emissions. We estimate that communities with "high" marginalization levels are on average exposed to 7,600 more tons of CO_2 emissions than the communities with "very low" marginalization levels during the 2016–2018 period. To the extent that these emissions are produced with co-pollutants, a cap and trade program that reduces CO_2 emissions is likely to benefit these communities in terms of air pollution exposure. We also found that communities with "high" marginalization levels are also exposed to higher NO_2 emissions; however, this is not statistically significant. Finally, with a 5% reduction of CO_2 emissions consistent with the program's target, we expect a decrease in NO_2 emissions for the most vulnerable populations. This could be likely translated to gains in co-pollutant reductions due to the ETS.

Environmental justice concerns have been part of climate policy implementation in other places of the world. These concerns have allowed the development of regulations that could potentially be implemented together with cap and trade to address disparities in air pollution exposure. For instance, California's AB 32 establishes that at least 25% of the revenues from cap and trade need to support disadvantaged communities and 5% of the revenues need to be used for developing projects in low-income communities. Moreover, the other revenue from cap and trade is used for grants to local environmental groups to implement projects such as community-owned air quality monitoring stations. Other emissions trading systems use part of their auction proceeds to mitigate electricity ratepayer effects (RGGI) financial support to mid and low-income households (EU-ETS), or directed towards funds that finance climate actions, including awareness raising (Québec) (Borghesi et al. 2016).

These actions might not apply to Mexico in the context of its emissions trading program. Nevertheless, analysing possible co-benefits of climate policy and its environmental justice implications is likely to be an important first step in achieving emission reductions with greater equality in terms of environmental exposure. This is especially relevant in the case of Mexico, given its large inequalities in income and along other important dimensions. The Ministry of Environment might benefit from using auction revenues in an environmentally progressive way. Additionally, it might find it optimal to introduce criteria for offsetting projects to be developed in environmentally disadvantaged communities, to relax their environmental burden.

There are other potential sources of environmental injustice that were not covered by this chapter, which could be exacerbated or reduced due to the Mexican ETS. The pass-through of carbon-related costs to consumers might affect disadvantaged communities heterogeneously, creating or alleviating energy expenditure gaps (Lyubich 2020). Additional sources related to environmental inequality are related to information gaps (Hausman and Stolper 2020), direct discrimination by demographics, firm location decisions, and housing decisions influenced by income inequality, among others.

Further research is needed to address climate justice concerns from this and other environmental policies. In the context of this chapter, research is needed to explore further relationships between CO_2, NO_2 and toxic contaminants, so the feasibility of GHG and co-pollutant reductions can be assessed. Improved data availability from the RENE would better inform the ETS policy and aid in developing pathways to the maximization of its potential co-benefits. However, the environmental justice dimension detailed in this chapter is a starting point to the evaluation of the distributional aspects of the ETS and could be a fruitful agenda for Mexican climate policy.

References

Altamirano M, Flamand L (eds) (2018) Desigualdades en México / 2018. Colegio de México, Ciudad de México

Arceo E, Hanna R, Oliva P (2016) Does the effect of pollution on infant mortality differ between developing and developed countries? evidence from Mexico City. Econ J 126(591):257–280

Banzhaf S, Ma L, Timmins C (2019) Environmental justice: the economics of race, place, and pollution. J Econ Perspect 33(1):185–208

Borghesi S, Montini M, Barreca A (2016) The European emission trading system and its followers: comparative analysis and linking perspectives. Springer, p 18

Burtraw D, Evans DA, Krupnick A, Palmer K, Toth R (2005) Economics of pollution trading for SO2 and NOx. Annu Rev Environ Resour 30:253–289

Cámara de Diputados (2018) Ley general de cambio climático. http://www.diputados.gob.mx/Ley esBiblio/pdf/LGCC_130718.pdf. Accessed 20 May 2020

CEEY (Centro de Estudios Espinosa Yglesias) (2019) Informe movilidad social en méxico 2019. Hacia la igualdad regional de oportunidades. https://ceey.org.mx/wp-content/uploads/2019/05/Informe-Movilidad-Social-en-M%C3%A9xico-2019.pdf. Accessed 20 May 2020

Chakraborti L, Margolis (2017) Do industries pollute more in poorer neighborhoods? Evidence from toxic releasing plants in Mexico. Econ Bull 37(2)

Currie J, Greenstone M, Moretti E (2011) Superfund cleanups and infant health. Am Econ Rev 101(3):435–441

Cushing L, Blaustein-Rejto D, Wander M, Pastor M, Sadd J, Zhu A, Morello-Frosch R (2018) Carbon trading, co-pollutants, and environmental equity: evidence from California's cap-and-trade program (2011–2015). PLoS medicine 15(7):e1002604

Dávila E, Kessel G, Levy S (2002) El sur también existe: un ensayo sobre el desarrollo regional de México. Economía Mexicana Nueva Época 11(2):205–260

Deryugina T, Heutel G, Miller NH, Molitor D, Reif J (2019) The mortality and medical costs of air pollution: evidence from changes in wind direction. Am Econ Rev 109(12):4178–4219

EPA (Environmental Protection Agency) (2018) Learn about environmental justice. https://www.epa.gov/environmentaljustice/learn-about-environmental-justice. Accessed 20 May 2020

Esquivel G (1999) Convergencia regional en México, 1940–1995. El Trimestre Económico 264(4):725–761

Fowlie M, Holland SP, Mansur ET (2012) What do emissions markets deliver and to whom? evidence from Southern California's NOx trading program. Am Econ Rev 102(2):965–993

Friedmann J, Fan Z, Tang K (2019) Low-Carbon heat solutions for heavy industry: sources, options, and costs today. Center on Global Energy Policy. https://www.energypolicy.columbia.edu/sites/default/files/file-uploads/LowCarbonHeat-CGEP_Report_100219-2_0.pdf. Accessed 20 May 2020

Grainger C, Ruangmas T (2018) Who wins from emissions trading? evidence from California. Environ Resour Econ 71(3):703–727

Grineski SE, Collins TW (2008) Exploring patterns of environmental injustice in the Global South: Maquiladoras in Ciudad Juárez. Mexico Popul Environ 29(6):247–270

Gutierrez E (2015) Air quality and infant mortality in Mexico: evidence from variation in pollution concentrations caused by the usage of small-scale power plants. J Popul Econ 28(4):1181–1207

Harris K (2019) Climate equity act—Kamala Harris. https://www.harris.senate.gov/imo/media/doc/CEA_background.pdf. Accessed 20 May 2020

Hausman C, Stolper S (2020) Inequality, information failures, and air pollution. National Bureau of Economic Research (No. w26682)

Hernandez-Cortes D, Meng KC (2020) Do environmental markets cause environmental injustice? evidence from California's carbon market. National Bureau of Economic Research (No. w27205)

Holland SP (2012) Spillovers from climate policy to other pollutants. In: Fullerton D, Wolfram C (eds), The design and implementation of US climate policy. National bureau of economic research conference report. University of Chicago Press, Chicago, pp 79–90

INECC (Instituto Nacional de Ecología y Cambio Climático) (2018) Costos de las contribuciones nacionalmente determinadas de México. http://cambioclimatico.gob.mx:8080/xmlui/bitstream/handle/publicaciones/40/723_2018_Costos_Contribuciones_Nacionalmente_Determinadas_Mexico_.pdf?sequence=1&isAllowed=y. Accessed 20 May 2020

Kaswan A (2008) Environmental justice and domestic climate change policy. Envtl l Rep News Analysis 38:10287

Lara-Valencia F, Harlow SD, Lemos MC, Denman CA (2009) Equity dimensions of hazardous waste generation in rapidly industrialising cities along the United States-Mexico border. J Environ Plan Manage 52(2):195–216

Liu L (2013) Geographic approaches to resolving environmental problems in search of the path to sustainability: the case of polluting plant relocation in China. Appl Geogr 45:138–146

Lome-Hurtado A, Touza-Montero J, White PC (2019) Environmental injustice in Mexico City: a spatial quantile approach. Expo Health 1–15

Lyubich E (2020) The race gap in residential energy expenditures. Energy Institute at Haas

Mahady JA, Octaviano C, Bolaños OSA, López ERR, Kammen DM, Castellanos S (2020) Mapping opportunities for transportation electrification to address social marginalization and air pollution challenges in Greater Mexico City. Environ Sci Technol 54(4):2103–2111

Meadows D, Vis P, Zapfel P (2020) The EU Emissions Trading System. In: Delbeke J, Vis P (eds) Towards a climate-neutral Europe. Routledge, London, pp 66–94

SEMARNAT, GIZ (2018) Distributing allowances in the Mexican emissions trading system: indicative allocation scenarios. In: Preparation of an emissions trading system in Mexico. https://www.gob.mx/cms/uploads/attachment/file/505767/Distributing_Allowances_in_the_Mexican_ETS.pdf. Accessed 20 May 2020

SEMARNAT (Secretaría de Medio Ambiente y Recursos Naturales) (2019a) Acuerdo por el que se establecen las bases preliminares del programa de prueba del sistema de comercio de emisiones. Diario Oficial de la Federación. http://www.dof.gob.mx/nota_detalle.php?codigo=5573934&fecha=01/10/2019. Accessed 20 May 2020

SEMARNAT (Secretaría de Medio Ambiente y Recursos Naturales) (2019b) Programa de prueba del sistema de comercio de emisiones en México. In: Preparación de un sistema de comercio de emisiones en México. https://www.gob.mx/cms/uploads/attachment/file/505746/Brochure_SCE-ESP.pdf. Accessed 20 May 2020

Walch RT (2018) The effect of California's carbon cap and trade program on co-pollutants and environmental justice: evidence from the electricity sector. Mimeo

Chapter 13
Blue Carbon in Emissions Markets: Challenges and Opportunities for Mexico

Antonina Ivanova Boncheva and Alfredo Bermudez-Contreras

Abstract Mangroves are ecosystems made up of trees or shrubs that develop in the intertidal zone and provide many vital environmental services for livelihoods in coastal areas. They are a habitat for the reproduction of several marine species. They afford protection from hurricanes, tides, sea-level rise and prevent the erosion of the coasts. Just one hectare of mangrove forest can hold up to 1,000 tons of carbon dioxide, more than tropical forests and jungles. Mexico is one of the countries with the greatest abundance of mangroves in the world, with more than 700,000 ha. Blue carbon can be a novel mechanism for promoting communication and cooperation between the investor, the government, the users, and beneficiaries of the environmental services of these ecosystems, creating public–private-social partnerships through mechanisms such as payment for environmental services, credits, or the voluntary carbon market. This chapter explores the possibilities of incorporating blue carbon in emissions markets. We explore the huge potential of Mexico's blue carbon to sequester CO_2. Then we analyse the new market instrument that allows countries to sell or transfer mitigation results internationally: The Sustainable Development Mechanism (SDM), established in the Paris Agreement. Secondly, we present the progress of the Commission for Environmental Cooperation (CEC) to standardize the methodologies to assess their stock and determine the magnitude of the blue carbon sinks. Thirdly, as an opportunity for Mexico, the collaboration with the California cap-and-trade program is analysed. We conclude that blue carbon is a very important mitigation tool to be included in the compensation schemes on regional and global levels. Additionally, mangrove protection is an excellent example of the mitigation-adaptation-sustainable development relationship, as well as fostering of governance by the inclusion of the coastal communities in decision-making and incomes.

A. Ivanova Boncheva (✉)
Department of Economics and Director of APEC Studies Center, Universidad Autonoma de Baja California Sur, La Paz, Mexico
e-mail: aivanova@uabcs.mx

A. Bermudez-Contreras
Autonomous University of Baja California Sur (UABCS), La Paz, Mexico
e-mail: abermudez@uabcs.mx

Keywords Blue carbon · Oceans · Emission trading · Sustainable development ·
International cooperation

Introduction

Blue carbon ecosystems are "the coastal ecosystems of mangroves, tidal marshes,
and seagrass meadows" (CI, IUCN and IOC-UNESCO 2019:2). Mangroves have an
enormous capacity for sucking up carbon dioxide and other greenhouse gases and
trapping them in flooded soils for millennia. In addition to their mitigation potential, they contribute to climate change adaptation stabilizing the coastal areas and
protecting them from sea-level rise, storms, and soil erosion (Ibid). Mangroves also
provide important ecosystem services as a habitat for the reproduction of different
marine organisms. Their social contribution is also very significant in providing food
and employment opportunities to coastal communities.

Governments, international actors (NGOs and academia), and local communities
around the world are supporting coastal wetland conservation as a part of the mitigation strategy. The initiatives have varying levels of private sector involvement and
different objectives, targets, and timelines. Some efforts focus on reducing emissions
from deforestation and degradation, while others focus on negative emissions through
the restoration of cleared or degraded landscapes. The United Nations Framework
Convention on Climate Change (UNFCCC) is beginning to include blue carbon in
the discussion of natural ecosystems. The existing REDD+ framework set-up under
decisions of the UNFCCC COP specifies modalities for Measuring, Reporting, and
Verifying (MRV) greenhouse gas emissions and removals (Park et al. 2013). Article
5 of the Paris Agreement explicitly calls for parties to take action to conserve and
enhance sinks and reservoirs of greenhouse gases, including forests, and encourages
countries to engage in cooperative approaches to this end. The explicit inclusion
of forest and mangrove conservation is potentially a "game changer" as it encourages countries to safeguard ecosystems for climate mitigation purposes (Grassi et al.
2017) and facilitates the access of developing countries with abundant forests and
mangroves to international carbon mitigation financing.

The objective of this chapter is to present Mexico's potential to involve blue
carbon in the emissions trading system. First, we present the stocks of blue carbon
in Mexico, the country with the greatest abundance of mangroves in the world, with
more than 700 thousand hectares. Second, the potential of blue carbon as carbon
storage is explored. Third, we analyse the opportunities that the new Sustainable
Development Mechanism (SDM) introduced by the Paris Agreement is presenting
for cap-and-trade and REDD+ mitigation options, especially for developing countries
like Mexico. Further, we explore Mexico's collaboration with the California cap-and-
trade program, as a possibility to introduce blue carbon in the emissions trade. A
case study of the Vizcaino Biosphere Reserve illustrates the country's likelihood of
entering the blue carbon emissions markets. We conclude that blue carbon is a great

area of opportunity for Mexico to perform mitigation strategies and participate in the regional and world cap-and-trade systems.

Blue Carbon in the World and Mexico

Climate Change and the Ocean

Total anthropogenic CO_2 emissions have been steadily and undoubtedly rising over the past decades and with them, the energy we trap on the planet. How emissions will behave in years to come depends on the decisions we make and the pathways we follow, as depicted in Fig. 13.1. According to the IPCC (2014), in excess of 90% of the energy accumulated in the climate system on Earth is in the ocean.

To make matters worse, alongside emissions growth and increased energy entrapment in the climate system, degradation of natural ecosystems that could serve as carbon sinks is also progressing, thus reducing their capacity to absorb CO_2 from the atmosphere. Fortunately, the mechanisms provided by forests to store carbon are well understood leading to the formulation of financial support schemes to promote their conservation in an effort to reverse the aforementioned trend. However, the carbon storage potential of ocean ecosystems, where 55% of biological carbon is captured (Nellemann et al. 2009), has not received enough attention in the fight against climate change.

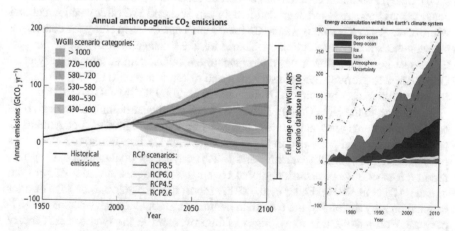

Fig. 13.1 Left: Annual anthropogenic CO_2 emissions history and IPCC scenarios. Right: Energy accumulation in the climate system. *Source* (IPCC 2014)

Blue Carbon

Coastal ecosystems can provide a wide range of services such as supporting fisheries, coastline protection from storms and sea-level rise, shoreline erosion prevention, water purification, biodiversity conservation, or providing food security for coastal communities, all of which are essential for climate change adaptation. Additionally, some of them can also work effectively to capture and store organic carbon acting as a carbon sink in the plants themselves and the sediments below. These include mangroves, salt marshes, and seagrass meadows. These ecosystems stretch through the land-sea interface covering supratidal (salt marshes), intertidal (mangroves), and shallow subtidal regions (seagrasses). In 2012, Pendleton et al. estimated the global extent of blue carbon ecosystems in the world to be 48.9 million hectares. The carbon stored in such ecosystems is referred to as blue carbon and accounts for perhaps as much as 71% of all carbon stored in ocean sediments (Ashok et al. 2019; CI et al. 2019; Nellemann et al. 2009; SEMARNAT 2017).

Blue carbon ecosystems are very fragile. Despite the enormous value found in the services they provide, economic development and human activities put them under sustained pressure. In fact, marine ecosystems are being lost at a faster rate than those based on land. The loss of these ecosystems due to unsustainable natural resource exploitation practices, poor watershed management, poor coastal development practices, and poor waste management is a serious threat for them and for the long list of services they provide in coastal regions, including carbon capture (uptake) and long-term carbon storage. Using a social cost of carbon of $41 per ton of CO_2 (2007 USD), Pendleton et al. (2012) estimated an annual global cost of blue carbon ecosystems conversion and degradation between $6.1 and $42 billion and noted that while damages to these ecosystems are located in a narrow strip along the coasts, the consequences are endured globally. Therefore, the management of marine ecosystems must be regarded as a desirable investment rather than an unnecessary cost. One method to promote their conservation and restoration would be their successful incorporation in carbon markets (Nellemann et al. 2009; OCM-NOAA 2018).

According to the Blue Carbon Initiative (www.thebluecarboninitiative.org), blue carbon ecosystems are found in all continents around the globe except Antarctica with the distribution presented in Fig. 13.2. The Blue Carbon Initiative reports loss rates of 1.5% per year for seagrasses and 1–2% per year for tidal marshes. They also report a loss of 30% of historical global coverage for seagrasses and 50% for tidal marshes (CI et al. 2019). Feller et al. (2017) report mangrove losses of 35% of their original area by the end of the twentieth century at global loss rates between 1–3% per year. Fortunately, this shows signs of improvement in the twenty-first century with loss rates of less than 1% per year and even as low as 0.16% per year between 2000 and 2012 (Feller et al. 2017; Hamilton and Casey 2016).

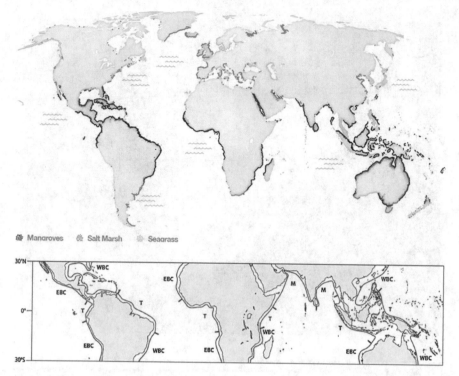

Fig. 13.2 Global distribution of blue carbon ecosystems (top) and mangroves alone (bottom). *Source* Top: Reproduced from CI et al. (2019). Bottom: Reproduced from Alongi et al. (2015)

Blue Carbon in Mexico

Mexico has considerable extensions of mangroves and seagrasses covering an estimated surface of 1.1 million hectares (Herrera-Silveira et al. 2020). Because of these ecosystems, Mexico has significant potential to capture and store blue carbon. Despite some protection provided by Mexican regulations, these ecosystems are under constant pressure from land-use changes for various purposes. Over the past 20 years, 24 Mt[1] CO$_2$ has been emitted in Mexico due to mangrove coverage loss (Herrera-Silveira et al. 2020). That's equivalent to 3% of the total emissions of Mexico in 2017 (INECC 2018). Financial schemes to support the conservation of blue carbon ecosystems in Mexico could help in maintaining a healthy stock of mangroves and seagrasses.

According to the most recent evaluation by Mexico's Biodiversity Commission (CONABIO), the country has a total mangrove surface of 775,555 ha (Valderrama-Landeros et al. 2017) distributed in five regions as shown in Fig. 13.3. The Yucatan Peninsula alone accounts for more than half of the total. Overall, this mangrove area

---

[1] 1 Mt equals 1 million metric tons.

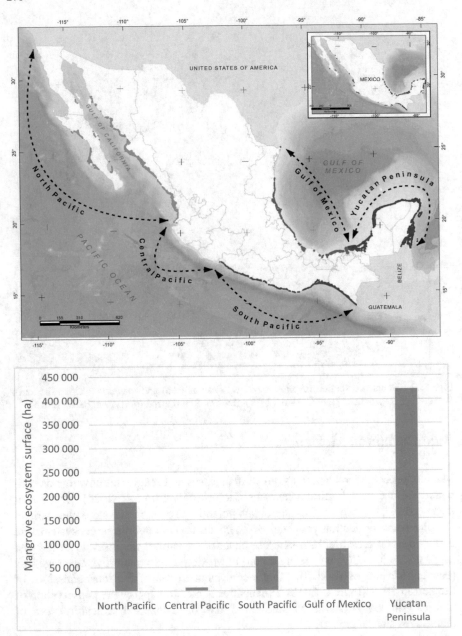

Fig. 13.3 Mangrove ecosystem surface by region in Mexico. *Source* Top: Reproduced from Herrera-Silveira et al. (2020). Bottom: Authors' elaboration with data from Valderrama-Landeros et al. (2017)

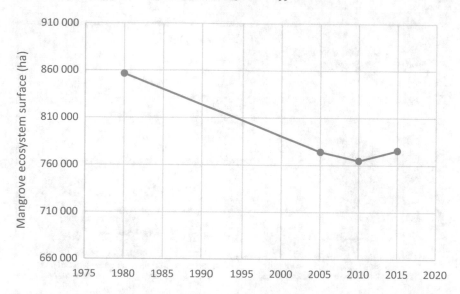

Fig. 13.4 Evolution of mangroves in Mexico. *Source* Authors' elaboration with data from Valderrama-Landeros et al. (2017). The value presented for 1980 corresponds to the aggregation of data for individual states during the 1970s and 1980s as presented by Valderrama-Landeros et al. (2017)

had a net loss of 9.4% between the 1970/1980 records and 2015, with a small recovery between 2010 and 2015 (Fig. 13.4). Hamilton et al. (2016) ranked Mexico in the top 10 countries with the most mangrove forest area but Mexico was also ranked in the top 10 countries with the highest annual total area of mangrove deforestation between 2000–2012 (Feller et al. 2017). Nevertheless, some steps are already being taken in the right direction. The "Adaptation and Blue Carbon" project is the first adaptation project in Mexico funded with national resources through the Climate Change Fund created by the General Law of Climate Change.[2]

Carbon Storage

Annually, between 235–450 Mt of carbon are captured and stored by blue carbon ecosystems around the world (Alongi 2014). The attention that these ecosystems and mangroves, in particular, have received as carbon sinks is due to their disproportionately high capacity to trap carbon in the long term in relation to the area they cover (see Fig. 13.5). However, as could be reasonably expected, not all mangrove plants around the world are the same nor are the conditions they grow in. Plant species, local climate, coastal geomorphology, fertility gradients, hydrodynamic types, and

---

[2] A video summarizing the experience in this project can be found in www.youtube.com/watch?v=zvDtxrizRws (04 May of 2020).

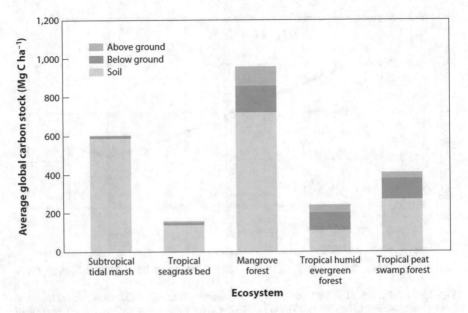

Fig. 13.5 Average global carbon stocks of various ecosystems. *Source* Reproduced from Alongi (2014)

even surface salinity are all factors that affect carbon storage capabilities per unit area (Herrera-Silveira et al. 2020; Ochoa-Gómez et al. 2019). Figure 13.6 presents

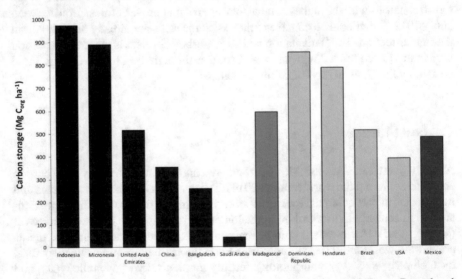

Fig. 13.6 Organic carbon stores per unit area in mangroves around the world. *Source* Reproduced from Herrera-Silveira et al. (2020)

Fig. 13.7 Mangrove distribution in VIBIRE. The Reserve is in a darker shade; mangrove ecosystems are in red. *Source* Authors' elaboration with data from Google Earth, CONANP (2000), and CONABIO (2015)

a comparison of unit carbon storage for mangroves worldwide. It is then clear that there is a wide range of carbon capture capacities for these ecosystems (Fig. 13.7).

Despite the different carbon storage capacity of mangroves in various places, in general, they maintain similar amounts of carbon in their living biomass as other vegetated ecosystems. However, a larger benefit lies in the carbon stored in the soil over which they are located and that they successfully create through various mechanisms. These soils are carbon-rich environments that can extend several metres deep where lack of oxygen and other factors constrain decomposition (Pendleton et al. 2012). This is at the same time why the conversion of mangroves to other land uses also poses an enormous risk as a climate change contributor. In converted or degraded mangrove regions, surface vegetation harm will not only result in a loss of atmospheric carbon uptake capacity but, more importantly, lead to the unlocking and decomposition of the organic carbon contained in these soils, ultimately resulting in pulses or spikes of greenhouse gases released back into the atmosphere. Estimates of emissions released to the atmosphere in the range of 150–1020 Mt CO_2 each year (central value: 450 Mt) have been reported due to the degradation and conversion of coastal ecosystems (Pendleton et al. 2012). To put this into perspective, Mexico's total greenhouse gas emissions in 2017 (734 Mt CO_2e, INECC 2018) are within that same range. Therefore, conserving and adequately managing mangroves to ensure the long-term permanence of those carbon stores is of paramount importance, firstly for their climatic benefits but also for the long list of valuable other services they provide.

Sustainable Development Mechanism (SDM) of the Paris Agreement

The 2015 Paris Agreement to the UNFCCC is at the centre of international cooperative efforts for climate change mitigation and adaptation in the post-2020 period. Although its legal form was heavily disputed in its four-year negotiating process (Dagnet et al. 2016). The overall purpose of international cooperation through the Paris Agreement is to enhance the implementation of the UNFCCC, including its objective of stabilizing atmospheric GHG concentrations "at a level that would prevent dangerous anthropogenic interference with the climate system" (UNFCCC 1992 Art. 2).

Article 6.1 of the Paris Agreement recognizes the role that cooperative approaches can play, on a voluntary basis, in implementing parties' Nationally Determined Contributions (NDCs) "in order to allow for higher ambition" in their mitigation actions and to promote sustainable development and environmental integrity. It lists a number of specific types of cooperative approaches that come within its ambit, including Internationally Transferred Mitigation Outcomes (ITMOs), a "mechanism to contribute to mitigation and support sustainable development", and a framework for non-market mechanisms (UNFCCC 2015; Morgan and Northrop 2017).

Article 6.2 suggests ITMOs can originate from a variety of sources including regional carbon markets or REDD+. While this provision, unlike similar provisions in the Kyoto Protocol, does not create an international carbon market, it enables parties to pursue this option should they choose to do so, for example, through the linking of domestic or regional carbon markets (Ivanova et al. 2020). Article 6.2 could also be implemented in other ways, including direct transfers between governments, linkage of mitigation policies across two or more parties, sectoral or activity crediting mechanisms, and other forms of cooperation involving public or private entities or both (Linn 2016). Wetlands International, together with the Australian Government, organized the event "Incorporating Blue Carbon into Nationally Determined Contributions under the Paris Agreement" at the UN Climate Change Conference, COP22, in Marrakesh (Ullman et al. 2013; Herr et al. 2015). In 2013, Mexico began to explore the options to include blue carbon in countries' NDCs (Pronatura Sur A.C. 2016).

Article 6.4 concerns the mitigation mechanism, referred to by some parties as the "sustainable development mechanism" or SDM. It is a mechanism that has as output GHG emissions reductions, which can be used by any party towards its NDC. However, the limit exists that emission reductions cannot be used by the host party if another party applies them to demonstrate the achievement of its NDC. Unlike the Clean Development Mechanism (CDM), there is no restriction specified regarding which parties can host mitigation projects and which parties can use the resulting emissions reductions towards their NDCs (Streck et al. 2016). The SDM will operate under the authority and guidance of the CMA[3] and is to be supervised by a body designated by the CMA in a similar fashion to the CDM.

The SDM also has a mission to foster sustainable development. The decision adopting the Paris Agreement specifies experience with Kyoto mechanisms like the CDM as a basis for the new mitigation mechanism (UNFCCC 2016). Compared to the CDM under the Kyoto Protocol, which had a climate-centric focus on measuring emissions reductions, the SDM has a more balanced focus on both climate and development objectives and a stronger political mandate to measure sustainable development impact and to verify that the impacts are "real, measurable, and long-term" (Olsen et al. 2018).

Blue Carbon in North America: Assessing the Role of Coastal Habitats in the Carbon Balance of the Subcontinent

As a member of the Commission for Environmental Cooperation (CEC),[4] Mexico is involved in various trilateral projects. The CEC's first blue carbon project is a tool that facilitates the inclusion of Mexican blue carbon in the emission markets. It contributes to the conservation and restoration of coastal habitats that capture and store CO_2. The project systematizes the information, mapping, and approaches necessary to fill gaps in our knowledge of carbon dynamics in carbon ecosystems as marshes, mangroves, and beds (CEC 2014, CCEA 2015).

The progress of the project is presented as follows:

- Establishment of a community of scientific practice around blue carbon in North America to promote cooperation and the exchange of knowledge among experts on the subject in the three countries.
- Integration of a common set of data on habitats that capture and store blue carbon. This dataset includes maps, carbon emission counts, and data on emission potential, uptake, and storage, as well as documented methods, information, and results. The information will be available in the Environmental Atlas of North America (King 2012).

---

[3] CMA is the short form for the group of the countries who have signed and ratified the Paris Agreement. The full name of this governing body is "Conference of the Parties serving as the meeting of the Parties to the Paris Agreement".

[4] Since 1994, Canada, Mexico, and the United States have collaborated in protecting North America's environment through the North American Agreement on Environmental Cooperation (NAAEC). Accordingly, the NAAEC established an intergovernmental organization—the Commission for Environmental Cooperation (CEC)—to support cooperation among the NAFTA partners to address environmental issues of continental concern, including the environmental challenges and opportunities presented by continent-wide free trade (CEC n.d.).

- Completion of the first step in formulating an internationally recognized methodology to include ecosystem conservation projects that capture and store blue carbon in voluntary carbon markets.
- New information and methods from different scientific studies that help fill gaps in our knowledge of carbon dynamics in habitats that capture and store blue carbon, including both healthy and disturbed sites (CEC 2016).
  The following products have been generated based on the results of the joint work:
- The first set of blue carbon storage habitat maps for North America showing the mapping of 47 776 km^2 of this habitat type conducted to date.
- A tri-national workshop with the blue carbon community of practice and one with experts in blue carbon, forest carbon, and land cover (WCMC 2016).
- Publication of the report on methodological criteria for offsetting greenhouse gas emissions in favour of intertidal wetland conservation and derived recommendations.
- Five research projects on coastal habitats that capture and store blue carbon are as follows:

(1) Observation of carbon accumulation indexes in coastal marshes and their response to sea-level rise.
(2) Levels of uptake and storage of blue carbon in northern marshes: evaluation of processes, reserves, and accumulation rates of the element in undisturbed, drained, and restored marshes.
(3) Carbon reserves in mangroves and marshes in the most extensive wetlands in Mesoamerica: the Centla swamps, Mexico.
(4) Carbon stocks in seagrass bed systems across a range of environmental conditions and seagrass types, with the aim of determining the amount of carbon deposited.
(5) Spatial variability of carbon storage within the marshes belonging to the United States National Estuarine Research Reserve System for (NERRS): comparison of methodologies and coastal regions (Ivanova 2019).

This cooperation mechanism is an important tool for Mexico to evaluate blue carbon availability in the country and to develop the necessary methodologies for comparative studies.

California Cap-and-Trade Program

California decided that a cap with a broad scope would be the most effective way to ensure that the state could meet its mitigation targets. At its launch in 2013, California's program covered all six greenhouse gases (GHG emissions) within the industrial and electricity sectors (EDF 2014). In 2015, the cap coverage expanded to transportation fuels and natural gas, bringing about 85% of state emissions under the cap. Emissions from imported electricity and fuel are included in the cap, though the cap does not cover emissions from sectors that are currently challenging to measure or regulate on a large scale, such as agriculture and fugitive emissions (EDF 2018).

In 2010, California and representatives of Acre, Brazil, and Chiapas, Mexico, signed a Memorandum of Understanding (MoU) that led to the establishment of a working group to provide guidance to California on fighting tropical deforestation and carbon pollution around the world through innovative policies that reduce Emissions from Deforestation and Degradation (REDD). The working group examined design elements, including legal and institutional aspects and social and environmental safeguards, to develop a jurisdictional scale REDD credit-trading system that could be used for compliance within California's carbon market (ROW 2013).

Chiapas has been developing a state-wide approach to REDD+, but it is at an earlier stage than Acre, Brazil (EDF 2015). Chiapas is identifying and beginning to assimilate the substantive and procedural elements needed to build a successful jurisdictional REDD+ program that will work within the Mexican context (Herrera Silveira and Teutli Hernández 2017). It also brings an important set of experiences regarding land tenure, indigenous rights, and participation, highlighting the critical importance of establishing a process that incorporates all stakeholders from the beginning in designing and building jurisdictional programs for REDD+ and low emissions development.

A very small amount (about 5%) of emissions reductions from California Commerce-included industries can be purchased through uncapped sources, including forestry ("green carbon") here. California has begun to expand this trade to neighbouring states and even to the international arena with highly forested countries with which it can offset its emissions, such as Brazil, Indonesia, or Mexico, and has already begun to work with the scheme of the certifying organization VCS (Lopomo et al. 2011). Currently, the California scheme is focused on qualifying blue carbon as a compensation category. It is yet to be decided whether this category will be applicable exclusively to bonds within California, or also to compensation from other countries included in California Commerce, such as Mexico. This may be a good opportunity for Mexico, as it adjusts very well to the requirements of this ETS and the scheme of working between states. In this way, the compensation scheme would come to adhere to state climate change strategies.

The Mexican Emissions Trading System Pilot Program

On 1 January 2020, the Ministry of the Environment and Natural Resources (SEMARNAT) launched the Emissions Trading System Pilot Program, which aims to promote the reduction of emissions at the lowest possible cost, and which will last for three years, in compliance with the Reform of the Mexican General Law on Climate Change.

During the pilot program, only facilities whose annual emissions are equal to or greater than 100,000 tons of direct carbon dioxide emissions will participate.

According to the Regulations of the National Emissions Registry, the following activities are considered:

(1)   In the energy sector: exploitation, production, transportation, and distribution of hydrocarbons and generation, transmission, and distribution of electricity.
(2)   For the industrial sector: automotive industry, cement and lime industry, chemical industry, food and beverage industry, glass industry, steel industry, metallurgical industry, and mining industry.

Blue carbon and forestry are sectors not regulated by the pilot program. However, the ETS can incentivize CO_2 reductions in unregulated sectors. Mitigation projects in these sectors can be submitted to inspection under approved protocols to guarantee the quality of their reductions. Approved projects can access offset credits to achieve the mitigation goal. It is a good opportunity to promote blue carbon, but the priority of offset credits is still the regulated sector (SEMARNAT and GIZ 2020).

Case Study: El Vizcaino Biosphere Reserve[5]

Baja California Sur is an arid state in Northwest Mexico. With more than 2,100 km (SPYDE 2015), the state has the longest coastline in the country, where blue carbon ecosystems can be found. According to CONABIO's most recent report (Valderrama-Landeros et al. 2017), as of 2015, the state had 26,579 ha of mangroves. Unlike other places in Mexico and the world, mangrove coverage in Baja California Sur has been very stable over the past four decades (26,724 ha in 1978), losing only 0.5%. Nevertheless, mangrove ecosystems in Baja California Sur are not static but rather the result of a dynamic equilibrium of losses and gains balancing out (López-Medellín et al. 2011; Watson and Corona 2018).

Natural Protected Areas (NPA) in Baja California Sur cover more than 40% of the state total surface of 73,308 km^2. The El Vizcaino Biosphere Reserve (VIBIRE) located in the north of the state is the largest NPA in the country with an area of 2.5 million hectares. The San Ignacio Lagoon in the Reserve is an important location for biodiversity including considerable mangrove extensions. CONABIO reported that in 1978 the Lagoon and neighbouring estuaries La Bocana and El Datil had a combined 3,672 ha of mangroves. By 2005, this was reduced to 3,607 ha but recovered by 2010 for a total of 3,799 ha.

An estimate of the total carbon stock in the San Ignacio Lagoon and neighbouring estuaries can be worked out using the organic carbon stock value reported by Herrera-Silveira et al. (2020) for mangroves in the North Pacific region in Mexico of 204.9 Mg C$_{org}$ ha^{-1}. This includes both above- and below-ground stocks and results

---

[5] We present this case study in the chapter to highlight the potential of the blue carbon in the emissions trade of Mexico. The results that presents the Vizcaino Biosphere Reserve are an example of the great possibilities that the inclusion of the blue carbon in the emission trade can represent for Mexico. The blue carbon could be included in the Californian cap-and trade system and as a non-regulated sector in the Mexican cap-and trade system.

in 0.778 Mt of organic carbon in the Lagoon region. If all this carbon were oxidized fully to form carbon dioxide, this would translate to 2.85 Mt CO_2 in total. To put this into perspective, this would be in the same order of magnitude as the reported greenhouse gas emissions for the whole state of Baja California Sur in all categories in 2010 of 4.16 Mt CO_2e (Ivanova and Gámez 2012). Finally, using the social cost of carbon[6] reported by Pendleton et al. (2012) of US 2007 $41 per ton of CO_2, the 0.778 Mt of organic carbon in the Lagoon would equate to 117 million dollars in global costs if these mangroves were converted, which as the data suggests is fortunately not happening. As compelling as these figures may be, without the adequate mechanisms in operation at the required scales, mangroves and their management around the world will not be valued for their superb carbon sequestration capacity in the long term and will continue to sustain pressure from economic development and human activities.

Conclusions

In the short and medium term, expansion of mangrove ecosystems provides many environmental services such as supporting fisheries, coastline protection from storms and sea-level rise, shoreline erosion prevention, water purification, biodiversity conservation, or providing food security for coastal communities, all of which are essential for climate change adaptation. However, in the long term, expansion and conservation of mangroves translate into enhanced carbon stores, which is of great value in the mitigation of climate change.

However, in its broadest understanding, mitigation activities—as well as climate change adaptation and conservation activities—can also include national capacity building or awareness-raising efforts (e.g. enabling stakeholders to use mangroves in a sustainable manner), support for institutional set-up, developing and implementing sectoral policies, enforcing changes in national legislation, and engaging stakeholders. The goals for mitigation are most prominently aligned with climate adaptation objectives, especially for nature-based solutions such as in agriculture, forestry, and the rural land-use sectors. Recognition of how climate change is likely to influence other development priorities, as access, justice, and allocation, may be a first step towards building cost-effective strategies and integrated, institutional capacity in Mexico to respond to climate change.

The present analysis found that the integration of the concept of blue carbon in Mexican public policy is an important area of opportunity. However, there are some challenges to be faced. In the first place, an official Mexican standard specifically regulating matters related to blue carbon must be drafted, proposed, and implemented. Others are property rights; the federation, state, and community areas of influence; and access to the international green funds. Currently, there are substantial shortcomings in the functioning of the institutional framework for sustainable

---

[6] Global economic cost of new atmospheric carbon (Pendleton et al. 2012).

development. Mitigating climate change, adapting to sea-level rise, and alleviating coastal communities' poverty can all be complementary.

The international commitments of Mexico involve both mitigation and adaptation actions, and in both cases, conservation and restoration of blue carbon ecosystems is the best solution in terms of cost-effectiveness (CEC 2017).

Mexico as part of the Paris Agreement can use ITMOs to achieve their NDCs but when engaging in this activity shall promote sustainable development, ensure environmental integrity, ensure transparency, including in governance, and apply "robust accounting" in accordance with CMA guidance to prevent double counting. Additionally, as a member of the CEC, Mexico participates in the regional assessment of blue carbon stocks and shares with methodologies for implementation and monitoring the U.S. and Canada. Under this scheme, the compensation scheme would come to adhere to state climate change strategies.

The best resource for this is the creation of a national regulated emissions trading system—similar to and supported by the California Emissions Trading System (ETS)—that includes blue carbon through the guidelines imposed by the specific Mexican official standard, and with a system of concessions to be able to assign carbon credits. Currently, the Mexican ETS is in the pilot phase and only considers the energy and industrial sectors as regulated. The creation of the ETS in Mexico is based, therefore, on the lessons learned and good practices carried out by international and national voluntary markets and implemented through a strategy at all levels of government, where financing and capacities flow from the international to the local and community level. Considering the great potential of Mexico to benefit from blue carbon in emissions trading, we strongly recommend including it among the regulated sectors during the next phase of ETS.

References

Alongi DM (2014) Carbon cycling and storage in mangrove forests, annual review of marine science, 6. Ann Rev PALO ALTO 1:195–219

Alongi DM, Mukhopadhyay SK (2015) Contribution of mangroves to coastal carbon cycling in low latitude seas. Agric Forest Meteorol 213:266–272

Ashok A, Cusack M, Saderne V, Krishnakumar PK, Rabaoui L, Qurban MA, Duarte CM, Agustí S (2019) Accelerated burial of petroleum hydrocarbons in Arabian Gulf blue carbon repositories. Sci Total Environ 669:205–212

Canadian Council on Ecological Areas (CCEA) (2015) Conservation Areas Reporting and Tracking System (CARTS). http://www.ccea.org/tools-resources/carts/

CI, IUCN & IOC-UNESCO (2019) The blue carbon initiative, viewed 14 May 2020. https://www.thebluecarboninitiative.org/

Commission for Environmental Cooperation (CEC) (2014) Cartografía y evaluación de hábitats que captan y almacenan carbono azul a fin de determinar con mayor precisión su capacidad para eliminar gases de efecto invernadero. http://www.cec.org/es/sites/default/files/documents/project_resources/project_accomplishments_2013-2014/blue_carbon_sp.pdf

CEC (n.d) Commission for environmental cooperation. http://www5.cec.org/

CEC (2016) Carbono azul en América del Norte: evaluación de la distribución de los lechos de pasto marino, marismas y manglares, y su papel como sumideros de carbono, Comisión para la Cooperación Ambiental, Montreal, Canadá, 58 pp.

CEC (2017) Análisis de las oportunidades para la integración del concepto de carbono azul en la política pública mexicana, Comisión para la Cooperación Ambiental, Montreal, Canadá, 102 pp.

CONABIO (2015), Distribución de los manglares en México en 2015 - Sistema Nacional de Información sobre Biodiversidad, viewed 20 August 2017. http://www.conabio.gob.mx/informacion/gis/?vns=gis_root/biodiv/monmang/manglegw

CONANP (2000) Buscador de datos por área natural protegida - Reserva de la Biósfera El Vizcaíno - Archivo kml para Google Earth, viewed 20 August 2017. http://sig.conanp.gob.mx/website/pagsig/

Dagnet Y, Waskow D, Elliott C, Northrop E, Thwaites J, Mogelgaard K, Krnjaic M, Levin K, Mcgray H (2016) Staying on track from Paris: advancing the key issues of the paris agreement. In: Working paper. World Resources Institute, Washington, DC. http://www.wri.org/ontrackfromparis

Environmental Defense Fund (EDF) (2014) Carbon Market California. A comprehensive analysis of the Golden State's Cap-and-Trade Program. http://www.edf.org/sites/default/files/content/ca-cap-and-trade_1yr_22_web.pdf

EDF (2015) Mexico: an emissions Trading Case Study. https://www.edf.org/sites/default/files/mexico-case-study-may2015.pdf

EDF (2018) California's cap-and-trade program step by step. https://www.edf.org/sites/default/files/californias-cap-and-trade-program-step-by-step.pdf

Feller IC, Friess DA, Krauss KW, Lewis RR (2017) The state of the world's mangroves in the 21st century under climate change. Hydrobiologia 803(1):1–12

Grassi G, House J, Dentener F, Federici S, den Elzen M, Penman J (2017) The key role of forests in meeting climate targets requires science for credible mitigation. Nat Clim Chang 7:220–226. https://doi.org/10.1038/nclimate3227

Hamilton SE, Casey D (2016) Creation of a high spatio-temporal resolution global database of continuous mangrove forest cover for the 21st century (CGMFC-21). Glob Ecol Biogeogr 25(6):729–738

Herr DT, Agardy, Benzaken D, Hicks F, Howard J, Landis E, Soles A, Vegh T (2015) Coastal "blue" carbon. A revised guide to supporting coastal wetland programs and projects using climate finance and other financial mechanisms. Gland, Switzerland: IUCN. https://doi.org/10.2305/IUCN.CH.2015.10.en

Herrera Silveira JA & Teutli Hernández C (2017) Carbono azul, manglares y politica pública, Elementos para Políticas Públicas, vol 1, # 1, CINVESTAV

Herrera-Silveira JA, Pech-Cardenas MA, Morales-Ojeda SM, Cinco-Castro S, Camacho-Rico A, Caamal Sosa JP, Mendoza-Martinez JE, Pech-Poot EY, Montero J, Teutli-Hernandez C (2020) Blue carbon of Mexico, carbon stocks and fluxes: a systematic review. Peer J 8:e8790

INECC (2018) National inventory of greenhouse gasses and substances 2017. Instituto Nacional de Ecología y Cambio Climático, viewed 7 April 2020. https://datos.abiertos.inecc.gob.mx/Datos_abiertos_INECC/Inventario_Nacional_de_Gases_de_Efecto_Invernadero/INEGyCEI_2017/INEGyCEI_1990-2017_IPCC_2006.xlsx

IPCC (2014) Climate change 2014: synthesys report, viewed 29 December 2014. http://www.ipcc.ch/report/ar5/f

Ivanova A, Gámez AE (eds) (2012) Plan estatal de acción ante el cambio climatico para Baja California Sur

Ivanova A (2019) Las Áreas Protegidas en Norteamérica: experiencias de la cooperación trilateral en protección del ambiente y acción climática. En:(Lucatello, S., coord.) Del TLCAN al T-MEC: la dimensión olvidada del medioambiente en América del Norte, Instituto Mora & Siglo XXI, pp 277–302

Ivanova A, Zia A, Ahmad P, Bastos-Lima M (2020) Climate mitigation policies and actions: access and allocation issues. Int Environ Agreements. https://doi.org/10.1007/s10784-020-09483-7

King L (2012) Including mangrove forests in REDD+, Climate and Development Knowledge Network, Nairobi

Linn A (2016) Next steps for the Paris agreement: when and how will the agreement enter into force? Issue brief. NYU School of Law, Guarini Center, New York, NY. http://guarinicenter.org/wp-content/uploads/2016/03/Paris-Entry-into-Force-Final-30-Mar-16.pdf

López-Medellín X, Ezcurra E, González-Abraham C, Hak J, Santiago LS, Sickman JO (2011) Oceanographic anomalies and sea-level rise drive mangroves inland in the Pacific coast of Mexico. J Veg Sci 22(1):143–151

Lopomo G, Marx LM, McAdams D, Murray B (2011) Carbon allowance auction design: an assessment of options for the United States. Rev Environ Econ Policy 5(1):25–43. https://doi.org/10.1093/reep/req024

Morgan J, Northrop E (2017) Will the Paris Agreement accelerate the pace of change? Wiley Interdiscip Rev Clim Change 8. https://doi.org/10.1002/wcc.471

Nellemann C, Corcoran E, Duarte CM, Valdés L, De Young C, Fonseca L, Grimsditch G (Eds) (2009) Blue carbon. A rapid response assessment. UNEP, FAO, IOC-UNESCO, viewed 24 May 2020. https://www.grida.no/publications/145

Ochoa-Gómez JG, Lluch-Cota SE, Rivera-Monroy VH, Lluch-Cota DB, Troyo-Diéguez E, Oechel W, Serviere-Zaragoza E (2019) Mangrove wetland productivity and carbon stocks in an arid zone of the Gulf of California (La Paz Bay, Mexico). For Ecol Manage 442:135–147

OCM-NOAA (2018) Reserves advance "blue carbon" approach to conserving wetlands, office for coastal management. National Oceanic and Atmospheric Administration, viewed 23 May 2020. https://coast.noaa.gov/states/stories/first-carbon-market-guidance-for-wetlands.html

Olsen KH, Arens C, Mersmann F (2018) Learning from CDM SD tool experience for Article 6.4 of the Paris agreement. Clim Policy 18:383–395. https://doi.org/10.1080/14693062.2016.1277686

Park MS, Choi ES, Youn YC (2013) REDD+ as an international cooperation strategy under the global climate change regime. For Sci Technol 9:213–224. https://doi.org/10.1080/21580103.2013.846875

Pendleton L, Donato DC, Murray BC, Crooks S, Jenkins WA, Sifleet S, Craft C, Fourquean JW, Kauffman JB, Marbà N, Megonigal P, Pidgeon E, Herr D, Gordon D, Baldera A (2012) 'Estimating global "blue carbon" emissions from conversion and degradation of vegetated coastal ecosystems. PLoS ONE 7(9)

Pronatura Sur AC (2016) Incluyendo a los manglares en la estrategia REDD+ de México: un enfoque integrado de colaboración entre el sector privado y social. Informe final

REDD Offset Working Group (ROW) (2013) California, Acre and Chiapas partnering to reduce emissions from tropical deforestation. https://ww3.arb.ca.gov/cc/capandtrade/sectorbasedoffsets/row-final-recommendations.pdf

SEMARNAT (2017) Understanding blue carbon, Secretaria de Medio Ambiente y Recursos Naturales, viewed 14 May 2020. https://www.gob.mx/cms/uploads/attachment/file/249430/Blue_carbon.pdf

SEMARNAT and GIZ (2020) Programa de Prueba del Sistema de Comercio de Emisiones en México. http://www.mexico2.com.mx/uploadsmexico/file/NOTA%20Sistema%20de%20Comercio%20de%20Emisiones%20en%20M%C3%A9xico%20v040719%203.pdf

SPYDE (2015) Información estratégica: Baja California Sur, Secretaría de Promoción y Desarrollo Económico del Gobierno del Estado de Baja California Sur (SPYDE)

Streck Ch, Keenlyside P, von Unger M (2016) The Paris agreement: a new beginning. J Eur Environ Plan Law 13:3–29 ©koninklijke brill nv, leiden. https://doi.org/10.1163/18760104-01301002

Ullman R, Bilbao Bastida V, Grimsditch G (2013) Including blue carbon in climate market mechanisms. Ocean Coast Manag 83:15–18

UNFCCC (1992) United Nations Framework Convention on Climate Change

UNFCCC (2015) Paris agreement

UNFCCC (2016) Decision 1/CP.21 Adoption of the Paris agreement.In: Report of the conference of the parties on its twenty-first session, held in Paris from 30 November to 13 December 2015, FCCC/CP/2015/10/Add.1

Valderrama-Landeros LH, Rodríguez-Zúñiga MT, Troche-Souza C, Velázquez-Salazar S, Villeda-Chávez E, Alcántara-Maya JA, Vázquez-Balderas B, Cruz-López MI, Ressl R (2017) Manglares de México: actualización y exploración de los datos del sistema de monitoreo 1970/1980–2015, Comisión Nacional para el Conocimiento y Uso de la Biodiversidad, Ciudad de México, viewed 23 May 2020. http://bioteca.biodiversidad.gob.mx/janium/Documentos/12889.pdf

Watson EB and Corona AH (2018) Assessment of blue carbon storage by Baja California (Mexico) tidal wetlands and evidence for wetland stability in the face of anthropogenic and climatic impacts. Sensors (Switzerland) 18(1)

World Conservation Monitoring Center (WCMC) (2016) Task force on other effective area-based conservation measures: co-chairs report of first international expert meeting (Cambridge)

Chapter 14
Relationship Between Emissions Trading System and the 2030 Agenda for Sustainable Development

Gustavo Sosa-Nunez

Abstract With the Paris Agreement and through Nationally Determined Contributions, nation-states have agreed to reduce their emissions of greenhouse gases. Some of them have approached this aspect by setting emission trading systems. In some cases, it is in the regional and sub-national levels where these types of developments are taking place. The relevance of this market-based instrument is increasing over time, to the point of being regarded as a cornerstone of climate change mitigation strategies, despite the lack of global agreement on the matter. The importance of emission trading systems, however, can be observed when assessing their relevance for achieving the 2030 Agenda for Sustainable Development. Implementing them can, and should, assist in reaching diverse targets of different Sustainable Development Goals. This is the case of the goals related to energy, economic growth, inclusive industrialization, sustainable cities, sustainable production and consumption patterns, marine and land life, as well as the climate itself. Then, the relevance of emission trading systems can be observed throughout the whole 2030 Agenda. It is thus in this context that this contribution aims to assess the manner in which this relationship takes place in the global fora and in Mexico. A key argument is that there should be the participation of a wider set of sectors and actors.

Keywords Emission trading system · Mexico · 2030 Agenda · Sustainable development goals

Introduction

The main objective of an Emission Trading System (ETS) is to reduce Greenhouse Gas (GHG) emissions in a cost-effective way without damaging the competitiveness of participating sectors and actors (European Commission 2015). The creation of this market-based instrument is aimed at complying with commitments made at the international level to tackle climate change, essentially the 2015 Paris Agreement, in

G. Sosa-Nunez (✉)
Instituto Mora-CONACYT, Mexico City, Mexico
e-mail: gsosa@institutomora.edu.mx

© The Author(s) 2022
S. Lucatello (ed.), *Towards an Emissions Trading System in Mexico: Rationale, Design and Connections With the Global Climate Agenda*, Springer Climate,
https://doi.org/10.1007/978-3-030-82759-5_14

which technological innovation and energy efficiency are essential. To achieve this, it is crucial to measure, report, and verify emissions reductions.

There is a disparity between different ETS according to their geographical location, the type of GHG emissions to consider, and participating sectors. However, they all share the same goal. Now it is Mexico's turn to develop its own ETS. The pilot programme currently in place offers insight about the road the country is planning to follow, and the different chapters conforming to this book provide accounts from different academic approaches about what is coming next, including expectations and areas of opportunity.

It is in this sense that the links between ETS and the 2030 Agenda should be explored. Even when ETS aims at reducing GHG emissions, a relationship with different Sustainable Development Goals (SDGs) exists beyond climate. The underlying rationale is that ETS could gain greater presence and importance in government circles if they were to acknowledge that, by implementing this type of market-based instrument, a better implementation of the 2030 Agenda could happen. Of course, this is an argument that should be explored further, as the content of this contribution shows only initial findings.

In this context, the first section presents a general background of ETS, stating their relevance, followed by a description of the 2030 Agenda, as well as the links that the author observes between SDGs and ETS. Afterwards, the case of Mexico is presented, commenting broadly on the ETS pilot programme currently in place, the national approach to the 2030 Agenda, and the links that are observable as things stand now. Lastly, there are conclusions commenting on the impact that the ETS pilot programme has on the 2030 Agenda, as well as suggestions on how to move forward when the pilot programme transitions to the operative phase.

ETS Background

There are many ETS across the world with different features. The one developed by the European Union (EU) stands out, as it is the largest market for trading Greenhouse Gases (GHGs) emissions and is its flagship climate change mitigation policy (Jones 2013). It is also the first multinational cap-and-trade system that has been seen as a prototype for a global scheme, as its participants—that is, member states—present significant disparities in economic circumstance, institutional development, and political will (Ellerman 2010). It includes the power sector, energy-intensive industry, and commercial aviation (European Commission 2020).

Another is the Regional Greenhouse Gas Initiative (RGGI), which is the first mandatory market-based programme in the United States to cap and reduce carbon dioxide (CO_2) emissions from the power sector. It is a cooperative initiative of the New England and Mid-Atlantic States (RGGI 2020). There is also the Western Climate Initiative (WCI), which is a collaboration of independent jurisdictions with GHG trading programmes between California, USA, and the Provinces of Québec and Nova Scotia, in Canada (WCI 2020).

Asia also has its cases. In Japan, the Tokyo Metropolitan Area is the country's first mandatory ETS and is linked to the Saitama ETS, in which participants are factories, buildings, and other facilities that consume large quantities of fossil fuels (ICAP 2020a). For its part, South Korea launched East Asia's first nationwide mandatory ETS and the second-largest carbon market after the EU ETS (ICAP 2020b). It includes direct GHG emissions and indirect emissions coming from electricity consumption. China has also recently implemented a national ETS, which builds on the experience of having carried seven regional pilot programmes (ETS-China 2020).

Down south, Australia also developed an ETS, in which credits could be obtained via a carbon farming initiative where farmers were paid to maintain their land as carbon sinks; although this scheme did not last due to political reasons (Atchison 2020). New Zealand also has its own ETS, which creates financial incentives for businesses to reduce their emissions and landowners to earn money by planting forests that become carbon sinks (New Zealand's Ministry for the Environment 2019). This case is also an interesting one because it includes the majority of economic sectors: forestry, waste, industrial processes, stationary energy, fossil fuels, and synthetic gases.

Beyond political jurisdictions, there is the Carbon Offsetting and Reduction Scheme for International Aviation (CORSIA), which is a global scheme for the global international aviation industry, and States adhere as members (ICAO 2020).

Newly created programmes can learn from these previous experiences. For example, the EU ETS experience has provided some insights into this market-based instrument, like considering the power sector and large industrial facilities as the first ones for early inclusion in the ETS. However, initial partial coverage should not "preclude a later, more comprehensive system, although the issue will be whether an initial partial approach makes it more difficult to arrive ultimately at comprehensive coverage" (Ellerman 2010, p. 94).

A further lesson is the possibility to move beyond the energy sector to include forests and farming. Another is that companies reducing emissions regard the carbon price as a competitive advantage; such price is too low to foster technological innovation, and companies focusing on reducing GHG emissions are reducing their costs (MexiCO$_2$ 2019).

Lastly, a further lesson is that oversupplying emissions allowances contributes to a low carbon price that, in turn, does not stimulate investment in emission reduction measures. If this situation does not happen, then a series of follow-up actions could result in a compelling approach to develop not only strategies for climate change mitigation, but also actions to comply with the 2030 Agenda for Sustainable Development. The resulting spillover effect could in turn help to consider tackling climate change as the cornerstone for worldwide sustainable development.

2030 Agenda for Sustainable Development

Since 2000, the United Nations (UN) have developed a series of goals aimed at improving humankind's living conditions. With the Millennium Development Goals (MDGs), which lasted up until 2015, there was the idea to tackle different problems. Out of eight, the seventh MDG emphasized guaranteeing environmental sustainability with no explicit reference to climate change. By the time the MDGs ended, it was acknowledged that climate change was undermining progress and that global CO_2 emissions were increasing at a rate of over 50% since 1990 (UN 2015a). This would locate environmental sustainability at the core of the post-2015 development agenda, which would come jointly with the Paris Agreement.

The 2030 Agenda for Sustainable Development was subscribed in 2015. Trying to overcome the pending issues that the MDGs left, 17 Sustainable Development Goals (SDGs) and 169 targets were issued, aiming at leaving no one behind. Issues would range from poverty, hunger, and health, to land and sea environments, including cities, employment, and economic growth.

Climate change is also considered in the 2030 Agenda. The preamble states the acknowledgement of the United Nations Framework Convention on Climate Change as "the primary international, intergovernmental forum for negotiating the global response to climate change" (UN 2015b, p. 8). For this, it is of utmost importance to accelerate the reduction of global GHG emissions and to adapt to the adverse impacts of climate change. To show commitment, the Agenda dedicates SDG13 to this topic.

However, despite the presence of climate change in the 2030 Agenda, it is through the United Nations Framework Convention on Climate Change (UNFCCC) and the Paris Agreement where efforts to tackle climate change are concentrated. The SDG13 of the 2030 Agenda includes a disclaimer stating this, hence diluting the impact that this SDG could have on climate change combat and the overall Agenda. The SDG Report issued by the United Nations (UN) in 2019 reiterates the importance of reducing GHG emissions, but makes no consideration about the type of instruments or mechanisms to use (UN 2019).

This does not come as a surprise, as the core objective of the Paris Agreement is to tackle climate change and foster the development and implementation of mitigation and adaptation strategies—where market-based instruments like ETS are included—while in the 2030 Agenda this topic is one of many to attend, despite its links with different SDGs besides SDG13. The existence of ETS could relate to this context, as this market-based instrument is related essentially to climate change mitigation. However, developing and implementing ETS could assist in achieving a wider range of issues for the Agenda beyond SDG13.

SDGs Link with ETS

The 2030 Agenda entails a wide range of issues, like ending poverty and hunger, ensuring healthy lives and quality education, achieving gender equality, accessing water and energy, promoting economic growth, conserving natural sea and inland resources, and promoting peaceful societies, among other objectives (see Appendix 1) (UN 2015b). Many of them can relate to the development and implementation of ETS. It is worth mentioning, though, that the following are broad comments aimed at pointing out links between ETS and the 2030 Agenda, and further research is needed to support these arguments. In fact, each SDG link with ETS would merit a chapter in itself.

The obvious link is with **SDG13**, as one of the strategies to tackle climate change relates to integrating mitigation measures into national policies. In this sense, setting ETS can be observed as part of a wider climate change policy of a given nation (**target 13.2**). This also relates to the improvement of education and human and institutional capacity (**target 13.3**), as well as the increase of capacity for planning and management in the least developed countries (**target 13.b**).

Besides SDG13, there are links between ETS and other SDGs, which can be observed in two different ways. The first is that some targets of different SDGs can be assisted when implementing ETS. The second is that ETS can be the result of work done around specific SDGs. That is to say, ETS can be either a cause or a consequence of approaches made on specific targets of some SDGs.

Energy is a key factor, which allows for considering **SDG7** as having a link. Participating companies looking to be energy efficient (**target 7.3**) while transitioning to renewable energy—for the sake of GHG emissions reduction—is central to any successful ETS (**target 7.2**). For this, research and investment in technology are required (**target 7.a**), to be able to expand infrastructure on the matter (**target 7.b**).

Being a market-based instrument, ETS should promote sustained, inclusive, and sustainable economic growth, as aimed by **SDG8**. Participating companies should achieve higher levels of economic productivity through innovation and technological upgrading to reduce GHG emissions (**target 8.2**), although this can be costly. If this happens, it could lead to decouple economic growth from environmental degradation, as proposed by the 10-Year Framework of Programmes on Sustainable Consumption and Production (10YFP) (**targets 8.4** and **12.1**, the latter coming from **SDG12**). If this is attainable, it may result from the development of sustainable industrialization, including the research and adoption of clean and environmentally sound technologies and industrial processes, as proposed by **SDG9** (**targets 9.2**, **9.4**, and **9.5**).

Transitioning to sustainable and environment-friendly industrialization would imply a positive impact in cities and human settlements where they are located, thus contributing to **SDG11**. For example, the inclusion of sustainable infrastructure, meaning buildings, could foster the reduction of GHG emissions (**target 11.3**) and, hence, improve air quality (**target 11.6**). Besides, if the transport sector participates in ETS, its sustainability could be improved (**target 11.2**).

Besides actions suggested by the 10YFP (**target 12.1**), the implementation of sustainable production patterns, as aimed by **SDG12**, could assist in reducing GHG emissions of those companies participating in ETS whose production is changing due to innovative technological upgrades. With this, there would be cases in which natural resources are used in a more sustainable and efficient way (**target 12.2**). Moreover, if cities and human settlements were to participate in ETS as contributing entities, through their governments and representatives, reducing food waste and post-harvest losses could mean a reduction of methane (CH_4) emissions, which would in turn need to be considered in the set of emissions considered by some ETS. Another aspect of **SDG12** that relates heavily with ETS is that, by aiming to reduce GHG emissions, inefficient fossil-fuel subsidies that encourage wasteful consumption could be removed (**target 12.c**).

On top of the previous comments, if a global ETS were to exist, or regional ones were to be fostered, regardless of the level of country development, some other SDGs could be related, although many of the cases would have to consider the geographical scope, the timeframe, political will, economic context, and citizenry commitment, among other variables.

One can assume that if a reduction in GHG emissions takes place, improvements in health and the **SDG3** would also be seen. This could imply a decrease in air pollution, hence reducing the number of deaths and illnesses happening due to this factor (**target 3.9**). This is because sectors contributing to GHG emissions are also sources of air pollution. Additionally, reducing CO_2, which is a GHG, can mean lower emissions of particulate matter, ozone, precursors, and other air pollutants (EPA 2020).

Other SDGs have an indirect link with the development of ETS, and they depend on considering forestry as a component of this type of market-based instrument. This would allow for considering links with **SDG15**, due to the importance of forests as inland ecosystems that can perform as carbon sinks. It should be the case, as "forestry can either be a source of significant emissions reductions or increasing emissions, depending on what incentive structures are put in place" (Brohé et al. 2009, p. 253). In this sense, their conservation, restoration, and sustainable use (**target 15.1**), together with halting deforestation and restoring degraded forests (**target 15.2**), could be actions going in the correct direction if included in ETS, besides being regarded as offset providers. For this, financial support would be necessary (**targets 15.a** and **15.b**). In this context, **SDG6** about water is intrinsically related, since forests, as water-related ecosystems, need to be protected and restored (**target 6.6**).

This is also the case of **SDG1**, aimed at tackling poverty. Granting equal rights and access to poor and vulnerable people to own land and natural resources (**target 1.4**) can encourage the participation of a wider set of actors in different ETS. Forest protection for ETS purposes would also help reach other SDG1 targets, like implementing programmes to reduce poverty in the least developed countries (**target 1.a**), creating related policy frameworks (**target 1.b**), and building resilience for the poor and reducing their exposure and vulnerability to climate-related extreme events (**target 1.5**).

In this context, it is likely that women take the leading role in plenty of cases, due in part to their ownership and control over land and natural resources (**target 5.a**), as stated by the **SDG5**, about gender.

If forests are to be included in ETS, agriculture should also be looked at. Agroforestry, forest farming, and mixed farming are examples of interventions in the agriculture sector that can assist to reduce deforestation emissions as well as reducing emissions in the agriculture sector (Carter et al. 2015). That is, although **SDG2** relates to ending hunger and achieving food security, it also focuses on the promotion of resilient and sustainable agricultural practices that help maintain ecosystems (**target 2.4**), which in turn could help to ensure access to food for local communities and a potential trade-related income (**targets 2.1** and **2.3**).

Would all ETS eventually include a wider range of actors? In developed and middle-income countries, it could be possible, which would lead one to think that not only local communities could participate through the forestry sector, but also small-scale industries and companies, who could promote development-oriented policies and access to financial services (**target 8.3**), as suggested by **SDG8**. The more the topic permeates in the society, the more people would be willing to participate, which would allow for establishing links with **SDG4**, about education. If learners acquire the knowledge and skills needed to promote sustainable development (**target 4.7**), an increase in public understanding of the topic could take place and, hence, wider participation coming from a broader set of actors would exist. This action could be assisted by an increase of qualified teachers (**target 4.c**) and by the expansion of scholarships to people interested in learning and deepening knowledge about ETS (**target 4.b**).

And what about **SDG14**? Links may appear if ETS were to move from a geographical approach to an emissions-based perspective in which estimated numbers from maritime transportation could be traded in an attempt to reduce marine pollution, regardless of location (**target 14.1**).

Continuing with potential links, if ETS were to be a global approach to reduce emissions, then it would be necessary to broaden and strengthen the participation of developing countries in institutions of global governance that would deal with the matter (**target 16.8**). A global carbon price could point in this direction, although this is far from certain for the near future. To make this happen, alliances to foster cooperation in technology, technology transfer, and innovation could assist, as proposed by **SDG17** (**targets 17.6**, **17.7**, and **17.8**). In turn, all countries would have access to similar information, thus enhancing policy coherence for sustainable development (**target 17.14**).

Then, as it is observable, many SDGs can relate to ETS, either as the cause of or as consequence from, implementing actions to reach specific targets (see Appendix 2 for a full list of ETS-related SDGs and targets listed in this section). Some links are noticeable, and others require long-term research to assess whether their connection is viable. However, the lifespan of the 2030 Agenda does not address this regard. In spite of this, some findings are conclusive for the current situation of Mexico.

The Case of Mexico

Continuing with the idea of pointing at potential links between ETS and different SDGs of the 2030 Agenda, this section aims to identify them for the case of Mexico. To do so, first, it is necessary to contextualize the topic; the reason for which a brief description of the ETS pilot programme recently put in place is provided. Subsequently, Mexico's work around the 2030 Agenda is commented upon, emphasizing on SDG indicators developed for the country and on its National Strategy for the Implementation of the 2030 Agenda. This allows identifying current SDG links with the establishment of the ETS pilot programme.

ETS Pilot Programme

Mexico has a legal and institutional framework focusing on tackling climate change. The cornerstone is the 2012 General Law on Climate Change, which was amended in 2018 to update the reduction percentages committed in both this legal instrument and the Nationally Determined Contribution (NDC) and it sets the context in which the ETS would exist (LGCC 2012). The federal government is responsible for creating, authorizing, and regulating emissions trading (Art. 7, Fraction IX), while also fostering the competitiveness of companies participating in such trading (Art. 7, Fraction XVIII).

The 2018 amendment entailed the gradual creation of an ETS (Art. 94) that includes a 36-month long pilot programme with no economic effects for participants (Transitory Art. 2). Preparations for Mexico's own ETS were supported by international exchange and experience coming from other parts of the world, as it was the case of California. Germany also assisted Mexico in setting up the National Emissions Registry (RENE), which is a database that collects information on GHG emissions from all major GHG emitting sectors, like energy, industry, transport, and others (IKI Alliance Mexico 2020). Such information would be essential for the ETS pilot programme.

The agreement setting the preliminary basis for the ETS pilot programme was published on 1 October 2019 (Gobierno de México 2019a). Lasting 36 months, the pilot programme began on 1 January 2020 and includes a transition phase towards the operative phase during the whole of 2022. The pilot programme is applicable to energy and industry sectors in cases when their yearly direct emissions surpass 100.000 tons of CO_2. Subsectors of the former relate to the production and supply chain of hydrocarbons and electricity. Subsectors of the latter compile a wide range of industries: automotive, cement, chemical, food processing, glass, iron and steel, metallurgical, mining, petrochemical, cellulose and paper, as well as those subsectors that produce emissions out of stationary sources (SEMARNAT 2019). It is worth pointing out that the ETS pilot programme considers only CO_2 emissions since it is the most emitted gas in the country (Gobierno de México 2019a).

The argument to select only energy and industry sectors is that they represent more than 90% of emissions reported by RENE. The agreement setting the pilot programme states that those sectors and GHG detailed in the LGCC will be added when the operative phase starts. This implies widening the scope of the ETS so that transportation, waste, agriculture, farming, forestry, and other land are considered in the future. It also implies that emissions other than CO_2 may be included.

Then, the potential inclusion of a broader range of sectors, once the operative phase starts, allows for consideration of the possibility of Mexico to set the links between the ETS and its approach to the 2030 Agenda. Being in the pilot phase implies a learning process that allows for focusing on areas of opportunities.

Mexico's Approach to the 2030 Agenda

In accordance with Mexico's usual performance in global affairs, this country adopted the 2030 Agenda at the beginning of 2016. However, the Agenda has not truly permeated public policies, despite the creation of an institutional framework to follow up on the matter and the promotion given to the topic by both governmental and non-governmental actors. Instead, there are broad references to it in governmental development programmes and public discourse, neither of which have the expected impact, mainly due to a lack of interest or indifference that rests on the immediacy of political preferences and government strategies, which focus essentially on social issues. Yet, in 2017, the Office of the Presidency formed a task force that, since then, has discussed the manner in which the Agenda can be implemented and measured.

Including members of the government, academia, private sector, and civil society, the task force assisted to develop a 2018 national strategy to begin the adoption of the Agenda. Nevertheless, it was not until 2019 that a National Strategy for the Implementation of the 2030 Agenda was issued. The task force would also support the National Institute of Statistics and Geography (INEGI) to develop indicators applicable to Mexico to measure SDGs performance.

Taking into account the targets that were identified as potentially having a link with ETS, which were shown in prior sections of this contribution, it is not surprising to argue that all of them could be applicable to Mexico, either explicitly or implicitly. Climate, energy, and industry appear as the obvious connections, but other SDGs are also related. Yet, if indicators to measure those SDG targets do not consider the involvement of ETS, little can be expected to promote the importance of this market-based instrument.

Of the 40 targets previously listed and that seem to relate to the implementation of ETS, indicators were developed in 29 cases (see Gobierno de México 2020). Unsurprisingly, they are mostly unrelated to ETS, except for two indicators of SDG7:

- Indicator 7.2.1.—Share of renewable energy in total final energy consumption.
- Indicator 9.4.1.—Total carbon dioxide emissions by Gross Domestic Product (GDP) by purchasing power parity.

These indicators could consider the inclusion of ETS—through GHG emissions reduction—as a means to contribute to their respective targets. For this, the meticulous analysis would be required to assess how to incorporate ETS contributions to the indicators, while at the same time keeping them in a side list to show the importance of this market-based instrument not only for the two related indicators, but also for showing their potential relevance for the overall Agenda. It is worth pointing out, however, that if forestry were to be included as a participating sector in the ETS once the operative phase starts, a further indicator could also be considered: indicator 15.1.1, measuring forest area as a proportion of the total area.

Maybe indicators for other SDGs could be developed to assert the importance of ETS for reducing emissions. Again, including forestry could have a spillover effect on other development areas. It would also widen the scope and number of actors and, as a result, it could increase the relevance of programme once its pilot phase ends.

With regard to the National Strategy for the Implementation of the 2030 Agenda, there is only one reference to the relevance of developing an ETS. It is presented as a way to transition to clean technologies that can assist in reducing GHG emissions from productive sectors and, hence, should be promoted and implemented (Gobierno de México 2019b). A further suggestion made in the National Strategy relates to the optimization of a carbon tax that goes beyond the mere fiscal collection, one that is properly destined for emissions reduction.

Comments/Conclusions

There are different aspects that come to mind when assessing the potential links that ETS can have with the 2030 Agenda SDGs, and the way this can be seen for Mexico. It is important to keep in mind, however, that the ideas expressed in this section require further research for both global and national levels.

To begin with, the development and implementation of ETS can have a spillover effect across many of the SDG targets, an issue that corroborates the transversal nature of the 2030 Agenda. For this to happen, it is essential that the formulation of SDG indicators includes a perspective on the ETS. This implies that whoever participates in the drafting has a clear understanding of the topic. At the same time, those in charge of preparing the ETS operative phase out of the pilot programme should also have the 2030 Agenda in mind. The addition of forests as carbon sinks and the people who manage them can be a turning point, the reason for which assessing the case of New Zealand should be interesting to observe and learn from the experience.

With regard to Mexico, both energy and transport sectors are key for reducing emissions: They are responsible for around three-quarters of the country's total GHG emissions. However, Mexico is not considering the transport sector in the pilot programme. Other sectors should also be included, namely forestry, domestic aviation, management, and public buildings. In addition, other GHG should be included to expand the scope of the emissions. In theory, this will happen with the operative phase, so a close follow-up will be required.

Considering environmental pollution—including GHG emissions—as economic externalities could assist not only in developing fairer market rules, but also to reach many of the goals of the 2030 Agenda. However, the slow pace and the voluntary approach that characterizes the ETS imply that any potential impact on this type of market instrument to mitigate climate change in Mexico will take place in the distant future. The key message is to foster investments in measures to mitigate emissions, but it is easier said than done. It may be possible that many commercial and industrial entities prefer to pay to pollute and innovate later. This is because it gives them the flexibility to reduce their emissions on their own terms (e.g. decide a timeframe for low carbon investment tailored to their business plan) and do so in the most cost-effective manner.

The aforementioned does not mean that ETS is the panacea to climate change, despite some governments and advocates claiming so. Market-based instruments can be criticized for monetizing the combat to climate change, although this is an ideological perspective that does not hold to the basic premises of the economic system in which we live. Furthermore, due to its market nature, ETS can mean that only those capable of investing in technology to reduce emissions or paying to pollute can persist.

Could ETS foster the creation and preservation of monopolies or oligopolies in specific productive sectors? The best-positioned industries could work together to sideline small competitors that cannot afford to reduce emissions or innovate at the speed, pace, and scope that a wealthier industry can. This is why it is important to broaden the scope of participant actors in ETS. The role of national governments is also up for discussion. They would be expected to regulate monopolies and oligopolies. Advocating for a bigger role from the state, either through incentives or regulations, depends on an environmentally conscious government that commits to the issue. Here, it is necessary to consider another set of variables, like the political party in power, policy preferences, and institutional structure, to cite a few.

Should the invisible hand be fully in charge of ruling this market-based instrument? How to set adequate carbon prices? There are many questions and few answers. Delayed decisions do not contribute to develop the urgent actions necessary to give ETS the importance that the global community is willing to give to them.

Then, it is worth noticing that the ETS is merely one of the many climate change mitigation measures that exist, and focusing on it due to its potential and market-driven perspective can be counterproductive if other measures are sidelined or reduced in importance. Therefore, there is the need to see the whole picture and include more sectors and actors. The consideration of the 2030 Agenda for Sustainable Development can be of assistance to get to this point.

Annex 1—Sustainable Development Goals of the 2030 Agenda

Goal 1. End poverty in all its forms everywhere.

Goal 2. End hunger, achieve food security and improved nutrition, and promote sustainable agriculture.

Goal 3. Ensure healthy lives and promote well-being for all at all ages.

Goal 4. Ensure inclusive and equitable quality education and promote lifelong learning opportunities for all.

Goal 5. Achieve gender equality and empower all women and girls.

Goal 6. Ensure availability and sustainable management of water and sanitation for all.

Goal 7. Ensure access to affordable, reliable, sustainable, and modern energy for all.

Goal 8. Promote sustained, inclusive, and sustainable economic growth; full and productive employment; and decent work for all.

Goal 9. Build resilient infrastructure, promote inclusive and sustainable industrialization, and foster innovation.

Goal 10. Reduce inequality within and among countries.

Goal 11. Make cities and human settlements inclusive, safe, resilient, and sustainable.

Goal 12. Ensure sustainable consumption and production patterns.

Goal 13. Take urgent action to combat climate change and its impacts.

Goal 14. Conserve and sustainably use the oceans, seas, and marine resources for sustainable development.

Goal 15. Protect, restore, and promote sustainable use of terrestrial ecosystems, sustainably manage forests, combat desertification, and halt and reverse land degradation and halt biodiversity loss.

Goal 16. Promote peaceful and inclusive societies for sustainable development; provide access to justice for all; and build effective, accountable, and inclusive institutions at all levels.

Goal 17. Strengthen the means of implementation and revitalize the Global Partnership for Sustainable Development.

Annex 2—ETS-Related SDGs and Targets

Goal 1. End poverty in all its forms everywhere

1.4 By 2030, ensure that all men and women, in particular the poor and the vulnerable, have equal rights to economic resources, as well as access to basic services, ownership and control over land and other forms of property, inheritance, natural resources, appropriate new technology, and financial services, including microfinance.

1.5 By 2030, build the resilience of the poor and those in vulnerable situations and reduce their exposure and vulnerability to climate-related extreme events and other economic, social, and environmental shocks and disasters.

1.a Ensure significant mobilization of resources from a variety of sources, including through enhanced development cooperation, in order to provide adequate and predictable means for developing countries, in particular least developed countries, to implement programmes and policies to end poverty in all its dimensions.

1.b Create sound policy frameworks at the national, regional, and international levels, based on pro-poor and gender-sensitive development strategies, to support accelerated investment in poverty eradication actions.

Goal 2. End hunger, achieve food security and improved nutrition, and promote sustainable agriculture

2.1 By 2030, end hunger and ensure access by all people, in particular the poor and people in vulnerable situations, including infants, to safe, nutritious, and sufficient food all year round.

2.3 By 2030, double the agricultural productivity and incomes of small-scale food producers, in particular women, indigenous peoples, family farmers, pastoralists, and fishers, including through secure and equal access to land, other productive resources and inputs, knowledge, financial services, markets, and opportunities for value addition and non-farm employment.

2.4 By 2030, ensure sustainable food production systems and implement resilient agricultural practices that increase productivity and production; that help maintain ecosystems; that strengthen capacity for adaptation to climate change, extreme weather, drought, flooding, and other disasters; and that progressively improve land and soil quality.

Goal 3. Ensure healthy lives and promote well-being for all at all ages

3.9 By 2030, substantially reduce the number of deaths and illnesses from hazardous chemicals and air, water, and soil pollution and contamination.

Goal 4. Ensure inclusive and equitable quality education and promote lifelong learning opportunities for all

4.7 By 2030, ensure that all learners acquire the knowledge and skills needed to promote sustainable development, including, among others, through education for sustainable development and sustainable lifestyles, human rights, gender equality, promotion of a culture of peace and non-violence, global citizenship, and appreciation of cultural diversity and of culture's contribution to sustainable development.

4.b By 2020, substantially expand globally the number of scholarships available to developing countries, in particular least developed countries, small island developing States, and African countries, for enrolment in higher education, including vocational training and information and communications technology, technical, engineering, and scientific programmes in developed countries and other developing countries.

4.c By 2030, substantially increase the supply of qualified teachers, including through international cooperation for teacher training in developing countries, especially least developed countries and small island developing States.

Goal 5. Achieve gender equality and empower all women and girls

5.a Undertake reforms to give women equal rights to economic resources, as well as access to ownership and control over land and other forms of property, financial services, inheritance, and natural resources, in accordance with national laws.

Goal 6. Ensure availability and sustainable management of water and sanitation for all

6.6 By 2020, protect and restore water-related ecosystems, including mountains, forests, wetlands, rivers, aquifers, and lakes.

Goal 7. Ensure access to affordable, reliable, sustainable, and modern energy for all

7.2 By 2030, increase substantially the share of renewable energy in the global energy mix.

7.3 By 2030, double the global rate of improvement in energy efficiency.

7.a By 2030, enhance international cooperation to facilitate access to clean energy research and technology, including renewable energy, energy efficiency, and advanced and cleaner fossil-fuel technology, and promote investment in energy infrastructure and clean energy technology.

7.b By 2030, expand infrastructure and upgrade technology for supplying modern and sustainable energy services for all in developing countries, in particular least developed countries, small island developing States, and landlocked developing countries, in accordance with their respective programmes of support.

Goal 8. Promote sustained, inclusive, and sustainable economic growth; full and productive employment; and decent work for all

8.2 Achieve higher levels of economic productivity through diversification, technological upgrading, and innovation, including through a focus on high-value added and labour-intensive sectors.

8.3 Promote development-oriented policies that support productive activities, decent job creation, entrepreneurship, creativity, and innovation, and encourage the formalization and growth of micro-, small-, and medium-sized enterprises, including through access to financial services.

8.4 Improve progressively, through 2030, global resource efficiency in consumption and production and endeavour to decouple economic growth from environmental degradation, in accordance with the 10-Year Framework of Programmes on Sustainable Consumption and Production, with developed countries taking the lead.

Goal 9. Build resilient infrastructure, promote inclusive and sustainable industrialization, and foster innovation

9.2 Promote inclusive and sustainable industrialization and, by 2030, significantly raise industry's share of employment and gross domestic product, in line with national circumstances, and double its share in the least developed countries.

9.4 By 2030, upgrade infrastructure and retrofit industries to make them sustainable, with increased resource-use efficiency and greater adoption of clean and environmentally sound technologies and industrial processes, with all countries taking action in accordance with their respective capabilities.

9.5 Enhance scientific research and upgrade the technological capabilities of industrial sectors in all countries, in particular developing countries, including, by 2030, encouraging innovation and substantially increasing the number of research and development workers per 1 million people and public and private research and development spending.

Goal 11. Make cities and human settlements inclusive, safe, resilient, and sustainable

11.2 By 2030, provide access to safe, affordable, accessible, and sustainable transport systems for all, improving road safety, notably by expanding public transport, with special attention to the needs of those in vulnerable situations, women, children, persons with disabilities, and older persons.

11.3 By 2030, enhance inclusive and sustainable urbanization and capacity for participatory, integrated, and sustainable human settlement planning and management in all countries.

11.6 By 2030, reduce the adverse per capita environmental impact of cities, including by paying special attention to air quality and municipal and other waste management.

Goal 12. Ensure sustainable consumption and production patterns

12.1 Implement the 10-Year Framework of Programmes on Sustainable Consumption and Production Patterns, all countries taking action, with developed countries taking the lead, taking into account the development and capabilities of developing countries.

12.2 By 2030, achieve the sustainable management and efficient use of natural resources.

12.c Rationalize inefficient fossil-fuel subsidies that encourage wasteful consumption by removing market distortions, in accordance with national circumstances, including by restructuring taxation and phasing out those harmful subsidies, where they exist, to reflect their environmental impacts, taking fully into account the specific needs and conditions of developing countries and minimizing the possible adverse impacts on their development in a manner that protects the poor and the affected communities.

Goal 13. Take urgent action to combat climate change and its impacts

13.2 Integrate climate change measures into national policies, strategies, and planning.

13.3 Improve education, awareness-raising, and human and institutional capacity on climate change mitigation, adaptation, impact reduction, and early warning.

13.b Promote mechanisms for raising capacity for effective climate change-related planning and management in the least developed countries and small island developing States, including focusing on women, youth, and local and marginalized communities.

Goal 14. Conserve and sustainably use the oceans, seas, and marine resources for sustainable development

14.1 By 2025, prevent and significantly reduce marine pollution of all kinds, in particular from land-based activities, including marine debris and nutrient pollution.

Goal 15. Protect, restore, and promote sustainable use of terrestrial ecosystems, sustainably manage forests, combat desertification, and halt and reverse land degradation and halt biodiversity loss

15.1 By 2020, ensure the conservation, restoration, and sustainable use of terrestrial and inland freshwater ecosystems and their services, in particular forests, wetlands, mountains, and drylands, in line with obligations under international agreements.

15.2 By 2020, promote the implementation of sustainable management of all types of forests, halt deforestation, restore degraded forests, and substantially increase afforestation and reforestation globally.

15.a Mobilize and significantly increase financial resources from all sources to conserve and sustainably use biodiversity and ecosystems.

15.b Mobilize significant resources from all sources and at all levels to finance sustainable forest management and provide adequate incentives to developing countries to advance such management, including for conservation and reforestation.

Goal 16. Promote peaceful and inclusive societies for sustainable development; provide access to justice for all; and build effective, accountable, and inclusive institutions at all levels

16.8 Broaden and strengthen the participation of developing countries in the institutions of global governance.

Goal 17. Strengthen the means of implementation and revitalize the Global Partnership for Sustainable Development

Technology

17.6 Enhance North-South, South-South, and triangular regional and international cooperation on and access to science, technology, and innovation and enhance knowledge sharing on mutually agreed terms, including through improved coordination among existing mechanisms, in particular at the United Nations level, and through a global technology facilitation mechanism.

17.7 Promote the development, transfer, dissemination, and diffusion of environmentally sound technologies to developing countries on favourable terms, including on concessional and preferential terms, as mutually agreed.

17.8 Fully operationalize the technology bank and science, technology and innovation capacity-building mechanism for least developed countries by 2017 and enhance the use of enabling technology, in particular information and communications technology.

Systemic issues—Policy and institutional coherence

17.14 Enhance policy coherence for sustainable development.

References

Atchison J (2020) Australia's ill-fated emissions trading system. Climate scorecard. 6 Mar 2020. https://www.climatescorecard.org/2020/03/australias-ill-fated-emissions-trading-system/ Accessed 8 June 2020

Brohé A, Eyre N, Howarth N (2009) Carbon markets: an international business guide. Earthscan, London

Carter S, Herold M, Rufino MC, Neumann K, Kooistra L, Verchot L (2015) Mitigation of agricultural emissions in the tropics: comparin forest land-sparing options at the national level. Biogeosciences 12:4809–4025. https://www.cifor.org/knowledge/publication/5630/. Accessed 7 June 2020

European Commission (2015) EU ETS handbook. https://ec.europa.eu/clima/sites/clima/files/docs/ets_handbook_en.pdf. Accessed 20 Aug 2020

Ellerman D (2010) The EU emission trading scheme: a prototype global system? In: Aldy JE, Stavins RN (eds) Post-Kyoto international climate policy: implementing architectures for agreement. Cambridge University Press, Cambridge, pp 88–118

EPA (United States Environmental Protection Agency) (2020) Air quality and climate change research. 11 Mar 2020. https://www.epa.gov/air-research/air-quality-and-climate-change-res earch. Accessed 7 June 2020

ETS-China (2020) Emission trading schemes in China. https://ets-china.org/ets-in-china/. Accessed 9 June 2020

European Commission (2020) EU Emissions Trading System (EU ETS). https://ec.europa.eu/clima/policies/ets_en. Accessed 10 June 2020

Gobierno de México (2020) Indicadores por objetivo y meta [Indicators by goal and target]. Objetivos de desarrollo sostenible. INEGI. http://agenda2030.mx/ODSopc.html?ti=T&goal=0&lang=es#/ind. Accessed 9 June 2020

Gobierno de México (2019a) Acuerdo por el que se establecen las bases preliminares del Programa de Prueba del Sistema de Comercio de Emisiones [Agreement setting the preliminary basis of the emissions trading system pilot program]. Diario Oficial de la Federación (DOF). 1 Oct 2019. https://dof.gob.mx/nota_detalle.php?codigo=5573934&fecha=01/10/2019 Accessed 10 June 2020

Gobierno de México (2019b) Estrategia nacional para la implementación de la agenda 2030 [National estrategy for the implementation of the 2030 agenda]. https://www.gob.mx/agenda 2030/documentos/estrategia-nacional-de-la-implementacion-de-la-agenda-2030-para-el-desarr ollo-sostenible-en-mexico. Accessed 10 June 2020

ICAO (International Civil Aviation Organization) (2020) Benefits for CORSIA participation. https://www.icao.int/environmental-protection/Pages/A39_CORSIA_FAQ5.aspx. Accessed 9 June 2020

ICAP (International Carbon Action Partnership) (2020a) Japan—Tokyo cap-and-trade program. Last update: 3 June 2020. https://icapcarbonaction.com/en/?option=com_etsmap&task=export&format=pdf&layout=list&systems%5B%5D=51. Accessed 9 June 2020

ICAP (International Carbon Action Partnership) (2020b) Korea emissions trading scheme. Last update: 3 June 2020. https://icapcarbonaction.com/en/?option=com_etsmap&task=export&for mat=pdf&layout=list&systems%5B%5D=47. Accessed 9 June 2020

IKI Alliance Mexico (2020) Preparation of an emission trading system (ETS) in Mexico. International Climate Initiative (IKI). GIZ-BMU. http://iki-alliance.mx/en/portafolio/preparation-of-an-emissions-trading-system-ets-in-mexico/. Accessed 2 June 2020

Jones CW (2013) A report on the European union emissions trading system (EU-ETS). Institute for Global Environmental Strategies. IGES working paper No. 2013-02

LGCC (Ley general de cambio climático) [General law on climate change] (2012) Diario oficial de la federación (DOF). 6 June 2012. Amendment date: 13 July 2018. http://www.diputados.gob.mx/LeyesBiblio/pdf/LGCC_130718.pdf. Accessed 12 May 2020

MexiCO$_2$ (2019) Nota Técnica. Sistema de comercio de emisiones en México [Technical sheet. Emissions trading system in Mexico]. MexiCO$_2$: Plataforma Mexicana de Carbono. July 2019. http://www.mexico2.com.mx/uploadsmexico/file/NOTA%20Sistema%20de%20C omercio%20de%20Emisiones%20en%20M%C3%A9xico%20v040719%203.pdf. Accessed 3 June 2020

New Zealand's Ministry for the Environment (2019) About the New Zealand emissions trading scheme. 18 Dec 2019. https://www.mfe.govt.nz/climate-change/new-zealand-emissions-trading-scheme/about-nz-ets. Accessed 9 June 2020

RGGI (Regional Greenhouse Gas Initiative) (2020) Elements of RGGI. https://www.rggi.org/pro gram-overview-and-design/elements. Accessed 10 June 2020

SEMARNAT (Secretaría de Medio Ambiente y Recursos Naturales) (2019) Programa de prueba del sistema de comercio de emisiones [ETS pilot program]. 27 Nov 2019. https://www.gob.mx/semarnat/acciones-y-programas/programa-de-prueba-del-sistema-de-comercio-de-emisiones-179414. Accessed 14 Apr 2020

UN (United Nations) (2015a) The millennium development goals report 2015. https://www.un.org/millenniumgoals/2015_MDG_Report/pdf/MDG%202015%20rev%20(July%201).pdf. Accessed 6 June 2020

UN (United Nations) (2015b) Transforming our world. The 2030 agenda for sustainable development. Resolution A/RES/70/1. 21 Oct 2015. https://www.un.org/ga/search/view_doc.asp?symbol=A/RES/70/1&Lang=E. Accessed 16 Apr 2020

UN (United Nations) (2019) Sustainable development goals report 2019. New York. https://unstats.un.org/sdgs/report/2019/. Accessed 10 June 2020

WCI (Western Climate Initiative) (2020) Greenhouse gas emissions trading: a cost-effective solution to climate change. https://wci-inc.org/. Accessed 10 June 2020